Advances in Intelligent Systems and Computing

Volume 294

Series editor

Janusz Kacprzyk, Polish Academy of Sciences, Warsaw, Poland
e-mail: kacprzyk@ibspan.waw.pl

For further volumes:
http://www.springer.com/series/11156

About this Series

The series "Advances in Intelligent Systems and Computing" contains publications on theory, applications, and design methods of Intelligent Systems and Intelligent Computing. Virtually all disciplines such as engineering, natural sciences, computer and information science, ICT, economics, business, e-commerce, environment, healthcare, life science are covered. The list of topics spans all the areas of modern intelligent systems and computing.

The publications within "Advances in Intelligent Systems and Computing" are primarily textbooks and proceedings of important conferences, symposia and congresses. They cover significant recent developments in the field, both of a foundational and applicable character. An important characteristic feature of the series is the short publication time and world-wide distribution. This permits a rapid and broad dissemination of research results.

Julio Sáez-Rodríguez · Miguel P. Rocha
Florentino Fdez-Riverola · Juan F. De Paz Santana
Editors

8th International Conference on Practical Applications of Computational Biology & Bioinformatics (PACBB 2014)

Springer

Editors
Julio Sáez-Rodríguez
European Bioinformatics Institute
Hinxton
United Kingdom

Miguel P. Rocha
University of Minho
Braga
Portugal

Florentino Fdez-Riverola
Department of Informatics
University of Vigo
Ourense
Spain

Juan F. De Paz Santana
Department of Computing Science
and control
University of Salamanca
Salamanca
Spain

ISSN 2194-5357 ISSN 2194-5365 (electronic)
ISBN 978-3-319-07580-8 ISBN 978-3-319-07581-5 (eBook)
DOI 10.1007/978-3-319-07581-5
Springer Cham Heidelberg New York Dordrecht London

Library of Congress Control Number: 2014939943

Printed on acid-free paper

Springer is part of Springer Science+Business Media (www.springer.com)

Preface

Biological and biomedical research are increasingly driven by experimental techniques that challenge our ability to analyse, process and extract meaningful knowledge from the underlying data. The impressive capabilities of next generation sequencing technologies, together with novel and ever evolving distinct types of omics data technologies, have put an increasingly complex set of challenges for the growing fields of Bioinformatics and Computational Biology. To address the multiple related tasks, for instance in biological modeling, there is the need to, more than ever, create multidisciplinary networks of collaborators, spanning computer scientists, mathematicians, biologists, doctors and many others.

The International Conference on Practical Applications of Computational Biology & Bioinformatics (PACBB) is an annual international meeting dedicated to emerging and challenging applied research in Bioinformatics and Computational Biology. Building on the success of previous events, the 8th edition of PACBB Conference will be held on 4–6 June 2014 in the University of Salamanca, Spain. In this occasion, special issues will be published by the Journal of Integrative Bioinformatics, the Journal of Computer Methods and Programs in Biomedicine and the Current Bioinformatics journal covering extended versions of selected articles.

This volume gathers the accepted contributions for the 8th edition of the PACBB Conference after being reviewed by different reviewers, from an international committee composed of 72 members from 15 countries. PACBB'14 technical program includes 34 papers from about 16 countries of origin, spanning many different sub-fields in Bioinformatics and Computational Biology.

Therefore, this event will strongly promote the interaction of researchers from diverse fields and distinct international research groups. The scientific content will be challenging and will promote the improvement of the valuable work that is being carried out by the participants. Also, it will promote the education of young scientists, in a post-graduate level, in an interdisciplinary field.

We would like to thank all the contributing authors and sponsors (Telefónica Digital, Indra, Ingeniería de Software Avanzado S.A, IBM, JCyL, IEEE Systems Man and Cybernetics Society Spain, AEPIA Asociación Española para la Inteligencia Artificial, APPIA Associação Portuguesa Para a Inteligência Artificial, CNRS Centre national de la recherche scientifique), AI*IA, as well as the members of the Program Committee and the Organizing Committee for their hard and highly valuable work and support. Their effort has helped to contribute to the success of the PACBB'14 event. PACBB'14 wouldn't exist without your assistance. This symposium is organized by the Bioinformatics, Intelligent System and Educational Technology Research Group (`http://bisite.usal.es/`) of the University of Salamanca and the Next Generation Computer System Group (`http://sing.ei.uvigo.es/`) of the University of Vigo.

Julio Sáez-Rodríguez
Miguel P. Rocha
PACBB'14 Programme Co-chairs

Florentino Fdez-Riverola
Juan F. De Paz Santana
PACBB'14 Organizing Co-chairs

Organization

General Co-chairs

Florentino Fdez-Riverola	University of Vigo, Spain
Juan F. De Paz	University of Salamanca, Spain
Julio Sáez-Rodríguez	European Bioinformatics Institute, United Kingdom
Miguel Rocha	University of Minho, Portugal

Program Committee

Alicia Troncoso	University Pablo de Olavide, Spain
Amparo Alonso	University of A Coruña, Spain
Ana Cristina Braga	University of Minho, Portugal
Anália Lourenço	University of Vigo, Spain
Armando Pinho	Universty of Aveiro, Portugal
Caludine Chaouiya	Gulbenkian Institute, Portugal
Camilo Lopez	Universidad Nacional de Colombia, Colombia
Carlos A.C. Bastos	University of Aveiro, Portugal
Daniel Glez-Peña	University of Vigo, Spain
Daniela Correia	CEB, University of Minho, Portugal
David Hoksza	Charles University in Prague, Czech Republic
Eva Lorenzo	University of Vigo, Spain
Fernanda Correia Barbosa	DETI/IEETA, University of Aveiro, Portugal
Fernando Díaz-Gómez	University of Valladolid, Spain
Fidel Cacheda	University of A Coruña -, Spain
Florencio Pazos	CNB, Spanish Council for Scientific Research, Spain
Francisco Torres-Avilés	Universidad de Santiago de Chile, Chile
Frank Klawonn-Ostafilia	University of Applied Sciences, Wolfenbuettel, Germany
Gonzalo Gómez-López	UBio/CNIO, Spanish National Cancer Research Centre, Spain

Gustavo Isaza	Universidad de Caldas, Colombia
Hagit Shatkay	University of Delaware, USA
Heri Ramampiaro	Norwegian University of Science and Technology, Norway
Hugo López-Fernández	University of Vigo, Spain
Hugo Miguel Santos	Universidade Nova de Lisboa, Portugal
Isabel C. Rocha	IBB/CEB, University of Minho, Portugal
Jiri Novak	Charles University in Prague, Czech Republic
João Rodrigues	University of Aveiro, Portugal
Joel P. Arrais	DEI/CISUC, University of Coimbra, Portugal
Jorge Ramirez	Universidad Nacional de Colombia, Colombia
Jorge Vieira	Institute for Molecular and Cell Biology, Portugal
José Antonio Castellanos Garzón	University of Valladolid, Spain
Jose Ignacio Requeno	University of Zaragoza, Spain
José Luis Capelo	Universidade Nova dc Lisboa, Portugal
José Luis Oliveira	Universty of Aveiro, Portugal
José Manuel Colom	University of Zaragoza, Spain
Juan Antonio García Ranea	University of Malaga, Spain
Julio R. Banga	IIM, Spanish Council for Scientific Research, Spain
Liliana Lopez-Kleine	Universidad Nacional de Colombia, Colombia
Loris Nanni	University of Bologna, Italy
Lourdes Borrajo	University of Vigo, Spain
Luis F. Castillo	Universidad de Caldas, Colombia
Luis Figueiredo	European Bioinformatics Institute, United Kingdom
Luis M. Rocha	Indiana University, USA
M Alamgir Hossain	Northumbria University at Newcastle, United Kingdom
Mª Araceli Sanchís de Miguel	University Carlos III of Madrid, Spain
Manuel Álvarez Díaz	University of A, Spain
Miguel Reboiro	University of Vigo, Spain
Mohammad Abdullah Al-Mamun	Northumbria University, United Kingdom
Mohd Saberi Mohamad	Universiti Teknologi Malaysia, Malaysia
Monica Borda	University of Cluj-Napoca, Romania
Narmer Galeano	Cenicafé, Colombia
Nuno Fonseca	CRACS/INESC, Porto, Portugal
Nuria Medina Medina	CITIC, University of Granada, Spain
Pierpaolo Vittorini	University of L'Aquila, Italy
Reyes Pavón	University of Vigo, Spain
Rita Ascenso	Polytecnic Institute of Leiria, Portugal
Rosalía Laza	University of Vigo, Spain

Rubén López-Cortés	Universidade Nova de Lisboa, Portugal
Rui Brito	University of Coimbra, Portugal
Rui C. Mendes	CCTC, University of Minho, Portugal
Rui Camacho	LIAAD/FEUP, Universty of Porto, Portugal
Rui Rijo	Polytecnic Institute of Leiria, Portugal
Sara C. Madeira	IST/INESC ID, Lisbon, Portugal
Sara P. Garcia	University of Aveiro, Portugal
Sérgio Deusdado	Polytecnic Institute of Bragança, Portugal
Sergio Matos	DETI/IEETA, University of Aveiro, Portugal
Silas Vilas Boias	Univerity of Auckland, New Zealand
Slim Hammadi	Ecole Centrale de Lille, France
Thierry Lecroq	Univeristy of Rouen, France
Tiago Resende	CEB, University of Minho, Portugal
Vera Afreixo	University of Aveiro, Portugal

Organising Committee

Juan M. Corchado	University of Salamanca, Spain
Javier Bajo	Polytechnic University of Madrid, Spain
Juan F. de Paz	University of Salamanca, Spain
Sara Rodríguez	University of Salamanca, Spain
Dante I. Tapia	University of Salamanca, Spain
Fernando de la Prieta Pintado	University of Salamanca, Spain
Davinia Carolina Zato Domínguez	University of Salamanca, Spain
Gabriel Villarrubia González	University of Salamanca, Spain
Alejandro Sánchez Yuste	University of Salamanca, Spain
Antonio Juan Sánchez Martín	University of Salamanca, Spain
Cristian I. Pinzón	University of Salamanca, Spain
Rosa Cano	University of Salamanca, Spain
Emilio S. Corchado	University of Salamanca, Spain
Eugenio Aguirre	University of Granada, Spain
Manuel P. Rubio	University of Salamanca, Spain
Belén Pérez Lancho	University of Salamanca, Spain
Angélica González Arrieta	University of Salamanca, Spain
Vivian F. López	University of Salamanca, Spain
Ana de Luís	University of Salamanca, Spain
Ana B. Gil	University of Salamanca, Spain
Mª Dolores Muñoz Vicente	University of Salamanca, Spain
Jesús García Herrero	University Carlos III of Madrid, Spain

Contents

Applications

Data Analysis and Mining

Proteins

Sequence Analysis

Systems Biology

Text Mining

Agent-Based Model for Phenotypic Prediction Using Genomic and Environmental Data

Sebastien Alameda[1], Carole Bernon[1], and Jean-Pierre Mano[2]

[1] Universite Paul Sabatier, Toulouse, France
[2] UPETEC, Toulouse, France

Abstract. One of the means to increase in-field crop yields is the use of software tools to predict future yield values using past in-field trials and plant genetics. The traditional, statistics-based approaches lack environmental data integration and are very sensitive to missing and/or noisy data. In this paper, we show how using a cooperative, adaptive Multi-Agent System can overcome the drawbacks of such algorithms. The system resolves the problem in an iterative way by a cooperation between the constraints, modelled as agents. Results show a good convergence of the algorithm. Complete tests to validate the provided solution quality are still in progress.

Keywords: Multi-Agent System, Adaptation, Self-organization, Phenotypic Prediction.

1 Introduction

Constant growth in global population, hence cereal consumption, increases the pressure on food processing industries to meet this increasing demand[1]. In this context, human and commercial necessity to produce more and more cereals implies the use of industrial processes that guarantee higher in-field yields. Amongst these processes, genomic breeding is a widely used set of techniques encompassing mathematical and software tools able to predict a crop yield based on genetics[2]. These tools, currently statistics-based, can be used to improve yield by choosing plant varieties with higher genetic potentials. The statistical methods traditionally used for these purposes lack the integration of environmental conditions in the predicting variables. Therefore, they have yet been unable to predict the yield variability of a crop depending on the weather - and other environmental parameters, such as the ground quality - it is exposed to.

To overcome this lack, we aim at building a system able to predict a yield value, depending on experimental conditions, by using cooperative, adaptive, self-organizing agent-based techniques. This system ought to be able to use raw data without any preprocessing. The experiments run on this system use data provided by seed companies, extracted from in-field maize experiments, which are both noisy and sparse. To validate this system, leave-one-out test cases will be executed to check its convergence.

J. Sáez-Rodríguez et al. (eds.), *8th International Conference on Practical Appl. of Comput. Biol. & Bioinform. (PACBB 2014)*, Advances in Intelligent Systems and Computing 294,
DOI: 10.1007/978-3-319-07581-5_1,

2 Problem Expression

2.1 Original Problem in Genomic Breeding

The problem is to predict the γ_i phenotype of an individual i ($i = 1..n$) knowing a $1 * p$ vector x_i of SNP genotypes on this individual. It is generally assumed that

$$\gamma_i = g(x_i) + e_i \tag{1}$$

with g being a function relating genotypes to phenotypes and e_i an error term to be minimized. The γ_i value found once the g function is computed is called the Genomic Estimated Breeding Value [3].

To find the actual value of the g function, i.e. to minimize the e_i terms, various methods can be used. Without drifting into too much detail, Random Regression Best Linear Unbiased Prediction (RR-BLUP) [4] offers good results in the context of biparental crosses, which is the case in maize breeding. This method, and the others used in plant breeding, have in common the goal to minimize, as said before, the error term and to find an accurate, global, expression of g.

Those global approaches pose the problem of the quality of the data involved in the predictions. For example it has been shown that the marker density - the size of the x_i vector related to the genome size of the considered species - needs to scale with population size and that the choice of the samples are of great importance in the accuracy of the results [2].

Furthermore, the accuracy of the prediction given by those models depends heavily on trait heritability. The more a trait is heritable, the more accurate the prediction [5]. This lack of accuracy in low-heritability traits may be explained by the influence of environmental parameters on those traits and by genomic-environmental interactions and brings the need for another problem expression able to integrate environmental data.

2.2 Problem Specification

The problem this paper addresses is the prediction of the γ yield value of a maize crop given a set x_i of n constraints on various genetical and environmental traits. The equation (1) becomes:

$$\gamma = g(x_i) + e \tag{2}$$

with g being a continuous function and e being the error term.

The assumed continuity property of g allows a local, exploratory search of the solution. In other terms, it removes the need of finding a global, search space wide definition for g. The means we offer to find a solution is to iteratively fetch relevant data on previously measured in-field tests from a database. To be deemed "relevant", a datum must match the constraints expressed by the x_i vector.

As discussed above, the relevant data $\{D_i\}$ are extracted from a database of past in-field trials on the basis of the constraints defined by the x_i parameters.

As the database typically holds more than a million of such data and can theoretically contain much more, for scalability purposes only a few of them is loaded in the memory at each iteration. Each datum D_i that constitutes the dataset is itself a set encompassing, for an observed phenotype, all phenotypic, environmental and genomic data related to this phenotype. In particular, the datum D_i holds a γ_i value for the phenotypic trait that is the goal of the prediction.

One of the challenges that the system must address is to cooperatively decide which constraints should be individually released or tightened, i.e. the tolerance to add to each constraint, in order to reach a satisfactory solution. Since a solution is defined as a dataset $\{D_i\}$, in the ideal case, all γ_i would be equal to one another (consistent solution) and the data set would contain a large number of data (trustworthy solution). Such a solution is deemed "satisfactory" when the system cooperatively decides that it cannot be improved anymore.

The solution satisfaction can then be expressed as a f_a function, aggregation of two functions:

- A function f_q that evaluates the quality of the solution as the range taken by the predicted values $\{\gamma_i\}$. The lower this range, the lower the value of $f_q(\{D_i\})$.
- A function f_t that evaluates the trust given to the solution provided. The more data D_i are implied in the solution, the lower the value of $f_t(\{D_i\})$.

With this definition, the goal of the prediction system is expressed as providing a solution $\{D_i\}$ as close as possible to the absolute minimum of f_a.

Linking back to the equation (2), $g(x_i)$ may then be defined as the average value of the $\{\gamma_i\}$ and e as a term bounded by the range of $\{\gamma_i\}$.

3 Solving Process

Agents are defined as autonomous entities able to perceive, make decisions and act upon their environment [6]. A system of those interconnected software agents is able to solve complex problems. The system used in order to solve this problem is based on the AMAS (Adaptive Multi-Agent System) theory [7], which provides a framework to create self-organizing, adaptive and cooperative software. The agents in the system, by modifying their local properties (adaptive) and the interactions between them (self-organizing), modify also the global function of the system.

3.1 The System and Its Environment

The AMAS considered here contains two different kinds of agents:

- n Constraint Agents, in charge of tightening or releasing the constraints defined in section 2.2. Each agent is responsible for one constraint. Each agent's goal is to minimize its estimation of the f_a function, calculated on the only basis of this agent's actions.

- A Problem Agent, in charge of evaluating the solution provided by the Constraint Agents and giving them a hint on the future actions they have to take in order to make the solution more satisfactory. Its goal is to minimize the actual f_a function.

3.2 Iterative Process

The resolution is iterative and the Fig.1 illustrates the way the system functions.

At each step, the Problem Agent (1) receives a data set $\{D_i\}$ and evaluates both values of $f_q(\{D_i\})$ and $f_t(\{D_i\})$. Every Constraint Agent (2) has three possible actions: tightening, releasing or leaving as is the constraint it is related to.

In order to decide amongst its possible actions, each Constraint Agent evaluates its influence on the solution quality f_q and the solution trust f_t by simulating the world state if it were to execute one or the other of its possible actions. Depending on this simulated state of the world, the agent chooses the most cooperative action to perform, that is the action that improves the value of the criterion the agent has the greatest influence on. This calculated influence gives the agent a hint on whether it should maintain as is, tighten or release the constraint it is related to.

The current restriction state of the constraints are aggregated (3) and used as a filter to find a new dataset $\{D_i\}$. This dataset consists of previously found data matching the new constraints and newly found data, also matching these new constraints, from the database (4). This way, the system simply ignores the missing data by including in the datasets only the existing, relevant data.

At each step, Each datum D_i in the database can be in one of these three states:

- Active: the datum is loaded into memory and, at each resolution step, gives a predicted value γ_i.
- Inactive: The datum was loaded into memory once but does not provide predicted values, as it does not match one of the current constraints.
- Existing: The datum exists in the database but has not currently been loaded into memory.

This model allows an iterative enrichment of the data pool. As the constraints become more precise regarding the problem to be solved, the Inactive + Active pool size tends to remain constant due to the fact that every datum matching the constraints has already been loaded into memory and no more data are loaded from the Existing data pool.

3.3 Convergence Measurement

The resolution ends when the dataset $\{D_i\}$ provided at the end of each resolution step is definitely stable. To guarantee this stability, two conditions must be met:

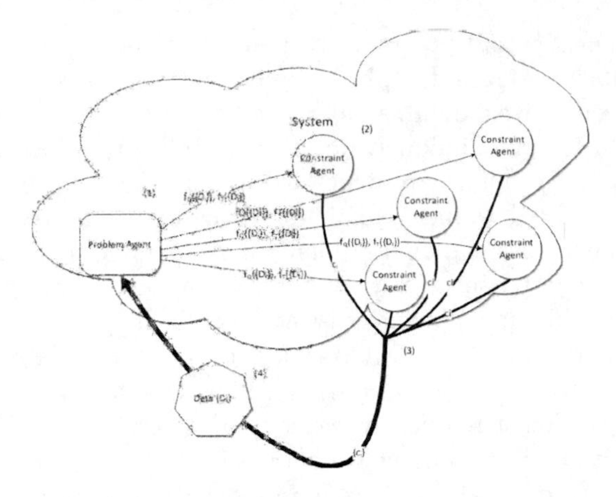

Fig. 1. A view of the system architecture exhibiting the information flow between the agents

- Every Constraint Agent estimates that the optimal (from its own point of view) action to take is to not modify its value.
- The Active + Inactive dataset size is stable, i.e. no more data are recruited from the database.

In those conditions, the system has reached a fixed point and the convergence process is complete. At this point, the data matching the constraints constitute the solution provided to the user.

4 Experiments and Results

As seen above, the convergence is characterized by the stability of the constraints and the stability of the Inactive + Active dataset size. The primary objective of the following experiments is to exhibit those two convergence conditions. The other objectives are to show that the convergence speed and the quantity of data used make this AMAS solution suitable for real-life use.

The experimental protocol set up is the random choice of several leave-one-out test cases. The data used are real-world in-field maize data, provided by seed companies that are partners of this research project.

4.1 Data Characterization

These data include:

- 300, 000 maize individuals with their pedigree and/or genomic data;
- 30, 000, 000 yield and other phenotypical data of in-field trials in the past years for these individuals;
- 150, 000 environmental (meteorological and pedological) data for these trials.

Those data are essentially sparse with respect to the various dependent variables in this problem. Indeed, the phenotypical measurements result from the interaction of a given maize individual, identified by its genomic data, and a specific environment, which can be uniquely determined by a given location and year, in which interfere the various environmental data specified above. If one considers for instance that these data measurements are arranged in a rectangular matrix, with individuals per rows and environments per columns, then the resulting matrix will be extremely sparse, i.e. with a high ratio of zero entries corresponding to unobserved data. This sparsity aspect is intrinsic to the problem, simply because it is infeasible to grow every year in every location all the existing maize individuals. With respect to the database considered here, in the case of the yield values (which is one of the most frequently collected data), the ratio of the number of measured values to the total number of entries in this matrix is less than 0.7 percent. In [8], the authors recall either techniques that try to input the missing data in some way, or methods that are designed to work without those missing input values, the first ones being sensitive to the ratio of observed to missing data, and the latter presenting some risk of overfitting. The AMAS method we consider here belongs to the second class of methods, and present the additional advantage that it does not suffer from overfitting issues, since the method itself aims at selecting a much denser subset of values that are relevant for a given problem.

As the data are provided by seed companies and protected by non-disclosure agreements, only raw estimations can be given for the size of the datasets. The total number of datasets present in the database is more than $1,000,000$. There are more than $50,000$ genomic, environmental and phenotypic variables (the n in 3.1), although only a limited number (10) of those variables were used as constraints in the following experiments.

4.2 Experiments

In the following figures, a sample of the most representative results are shown.

Figure 2 shows the convergence speed of the tolerance of a single constraint upon various experiments. It exhibits that a limited number of steps is needed to reach a fixed point, according to the constraints strength. The tolerance converges to different values due to the fact that this particular constraint may be of more or less importance depending on the problem. It can be seen that the tolerance evolves by stages. This pattern can be explained by the fact that the Constraint Agent tightens its constraint only if the number of Inactive+Active data still matching the new constraint is sufficient. As this number steadily increases over time, the constraint can be tightened only when a certain threshold is reached. For example, for Experiment 2, the tolerance remains constant from step 105, which means that from this step on, the Constraint Agent related to this constraint decides at each iteration to leave the tolerance as is. However, the other constraints –not shown in this figure– are still able to adjust their tolerance. To see when the fixed point is actually reached, the analysis presented in Fig.3 is necessary.

Figure 3 shows the total number of data used against the simulation time, in iteration steps. For example, for experiment 2, the fixed point is reached at 108 steps. Those results exhibit that less than 1% of the database is needed for the system to reach its fixed point and return a prediction to the user in less than 200 steps. An experiment runs in about 45 minutes, however we estimate that more than 75% of this time is consumed by database accesses. More precise measurements have still to be made.

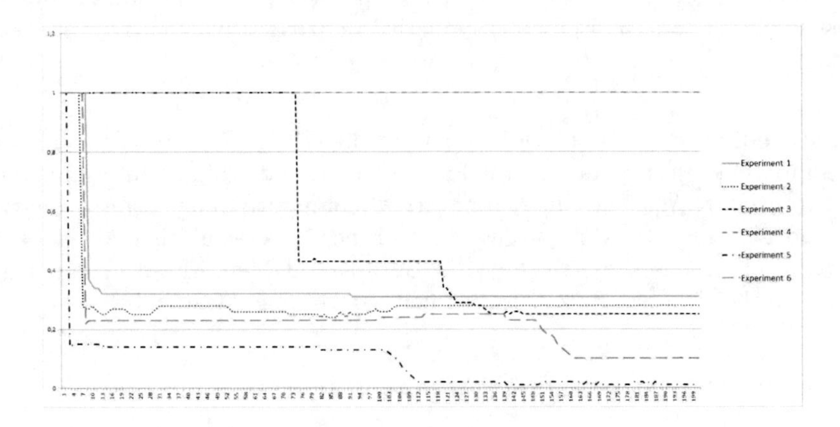

Fig. 2. Convergence of a single constraint upon various experiments

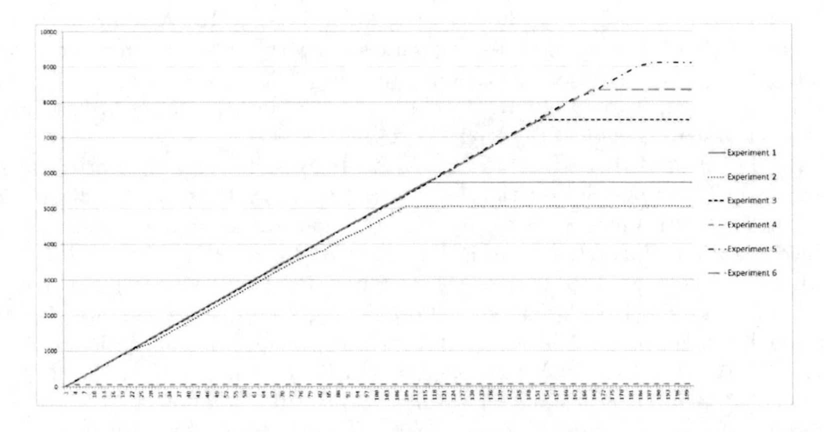

Fig. 3. Convergence of the Inactive+Active dataset size upon various experiments

5 Conclusion

In this paper, an Adaptive Multi-Agent System was presented to overcome the lack of traditional statistical approaches in phenotypic prediction. The system solves the problem using cooperation between agents, which are responsible for the various genomic and environmental constraints. Experiments show how the system converges towards a solution despite the high sparsity of the data involved. Since only the relevant datapoints are explored based on a very small fraction of the entire database, the system is not sensitive to missing data. The system convergence is characterized by the convergence of constraints tolerance and the stability of the data pool.

Current and future works are aimed at validating the solution by different cross-validation tests as well as systematic comparison with current statistical methods.

Acknowledgements. This work is part of the GBds (Genomic Breeding decision support) project funded by the French FUI (Fonds Unique Interministeriel) and approved by Agri Sud-Ouest Innovation (competitive cluster for the agriculture and food industries in southwestern France). We would like also to thank our partners in this project (Ragt 2n, Euralis and Meteo France) and Daniel Ruiz from the IRIT laboratory for his help concerning data analysis.

References

[1] Food, of the United Nations, A.O.: State of food and agriculture (2013)
[2] Lorenz, A.J., Chao, S., Asoro, F.G., Heffner, E.L., Hayashi, T., Iwata, H., Smith, K.P., Sorrells, M.E., Jannink, J.L.: Genomic selection in plant breeding: Knowledge and prospects. Advances in Agronomy 110, 77–121 (2011)
[3] Moser, G., Tier, B., Crump, R., Khatkar, M., Raadsma, H.: A comparison of five methods to predict genomic breeding values of dairy bulls from genome-wide snp markers. Genetics Selection Evolution 41(1), 56 (2009)
[4] Whittaker, J.C., Thompson, R., Denham, M.C.: Marker-assisted selection using ridge regression. Genetical Research 75, 249–252 (2000)
[5] Hayes, B.J., Visscher, P.M., Goddard, M.E.: Increased accuracy of artificial selection by using the realized relationship matrix. Genetics Research 91, 47–60 (2009)
[6] Ferber, J.: Multi-Agent Systems: An Introduction to Distributed Artificial Intelligence. Addison-Wesley Longman Publishing Co., Inc. (1999)
[7] Capera, D., George, J.P., Gleizes, M.P., Glize, P.: The AMAS Theory for Complex Problem Solving Based on Self-organizing Cooperative Agents. In: International Workshop on Theory And Practice of Open Computational Systems (TAPOCS@WETICE 2003), pp. 389–394. IEEE Computer Society (2003)
[8] Ilin, A., Raiko, T.: Practical approaches to principal component analysis in the presence of missing values. J. Mach. Learn. Res. 11, 1957–2000 (2010)

NAPROC-13: A Carbon NMR Web Database for the Structural Elucidation of Natural Products and Food Phytochemicals*

José Luis López-Pérez[1], Roberto Theron[2], Esther del Olmo[1], Beatriz Santos-Buitrago[3], José Francisco Adserias[4], Carlos Estévez[4], Carlos García Cuadrado[4], David Eguiluz López[4], and Gustavo Santos-García[5]

[1] Departamento de Química Farmacéutica – IBSAL – CIETUS, Universidad de Salamanca, Spain
[2] Dpto. Informática y Automática, Universidad de Salamanca, Spain
[3] School of Computing, University of the West of Scotland, UK
[4] Fundación General Universidad de Salamanca, Spain
[5] Computing Center, Universidad de Salamanca, Spain

Abstract. This paper describes the characteristics and functionalities of the web-based database NAPROC-13 (`http://c13.usal.es`). It contains Carbon NMR spectral data from more than 21.000 Natural Products and related derivates. A considerable number of structures included in the database have been revised and corrected from the original publications considering subsequent published revisions. It provides tools that facilitate the structural identification of natural compounds even before their purification. This database allows for flexible searches by chemical structure, substructure of structures as well as spectral features, chemical shifts and multiplicities. Searches for names, formulas, molecular weights, family, type and group of compound according to the IUPAC classification are also implemented. It supports a wide range of searches, from simple text matching to complex boolean queries. These capabilities are used together with visual interactive tools, which enable the structural elucidation of known and unknown compounds by comparison of their ^{13}C NMR data.

Keywords: structural elucidation, carbon NMR spectral database, natural compounds, chemoinformatics, bioinformatics, food phytochemicals, SMILES code.

1 Introduction

Chemoinformatics is the application of informatics methods to chemical problems [7]. All major areas of chemistry can profit from the use of information technology and management, since both a deep chemical knowledge and the

* Financial support came from the Ministerio de Educación y Ciencia, project TIN2006-06313 and the Junta de Castilla y León, project SA221U13. Research also supported by Spanish project Strongsoft TIN2012-39391-C04-04.

J. Sáez-Rodríguez et al. (eds.), *8th International Conference on Practical Appl. of Comput. Biol. & Bioinform. (PACBB 2014)*, Advances in Intelligent Systems and Computing 294,
DOI: 10.1007/978-3-319-07581-5_2,

processing of a huge amount of information are needed. Natural Products (NPs) structure elucidation requires spectroscopic experiments. The results of these spectroscopic experiments need to be compared with those of the previously described compounds. This methodology provides highly interesting challenges for chemoinformatics practitioners.

NPs from microbial, plant, marine, or even mammalian sources have traditionally been a major drug source and continue to play a significant role in today's drug discovery environments [10]. In fact, in some therapeutic areas, for example, oncology, the majority of currently available drugs are derived from NPs. However, NPs have not always been as popular in drug discovery research as one might expect, since in the NPs research, tedious purifications are needed in order to isolate the constituents. These procedures are often performed with the main purpose of structure identification or elucidation. Because of that, ultra-high throughput screening and large-scale combinatorial synthetic methods have been the major methods employed in drug discovery [19]. Yet if the structures of natural extract constituents could be known in advance, the isolation efforts could be focused on truly novel and interesting components, avoiding re-isolation of known or trivial constituents and in this way increasing the productivity [4]. Furthermore, it is generally known that the intrinsic diversity of NPs exceeds the degree of molecular diversity that can be created by synthetic means, and the vast majority of biodiversity is yet to be explored [10]. At present, it is unanimously assumed that the size of a chemical library is not a key issue for successful developmental leads and that molecular diversity, biological functionality and "drug likeness" are decisive factors for drug discovery processes [10]. For this reason, the natural products-based drug discovery is on the rise again.

Some chemoinformatics methods include predictive classification, regression and clustering of molecules and their properties. In order to develop these statistical and machine learning methods the need for large and well-annotated datasets has been already pointed out. These datasets need to be organized in rapidly searchable databases to facilitate the development of computational methods that rapidly extract or predict useful information for each molecule [3]. The progressive improvement of analytical techniques for structural elucidation makes today's structural identification more reliable and it permits the correction of structures of a large number of previously published compounds.

NPs databases are of high priority and importance for structure search, matching and identification [9]. In this paper, we present a web-based spectral database that facilitates the structural identification of the natural compounds previous to their purification.

2 ^{13}C NMR Spectroscopy: A Power Technique for Structural Elucidation of Natural Compounds

For the elucidation of natural compounds, ^{13}C NMR spectroscopy is the most powerful tool. This is largely due to the well-known and exquisite dependence of the ^{13}C chemical shift of each carbon atom on its local chemical environment and its number of attached protons. Furthermore, the highly resolved spectra, provided by a large chemical shift range and narrow peak width, could by easily converted into a highly reduced numerical lists of chemical shift positions with minimal loss of information. ^{13}C NMR spectroscopy can also provide the molecular formula. The analysis of spectral data for the determination of unknown compound structure remains a usual but a laborious task in chemical practice.

Since the advent of computers many efforts have been directed toward facilitating the solution to this problem [7]. Libraries of such spectral lists of data are common for synthetic organic compounds and are an invaluable tool for confirming the identity of known compounds [17]. However, the methods for structure elucidation of compounds apart from a database have not been exhaustively studied. In the field of NPs, where hundreds of thousands compounds have been reported in the literature, most compounds are absent from commercially available spectral libraries.

Once a researcher in NPs isolates and purifies a compound, he needs to know the compound's structure, the skeleton and if it has been previously described. If a database of NPs and their NMR spectral data are available, searching databases will allow for quick identifications by means of comparison of the new compound with the NMR spectrum of the registered compounds or with other related compounds. This search provides insight into the structural elucidation of unknown compounds.

NAPROC-13 has many search facilities and a set-up that allows comparative studies of related compounds. At present, new search tools are being developed and the data input methods are being improved so as to allow researchers from different institutions to introduce the information over the Net. The aim of this database is to help identify and elucidate the structure of hypothetical new compounds, by comparing their ^{13}C NMR data with those of already published related compounds.

2.1 NMR Databases for Phytochemicals

Mass spectrometry and NMR spectroscopy allow the efficient identification of phytochemicals and of other NPs. Because of the large spectral dispersion, the relative chemical shift invariance, and the simplicity of ^{13}C NMR spectra, most analytical chemists prefer to use ^{13}C NMR for the identification of phytochemicals, phytochemical metabolites, and other NPs. NAPROC-13, which is a ^{13}C NMR database of NPs, probably represents one of the richest NMR resource for phytochemists and phytochemical databases [18]. Along with NAPROC-13, NMRShiftDB2 [20] (`http://www.nmrshiftdb.org`) is another open web database for organic structures and their NMR spectra; unfortunately, it does not contain too many NPs. Other noteworthy NMR databases are: HMDB, HMDB, MMCD, BMRB, SDBS, and HaveItAll CNMR-HNMR.

The Human Metabolome Database [25] (HMDB, `http://www.hmdb.ca`) is a freely available electronic database containing detailed information about small molecule metabolites found in the human body. This database contains >40.000 metabolite entries. It contains experimental ^{1}H and ^{13}C NMR data (and assignments) for 790 compounds. Additionally, predicted ^{1}H and ^{13}C NMR spectra have been generated for 3.100 compounds.

Spectral Database for Organic Compounds (SDBS, http://sdbs.db.aist.go.jp) is an integrated spectral database system for 34.000 organic compounds, which includes 6 different types of spectra (an electron impact Mass spectrum EI-MS, a Fourier transform infrared spectrum FT-IR, a ^{1}H NMR spectrum, a ^{13}C NMR spectrum, a laser Raman spectrum, and an electron spin resonance ESR spectrum) under a directory of the compounds.

HaveItAll CNMR-HNMR Library (http://www.bio-rad.com) access over 500.000 high-quality ^{13}C NMR and 75.000 ^{1}H NMR spectra. It offers access to high-quality NMR spectral reference data for reliable identification and NMR prediction.

3 NAPROC-13: Database and Web Application

The structural elucidation of natural compounds poses a great challenge because of its great structural diversity and complexity [16]. For this reason, we are developing a database accessible through a standard browser (http://c13.usal.es). It provides the retrieval of natural compounds structures with ^{13}C NMR spectral data related to the query. At present it contains the structures of more than 21.000 compounds with their ^{13}C NMR information.

MySQL (http://www.mysql.com) has been chosen to develop NAPROC-13 for its high reliability and good performance; it is a fast, robust multithread, multiuser database. MySQL is an open-source relational database manager, based on SQL (Structured Query Language). We use the open-source Apache Tomcat web server and JavaServer Pages (JSP) technology to create dynamically web pages. By means of proper JSP programming, we bring about the communication between applets and the database. As for the interactive visualization tools, Java applets have also been integrated in this application.

There is a widespread belief that publicly funded scientific data must be freely available to the public [18]. Open accessibility has many benefits, not the least of which is increased visibility. Our database makes freely available resources that can be easily accessed over the Internet without passwords or logins.

Our aim was to design a reliable database. Data acquiring are fully and properly provided with references, data sources, and citations. References ensure that the data can be reproduced and allow users to investigate the data sources for further information. Structures and spectral data collected in the database proceeds from books, journals and our measurements. They are mainly compiled from papers in the following research journals: *Journal of Natural Products, Phytochemistry, Planta Medica, Chemical & Pharmaceutical Bulletin, Chemistry of Natural Compounds, Helvetica Chimica Acta*, and *Magnetic Resonance in Chemistry*.

NAPROC-13 is continually expanded and updated in order to enhance the database's querying capabilities, design, and layout. User-friendliness has been another important factor. NAPROC-13 interface allows for complex queries, which can be performed through simple pull-down menus or clickable boxes using plain language. Web capabilities of HTML language enables a high degree of interactivity.

4 Reliability of Structural and Spectroscopic Data from NAPROC-13

Over the course of the past four decades, the structural elucidation of unknown NPs has undergone a tremendous revolution. A battery of advanced spectroscopic methods, such as multidimensional NMR spectroscopy, high-resolution mass spectrometry or X-ray crystallography are available today for structural resolution of highly complex natural compounds.

Despite the available structural strategies and elucidation methods and despite the progress made in recent decades, constant revisions of structures of NPs are published in the literature. These revisions are not mere inversion of stereocenters, but they may entail profound changes in their chemical constitution. For example, more than thousands of articles on structural revisions published in the best journals cover virtually all types of compounds, steroids, terpenoids, alkaloids, aromatic systems, etc., regardless of the size of the molecule.

Often a structure obtained by X-ray diffraction is considered irrefutable proof of its structure. However, we can find examples in which the position of 2 heteroatoms has been changed. Another method to confirm the structure of a compound is by means of its total synthesis. In some of the synthesized compounds, we can observe a discrepancy between the natural product and its synthetic data, which means that the proposed structure for the natural compound is not correct. Although in most cases, the structure can be fixed, in others, the ambiguity persists since the NMR data of the synthesized compound is different from the structural proposal of the natural product and, hence, the actual structure remains unknown. This is due to the enormous structural complexity of the isolated compounds and the small quantities of sample available.

In the field of NPs, the structural assignment is often based on the structures of the related compounds. Thus, if the wrong structure of a substance is taken as a model, errors are continuously replicated. This problem can be avoided, if reliable NPs spectroscopic data is entered into a database such as NAPROC-13 and is used as reference. In this way, we can avoid some errors in publications. Let's consider the following example: the same compound was independently isolated by two research teams who propose different structures and names for the same spectroscopic data of an identical compound. Later, both structural proposals are proven to be incorrect (see Figure 1) Access to the spectroscopic data could help in assigning new compounds of a family and facilitate the process of structural reallocation.

Incorrect NPs assignments not only make the determination of the biosynthetic pathway more difficult, but may have costs in terms of time and money. Imagine that an interesting product is isolated from a pharmacological point of view. Current strategy synthesizes NPs and their closely related analogues.

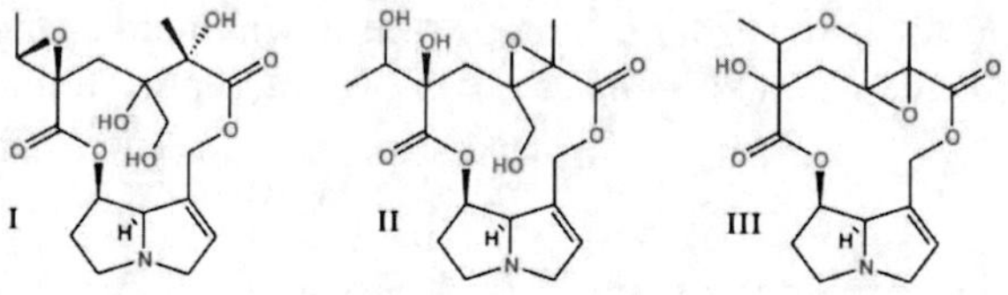

Fig. 1. I: Proposed erroneous structure in [23]; II: Proposed erroneous structure in [1]; III: Corrected structure of adenifoline [26]

Obviously, if the structure is not correct, we synthesize another compound different from the one we are interested in.

A database is as useful as the data it contains. Curators spend a considerable amount of time acquiring data in order to keep the database relevant. Data acquisition and data entry are not automated, but data is manually searched, read, assessed, entered, and validated. NAPROC-13 prioritizes the introduction of those compounds whose structures have been reviewed in recent literature. Since a database of this nature grows, manual transcription errors and those present in the literature are inevitable, We have developed some scripts to detect obvious chemical shift errors, such as shifts greater than 240.0 ppm, as well as errors based on a few simple rules regarding proper ranges of chemical shift ranges for several easily identifiable functional groups. Thus the data presented in NAPROC-13 has greater reliability when being considered as a pattern.

4.1 Database Design

Numbering system of the well-known *Dictionary of Natural Products* (`http://dnp.chemnetbase.com`, Chapmann & Hall/CRC Press) has been applied to each family skeleton in NAPROC-13. Numbering homogeneity within the same family compounds enables the comparison of spectral data for a variety of related structures.

NAPROC-13 contains a wide diversity of data types. It collects a rich mixture of text, numbers, charts, and graphs. The basic database schema is relationally organized and the molecular structures are defined and stored in the database with SMILES code (Simplified Molecular Input Line Entry Specification) [24]. This format of structural specification, that uses one line notation, is designed to share chemical structure information over the Internet [8]. For example, SMILES code for Melatonin ($C_{13}H_{16}N_2O_2$) is "`CC(=O)NCCC1=CNc2c1cc(OC)cc2`".

Substructural searches are performed by SMARTS specification (SMiles ARbitrary Target Specification). This is a language for specifying substructural patterns in molecules from the SMILES code. The SMARTS line notation is expressive and allows extremely precise and transparent substructural specification and atom typing. It uses logical operators that allow choosing all-purpose atoms, groups of alternative atoms, donor and acceptor groups of hydrogen bonds or lipophilic atoms. For example, SMARTS specification for Hydrazine (H_2NNH_2) is "`[NX3][NX3]`" and for an Oxygen in $-O-C=N-$ is "`[$([OX2]C=N)]`".

Evidently the type of patterns and notations that are used, both SMARTS and SMILES, are too complex to be interpreted by organic chemists without specific training in this area. For this reason, we use a tool able to convert these notations into a graph that represents a substructure that will act as question.

The spectral ^{13}C NMR data, in form of a numerical list of chemical shift and their multiplicity, is always associated with each compound structure. A script calculates and represents the ^{13}C NMR spectra of the selected compound in a very similar way to the experimentally obtained data, and shows the decoupled proton (broad band) and the DEPTs (Distortionless Enhancement by Polarization Transfer). Figure 2 displays the ^{13}C NMR spectrum calculated for the substance of a compound found by a search. Another script calculates and represents the signals corresponding to the deuterated solvent used in the experiment.

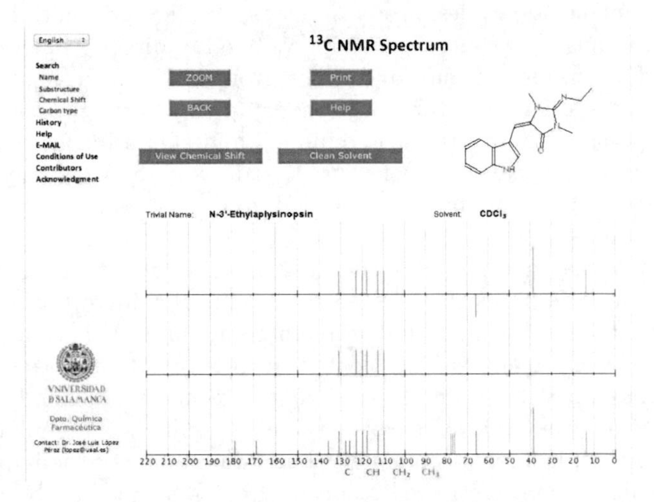

Fig. 2. Chart of the ^{13}C NMR spectra of a chemical compound. Multiplicities of the carbons are codified by colors.

5 Queries in NAPROC-13

NAPROC-13 allows for flexible searches by chemical structure, substructure of structures as well as spectral features, chemical shifts and multiplicities [21]. Searches for names, formulas, molecular weights, family, type and group of compound according to the IUPAC classification and other parameters are also included. NAPROC-13 database supports a wide range of searches, from simple text matching to complex boolean queries.

This database offers several alternatives of the chemical shift search process. The multiplicity for each chemical shift is always required and this constitutes a useful search restriction. The search can be undertaken for one specific position in the molecule. The system permits to formulate the enquiry with the required number of carbons, by one carbon or more, up to the totality of the carbons of the compound. There is a default established deviation (+/-1 ppm) for all chemical shifts, but the user can specify a particular deviation for every carbon. It is important to be able to repeat the search with different deviations and to select the search that provides the best results. If the deviation is too small, it may occur that an interesting compound will not be selected. In this way, a reasonable and manageable number of compounds can be obtained. Even a search based only on the most significant carbons of the studied compound ^{13}C NMR spectrum will lead to the identification of the family they belong to.

Moreover, users can address questions in a graphic form to the database using JME Molecular Editor, a structure editor that enables the user to draw several fragments that may not be related to each other. JME has a palette that speeds up the creation of structures and uses IUPAC recommendations to depict the stereochemistry. By using this palette it is possible to add preformed substructures, i.e., different size cycles, aromatic rings, simple and multiple bonds, frequently used atoms. The control panel allows to enter directly functional groups, i.e., carboxyl acids, nitro groups and other groups. The facilities of this applet rapidly generates a new structure and speeds up the search process.

It is also possible to undertake a combined and simultaneous search by substructure and by chemical shifts, a feature that undoubtedly enhances the search capacity and increases the possibilities of finding compounds related with the problem substance.

The iterative search is probably the most genuine search of this application. The user can include in his search from one chemical shift to the totality of the signals of the ^{13}C NMR spectrum problem compound. This tool will initially carry out a search of all the entered chemical shifts. If it does not find any compound that does not fulfill the full requirements, it will undertake a new iterative search by all the shifts except one. It will perform all the possible combinations until it finds a compound that fulfills some of the requirements.

The matching records retrieved resulting from a search can be displayed in the Results pane in the form of molecular structure. The chemical shifts of the matching records can be viewed in tables or in the compound structures by clicking the δ (ppm) in tables/structures buttons. Properties pane provides the details of a particular record. Spectrum pane shows spectrum graphically.

5.1 Interactive Visual Analytical Tool

As stated above, NAPROC-13 features a built-in visual analytical tool. It is a highly interactive interface integrated by four linked views: ^{13}C NMR spectrum, structure, parallel coordinates plot, and taxonomic information (see Figure 3).

The main advantage of this approach is that the user can deal with a great number of compounds that have matched a particular search. Thanks to interaction, a user can explore this result set, focusing on particular details of a given compound (name-family-type-group, structure, spectrum) while maintaining the context, i.e. the characteristics of the rest of the compounds in the result set. Parallel coordinates provide a way of representing any number of dimensions in the 2D screen space [11]. Each compound is drawn as a polyline passing through

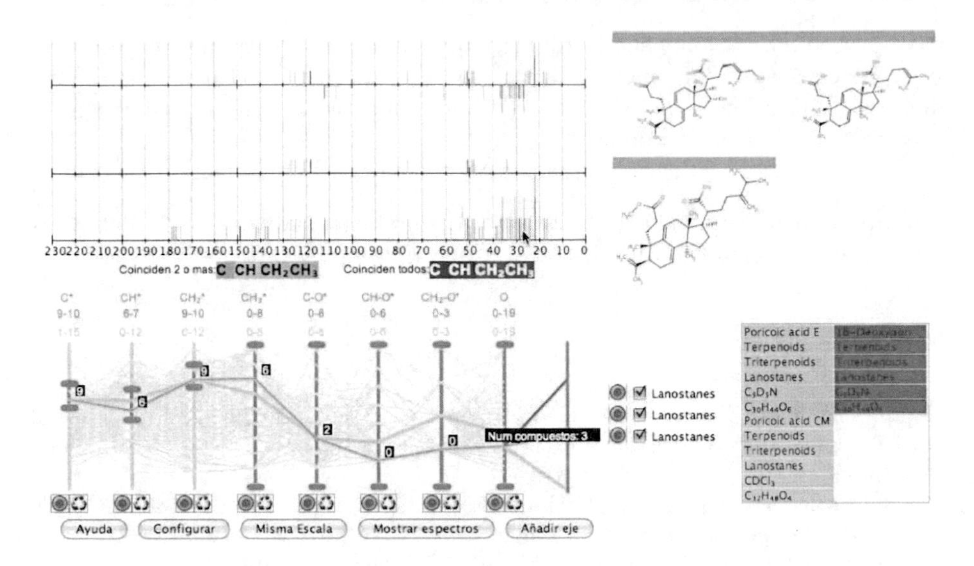

Fig. 3. Visual interactive exploration of results

parallel axes, which represent the number of particular elements or groups. Thus, it is possible to discover patterns, i.e. a number of polylines (compounds) with similar shapes.

A similar, although visually different, approach is taken with the representation of the spectra. Initially, all the compound spectra are shown as overlapped. Thus a global pattern can be discovered in the result set.

The user can interact with any of the four views and, as a result, the other three views will change accordingly. For instance, the expert may select any number of polylines and the corresponding spectra will appear as overlapped, and their structures are shown in order to facilitate their comparison.

Further inspection can be achieved by filtering the result set according to different criteria (e.g., a range in the number of occurrences of an element or a particular area in the spectrum). The filtered data is visually maintained in the background in order to keep always the context of the exploration.

All these features foster the discovery of knowledge and provide insight into the vast number of compounds included in the database.

6 Conclusions and Further Work

The development of a comprehensive and qualified open access NPs database will be able not only to considerably facilitate the dereplication process but also to accelerate the identification of unknown compounds.

Currently, we are working in order to increase the number of stored compounds and to add new search methods, such as the hot spot search, a powerful search for chemical shifts of carbons of one area of the molecule. We also intend to improve information visualization techniques that give more insight into analysis processes as well as include supervised and unsupervised machine learning methods conducive to interesting predictions for the assignment of the NMR spectral data of new compounds.

Acknowledgments. The authors wish to thank the courtesy of Dr Peter Ertl for consenting to the non-profit use of JME Molecular Editor.

References

1. Bohlmann, F., Zdero, C., Jakupovic, J., Grenz, M., et al.: Further Pyrrolizidine alkaloids and furoeremophilanes from Senecio species. Phytochemistry 25, 1151–1159 (1986)
2. Breitmaier, E., Woelter, W.: Carbon-13 NMR spectroscopy. High-resolution methods and applications in Organic Chemistry. VCH Publ., Weinheim (1987)
3. Chen, J., Swamidass, S.J., Dou, Y., et al.: Chemdb: a public database of small molecules and related chemoinformatics resources. Bioinformatics 21, 4122–4139 (2005)
4. Clarkson, C., Stærk, D., Hansen, S.H., Smith, P.J., Jaroszewski, J.W.: Discovering new natural products directly from crude extracts by HPLC-SPE-NMR: chinane diterpenes in Harpagophytum procumbens. J. Nat. Prod. 69, 527–530 (2006)
5. Cui, Q., Lewis, I.A., Hegeman, A.D., Anderson, M.E., et al.: Metabolite identification via the Madison Metabolomics Consortium Database. Nat. Biotechnol. 26, 162–164 (2008)
6. Ertl, P., Jacob, O.: WWW-based chemical information system. J. Mol. Struct. Theochem. 419, 113–120 (1997)
7. Gasteiger, J.: Chemoinformatics: a new field with a long tradition. Anal. Bioanal. Chem. 384, 57–64 (2006)
8. Grabley, S., Thiericke, R.: Bioactive agents from natural sources: trends in discovery and application. Adv. Biochem. Eng. Biotechnol. 64, 101–154 (1999)
9. Halabalaki, M., Vougogiannopoulou, K., Mikros, E., Skaltsounis, A.L.: Recent advances and new strategies in the NMR-based identification of natural products. Curr. Opin. Biotechnol. 25, 1–7 (2014)
10. Harvey, A.: Strategies for discovering drugs from previously unexplored natural products. Drug. Discov. Today 5, 294–300 (2000)
11. Inselberg, A.: The plane with parallel coordinates. The Visual Computer 1, 69–91 (1985)
12. Kochev, N., Monev, V., Bangov, I.: Searching chemical structures. In: Chemoinformatics: a Textbook, pp. 291–318. Wiley-VCH (2003)
13. Lei, J., Zhou, J.: J. Chem. Inf. Comput. Sci. 42, 742–748 (2002)
14. López-Pérez, J.L., Theron, R., del Olmo, E., Díez, D., Vaquero, M., Adserias, J.F.: Application of chemoinformatics to the structural elucidation of natural compounds. In: Corchado, E., Yin, H., Botti, V., Fyfe, C. (eds.) IDEAL 2006. LNCS, vol. 4224, pp. 1150–1157. Springer, Heidelberg (2006)
15. López-Pérez, J.L., Theron, R., Olmo, E., del Díaz, D.: NAPROC-13: a database for the dereplication of natural product mixtures in bioassay-guided protocols. Bioinformatics 23, 3256–3257 (2007)

16. Peláez, R., Theron, R., García, C.A., López-Pérez, J.L., et al.: Design of new chemoinformatic tools for the analysis of virtual screening studies: Application to tubulin inhibitors. In: Corchado, J.M., De Paz, J.F., Rocha, M.P., Riverola, F.F. (eds.) IWPACBB 2008. ASC, vol. 49, pp. 189–196. Springer, Heidelberg (2008)
17. Robien, W.: NMR data correlation with chemical structure. In: von Ragu Schleyer, P., Allinger, N.L., Clark, T., Gasteiger, J., Kollman, P.A., Schaefer, H.F. (eds.) Encyclopedia of Computational Chemistry, vol. 3, pp. 1845–1857. Wiley, New York (1998)
18. Scalbert, A., Andres-Lacueva, C., Arita, M., Kroon, P., et al.: Databases on food phytochemicals and their health-promoting effects. J. Agric. Food Chem. 59, 4331 (2011)
19. Shoichet, B.K.: Virtual screening of chemical libraries. Nature 432, 862–865 (2004)
20. Steinbeck, C., Kuhn, S.: NMRShiftDB compound identification and structure elucidation support through a free community-built web database. Phytochemistry 65, 2711–2717 (2004)
21. Theron, R., Olmo, E., del, D.D., Vaquero, M., Adserias, J.F., López-Pérez, J.L.: NATPRO-C13: an interactive tool for the structural elucidation of natural compounds. In: Corchado, E., Corchado, J.M., Abraham, A. (eds.) Innovations Hybrid Intelligent Systems. ASC, vol. 44, pp. 401–410. Springer, Heidelberg (2008)
22. Ulrich, E.L., Akutsu, H., Doreleijers, J.F., Harano, Y., Ioannidis, Y.E., Lin, J., et al.: BioMagResBank. Nucleic Acids Res. 36, D402–D408 (2008)
23. Urones, J.G., Barcala, P.B., Marcos, I.S., Moro, R.F., Esteban, M.L., et al.: Pyrrolizidine alkaloids from Senecio gallicus and S. adonidifolius. Phytochemistry 27, 1507–1510 (1988)
24. Weininger, D.: Smiles, a chemical language and information system. 1. Introduction to methodology and encoding rules. J. Chem. Inf. Comput. Sci. 28, 31–36 (1988)
25. Wishart, D.S., Jewison, T., Guo, A.C., Wilson, M., Knox, C., et al.: HMDB 3.0—The Human Metabolome Database in 2013. Nucleic Acids Res. 41(D1), D801–D807 (2013)
26. Witte, L., Ernst, L., Wray, V., Hartmann, T.: Revised structure of the main alkaloid of Senecio adonidifolius. Phytochemistry 31, 1027–1028 (1992)

Platform Image Processing Applied to the Study of Retinal Vessels

Pablo Chamoso[1], Luis García-Ortiz[2], José I. Recio-Rodríguez[2], and Manuel A. Gómez-Marcos[2]

[1] Computers and Automation Department, University of Salamanca, Salamanca, Spain
[2] Primary care Research unit La Alamedilla, Sacyl, IBSAL, Salamanca, Spain
{chamoso,lgarciao,donrecio,magomez}@usal.es

Abstract. Recent studies have found retinal vessel caliber to be related to the risk of hypertension, left ventricular hypertrophy, metabolic syndrome, stroke and others coronary artery diseases. The vascular system in the human retina is easily perceived in its natural living state by the use of a retinal camera. Nowadays, there is general experimental agreement on the analysis of the patterns of the retinal blood vessels in the normal human retina. The development of automated tools designed to improve performance and decrease interobserver variability, therefore, appears necessary. This paper presents a study focused on developing a technological platform specialized in assessing retinal vessel caliber and describing the relationship of the results obtained to cardiovascular risk.

Keywords: arteriolar–venular ratio, arterial stiffness, cardiovascular disease, AI algorithms, pattern recognition, image analysis, expert knowledge.

1 Introduction and Background

Retinal images have an orange form, varying with the skin color and age of the patient. Fundoscopy provides important information, as it enables detecting diseases of the eyes, which is also the only area of the body where small blood vessels can be studied with relative ease. There are many systemic diseases (hypertension, diabetes, atherosclerosis, left ventricular hypertrophy, metabolic syndrome, stroke, and coronary artery disease) that affect vessels of this size in a relatively slow and silent way. It is, however, frequently impossible to directly assess the extent of this damage during a physical examination, as the affected organs, e.g. kidneys, are well hidden. Evaluation of the retina provides an opportunity to directly visualize these functions. Based on this information, expert clinicians can make educated guesses as to what is occurring elsewhere in the body. Image processing techniques are growing in prominence in all fields of medical science. Automatic detection of parameters from retinal images is an important problem since they are associated with the risk of diseases such as those named above [29][31][33]. The cataloging of key features such as the optic disc, fovea and the retinal vessels as reference matches is a requirement to systems being able to achieve more complex responsibilities that identify pathological

J. Sáez-Rodríguez et al. (eds.), *8th International Conference on Practical Appl. of Comput. Biol. & Bioinform. (PACBB 2014)*, Advances in Intelligent Systems and Computing 294,
DOI: 10.1007/978-3-319-07581-5_3,

entities. There are a lot of techniques for identifying these structures in retinal photographs. The most studied areas in this field can be classified into three groups [21]: (i) The *location of the optic disc*, which is necessary for measuring distances in retinal images, and for identifying changes within the optic disc region due to disease. Techniques such as analysis of intensity pixels with a high grey-scale value [18][11] or principal component analysis (PCA) [19] are used for locating the disk. Other authors [16] use the Hough transform (a general technique for identifying the locations and orientations of certain types of shapes within a digital image [16]) to locate the optic disc. A ''fuzzy convergence'' algorithm is another technique used for this purpose [12]. (ii) The *detection of the fovea*, habitually chosen as the position of maximum correlation between a model template and the intensity image [19]. (iii) The *segmentation of the vasculature form retinal images*, that is, the representation of the blood vessels and their connections by segments or similar structures. There are many techniques to accomplish this, the most significant of which are: (i) matched filters, which typically have a Gaussian or a Gaussian derivative profile [3] [13] [20] [14]; (ii) vessel tracking, whereby vessel center locations are automatically sought over each cross-section of a vessel along the vessels longitudinal axis, having been given a starting and end point [28]; (iii) neural networks, which employ mathematical ''weights'' to decide the probability of input data belonging to a particular output [1]; and (iv) morphological processing, which uses characteristics of the vasculature shape that are known a priori, such as being piecewise linear and connected [12].

Present scientific literature includes much research focused on automating the analysis of retinal images [25] [15] [4][10][12]. In this paper, we propose a novel image processing platform to study the structural properties of vessels, arteries and veins that are observed with a red-free fundus camera in the normal human eye. The platform, called Altair "*Automatic image analyzer to assess retinal vessel caliber*" [30], employs analytical methods and AI (Artificial Intelligence) algorithms to detect retinal parameters of interest. The sequence of algorithms represents a new methodology to determine the properties of retinal veins and arteries. The platform does not require user initialization, it is robust to the changes in the appearance of retinal fundus images typically encountered in clinical environments, and it is intended as a unified tool to link all the methods needed to automate all processes of measurement on the retinas. Section 2 introduces the platform and its most important characteristics, showing some of the relevant techniques and results. Finally, some conclusions and results are presented in section 3.

2 Description of the Platform

The platform facilitates the study of structural properties of vessels, arteries and veins that are observed with a red-free fundus camera in the normal human eye. The retina is the only human location where blood vessels can be directly visualized non-invasively by the use of a retinal camera. Figure 1 displays a retinal image in which branching blood vessels are shown. The bigger, darker ones are the veins and the smaller, brighter red structures the arteries. Changes in the appearance of the arteries as well as alterations in the arterial-venous crossing pattern (av index) occur with atherosclerosis and hypertension. These vessels are more obvious in the superior and inferior aspects of the retina, with relative sparing of the temporal and medial regions. to the use of the platform allows expert clinicians to observe the parameters measured by regions or quadrants to discriminate the information that they do not consider relevant.

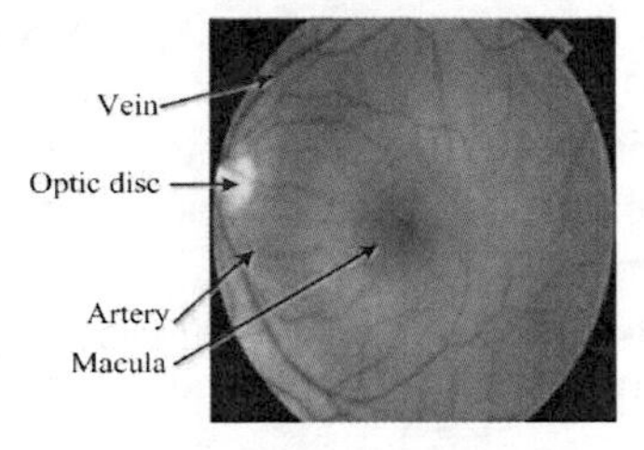

Fig. 1. A retinograph usually takes three images of each eye; this image corresponds to the photograph with the disc on one side. The nose is on the left side.

Different analytical methods and AI algorithms are used to determine the scaling properties of real objects, yielding different measures of the fractal dimension, length and area of retinal veins and arteries. The main objective is to relate the level of cardiovascular risk in patients to everything that can be observed in the retinas. In this work we are interested in obtaining as much information as possible from the images obtained, and have focused on the following: (i) Index Artery / Vein: represents a relationship between the thickness of arteries and veins. (ii) Area occupied by the veins and arteries. (iii) Distribution of the capillary: according to the blood distribution, the color distribution of the capillaries varies.

Based on the values for area, length and position of the vascular system in healthy patients, we expect to determine ranges of normalcy within the population for their subsequent application to subjects affected by various diseases.

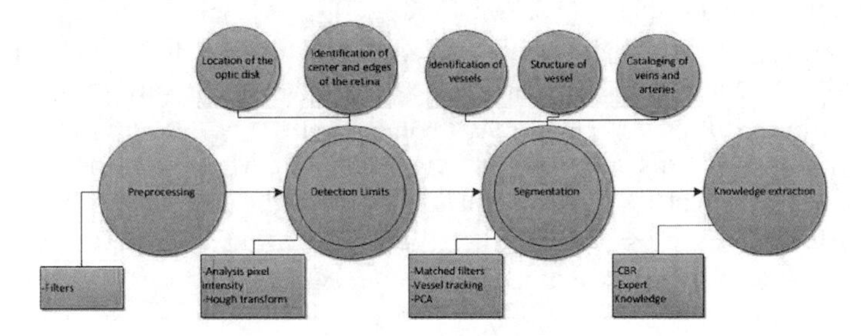

Fig. 2. Outline of the platform

The next subsections describe the main phases through which the retinal image passes within the platform. Figure 2 shows the diagram that represents these phases. The circles represent modules in the platform; some of them are divided and contain different submodules. The original image passes through each one of the modules (preprocessing, detection, segmentation and extraction of knowledge) which use different techniques and algorithms to obtain the desired image information. The main techniques are represented in the squares. This sequence of steps is a methodology that is explained in the following section, also showing examples of the results obtained.

2.1 Phases

The methodology used to obtain the functionality of the platform may be divided into two phases. Firstly, a phase called "digitization of the retina", in which the different parts of the eye image are identified. Here a data structure is created, which makes it possible to represent and process the retina without requiring the original image. This phase includes modules of preprocessing, detection and segmentation. Secondly, a phase of "measurements" in which we work with retinas that have been previously identified. This phase includes extraction of knowledge and manual correction, or expert knowledge, if necessary.

This paper focuses on the first phase, which is in charge of creating and identifying all the elements of interest of the retina. To carry out these phases, the following steps are necessary.

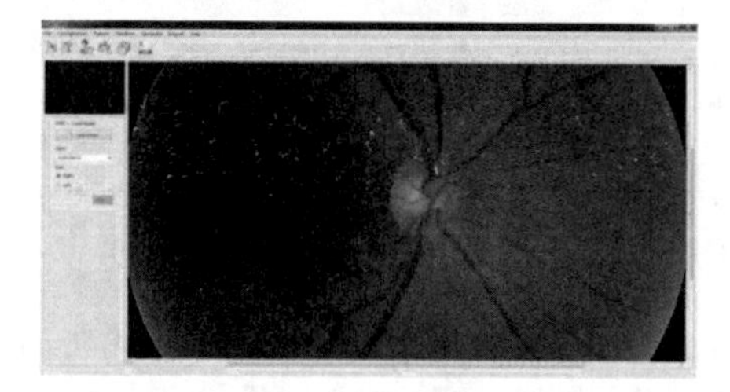

Fig. 3. Load image and find macula. First step in the platform.

Fig. 4. Identify the papilla in the retinal image

2.1.1 Preprocessing

During the testing, a retinography was performed using a Topcon TRC NW 200 nonmydriatic retinal camera (Topcon Europe B.C., Capelle a/d Ijssel, The Netherlands), obtaining nasal and temporal images centered on the disk (Figure 1). The nasal image with the centered disk is loaded into the platform software through the preprocessing module (Figure 3). The preprocessing or filtering module reduces noise, improves contrast, sharpens edges or corrects blurriness. Moreover, the platform is able to automatically detect the type of image (left or right eye) and find the macula by using dynamic thresholds, binary filters and comparing both sides. Some of these actions can be carried out at the hardware level, which is to say with the features included with the camera.

2.1.2 Detection Limits

This module is in charge of locating the disk and identifying the center, edges and regions of the retina (Figure 4). The aim is to construct a data structure that identifies each part of the retina based on the matrices of colors representing the images obtained (Figure 1). In this phase, image processing techniques were used to detect intensity based on the boundaries of the structures [12][4]. The identification of the papilla is important since it serves as the starting point for the detection and identification of the different blood vessels.

This phase identifies the boundaries and the retinal papilla from an RGB image of the retina. The following values are returned: C_r is the center of the retina, which identifies the vector with coordinates *x, y* of the center of the retina. C_p is the center of the disc, which identifies the vector with the coordinates *x, y* of the center of the papilla. R_r, is the radius of the retina. R_p, is the radius of the papilla. As an example, a sequence of output values in this phase is shown in the following table and figure:

Table 1. Sequence of output values in detection modules (pixel)

C_r	C_p	R_r	R_p
1012,44 ; 774,13	**1035,98 ; 734,11**	**692,68**	**111,76**
	1104,87 ; 562,52		108,92
	915,38 ; 736,77		122,15
	900,27 ; 658,74		101,95

In order to identify the limits, and in particular to identify the circumferences, it became necessary to carry out a process of image segmentation. Segmentation is the process that divides an image into regions or objects whose pixels have similar attributes. Each segmented region typically has a physical significance within the image. It is one of the most important processes in an automated vision system because it makes it possible to extract the objects from the image for subsequent description and recognition. Segmentation techniques can be divided into three main groups: techniques based on the detection of edges or borders [16], thresholding techniques [18], and techniques based on clustering of pixels [11]. After analyzing the possibilities, we chose one of the techniques from the first group that provided the best results and that, in this case, uses an optimization of the Hough transform [16]. This technique is very robust against noise and the existence of gaps in the border of the object. It is used to detect different shapes in digital images. When applying the Hough transform to an image, it is first necessary to obtain a binary image of the pixels that form part of the limits of the object (applying edge detection). The aim of the Hough transform is to find aligned points that may exist in the image to form a desired shape. For example, to identify line points that satisfy the equation of the line: ($\rho = x \cdot cos\ \theta + sen\ \theta$, in polar coordinate). In our case, we looked for points that verify the equation of the circle: (i) in polar coordinate system: $r^2 - 2sr \cdot cos\ (\theta - \alpha) + s^2 = c^2$, where (s, α) is the center and c the radius; (ii) in Cartesian coordinate system: $(x-a)^2 + (y-b)^2 = r^2$, where (a,b) is the center and r the radius.

The algorithm is not computationally heavy, as it does not check all radius, or all possible centers, only the candidate values. The candidate centers are those defined in a near portion of the retina, and the radius is approximately one sixth the radius of the retina. To measure the approximate diameter of the retina, the algorithm calculates the average color of the image column: diameter of the retina is the length that has a non-zero value (black).

Identifying the papilla is a necessary step because it provides a starting point for other stages of segmentation and serves as a reference point for some typical measurements. Typically the correct result is the circumference of the higher value in the accumulator (over 70% of cases). In almost 100% of the cases, the correct identification can be found among the 3 greatest values found by the accumulator. In this phase it is divided into image regions (squares), for further handling by experts.

2.1.3 Segmentation of the Vasculature from Retinal Images

The ultimate goal is to identify each blood vessel as a series of points that define the path of the vessel. Each of these points will be assigned a certain thickness. Moreover, it will be necessary to distinguish whether a particular blood vessel is a vein or an artery. AI algorithms responsible for identifying veins and arteries must perform a series of sweeps in search of "key points". Algorithms based on matched filters[3] [13] [20] [14], vessel tracking [25] and PCA [19], among others, are used for obtaining the proximity points between objects (veins, arteries, capillaries), the structures retinal structures or assemblies, branching patterns, etc. These algorithms work with transformations of the original image of the retina obtained from the previous step. Three steps are necessary within this module: (i) identification of vessels; (ii) definition of the structure of vessel; (iii) cataloging of veins and arteries.

- Identification of vessels: In this step the blood vessels are identified in the image by thresholding techniques. Their purpose is to remove pixels where the structuring element does not enter, in this case the blood vessels. The image on the retina is blurred to keep an image similar to the background. This image is used as a threshold so that the pixels of the original image will be treated as vessels if their intensity reaches 90% of the background intensity.
- Structure of vessel: This phase defines the tree forming blood vessels. Various techniques are used in conjunction with the following steps:
- Cataloging of veins and arteries: To detect whether a vessel is vein or artery, the main branch is taken of the vessel. The platform carries out an automatic differentiation between veins and arteries by comparing the detected vessels and the nearest background colors (R and G layers). It is possible to make manual changes if necessary for vessels that have been incorrectly identified.

The following images show the output of this phase. At the end of this stage the entire arterio-venous tree is stored in a structured way, making it possible to know not only if a vessel passes through a point or not, but through which point each vessel passes, which one is the parent vessel, etc.

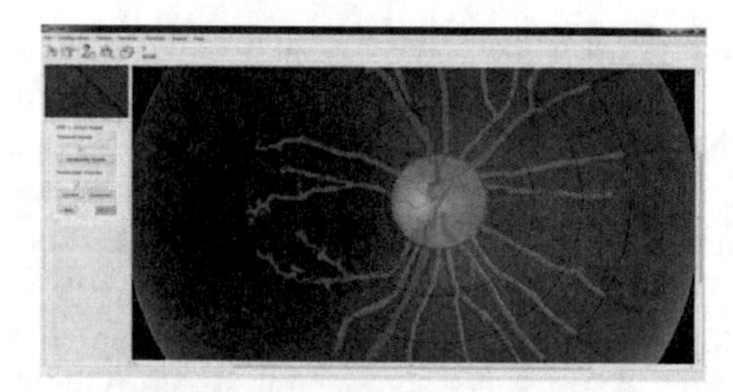

Fig. 5. Structure of the vessels

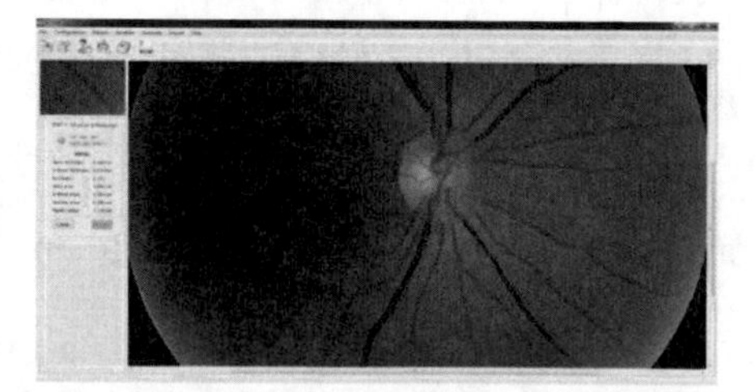

Fig. 6. Arteries and veins detection

2.1.4 Knowledge Extraction

The system provides statistical methods to facilitate the analysis of the data, and allows analyzing the relationship between vein thickness, artery thickness, AV index, artery area and vascular area with different pathologies. The analysis of different pathologies is performed in an assisted way and, so that it is not necessary to use additional software. Besides, the system incorporates traditional statistical functionality to analyze continuous variables. The ANOVA parametric test [6] and the Kruskal-Wallis non-parametric test [24] are provided to analyze the influence of a continuous variable with respect to a categorical variable.

The system does not define categorical variables, but it allows creating intervals from continuous variables. Chi squared [17] is provided to perform dependency analysis methods. When the expected frequencies are less than 5, the result may not be correct so a Yates correction is applied to try to mitigate this effect. Finally, a Fisher exact test [32] is applied when working with a small size sample and it is not possible to guarantee that 80% of the data in a contingency table have a value greater than 5.

A case based reasoning system is incorporated to make predictions about categorical variables from the system variables. The system allows grouping the cases by applying EM [2]. EM was selected because it allows an automatic establishment of the number of clusters. The most similar cluster to the new case is recovered during retrieve phase of the CBR cycle by applying the nearest neighbor technique. If the memory of cases was not structured in clusters, then there is a single cluster. During the retrieve phase, the system retrieves not only the most similar cluster, but also the classifier associated with the cluster. More especifically, C4.5 algorithm [22] is used because it makes easy to interpret the results over alternatives SVM , Bayesian networks, bagging, etc. In the reuse phase, a C4.5 decision tree is used to perform the classification of the patient according to the variable taken into consideration. Then, in the review phase, the system obtains information about the decision tree and the kappa index in order to interpret and analyze the efficiency of the classifier. Finally, in the retain phase, the user determines if the new case should be stored. If that is the case, the clusters and trees are rebuilt.

3 Results and Conclusions

In this work, we have assessed the performance of our platform using retinal images [30] acquired from Primary Care Research Unit La Alamedilla, SACYL, IBSAL, Salamanca, Spain. The images were obtained using a TopCon TRC-NW6S Non-Mydriatic Retinal Camera.

Figure 7 shows the tests performed using 10 retinal images. No difference was found between values in terms of age, sex, cardiovascular risk factors, or drug use. The figure shows: Area veins and arteries, AV index (AV), Veins P (VP) = number of veins around the papilla, Veins A (VA)= number of veins that cross the corona outlined with radius=2*R_p. R_p is the radio of the papilla, Veins B (VB)= number of veins that cross the corona outlined with radius=3*R_p, same values for arteries.

The values for the arteries are the same. It is possible to observe the measurement of the values for veins and arteries (thickness, area) are similar between different retinas (in this case no retinal images of sick patients were introduced).

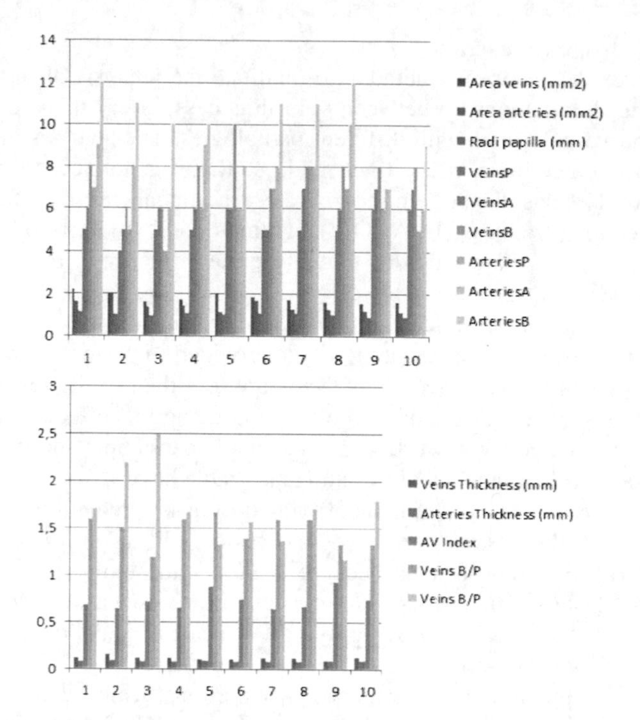

Fig. 7. Relations between the parameters obtained by the platform

Parameters like the veins in the papilla and AV index are the most fluctuating. Due to the lack of a common database and a reliable way to measure performance, it is difficult to compare our platform to those previously reported in the literature. Although some authors report algorithms and methods [25] [15] [4][10] [12] that performed in a manner similar to that of our platform, these results may not be comparable, since these methods are tested separately and were assessed using different databases. Since automation has been valid and verified, our next step is to compare the values obtained with significant medical values in our database including the case based reasoning system proposed in previous section.

The platform is intended to be used as a unified tool to link all the methods needed to automate all processes of measurement on the retinas. It uses the latest computer techniques both statistical and medical. In a research context, the platform offers the potential to examine a large number of images with time and cost savings and offer more objective measurements than current observer-driven techniques. Advantages in a clinical context include the potential to perform large numbers of automated screening for conditions such as risk of hypertension, left ventricular hypertrophy, metabolic syndrome, stroke, and coronary artery disease, which in turn reduces the workload required from medical staff. As a future line of study in this point, the next step would be to analyze the significance of the measurements obtained with regard to their meaning in a medical context. That is, to describe the relationship of the results obtained to the risk of cardiovascular disease estimated with the Framingham or similar scale and markers of cardiovascular target organ damage.

Acknowledgements. This work has been carried out by the project *Sociedades Humano-Agente: Inmersión, Adaptación y Simulación.* TIN2012-36586-C03-03. Ministerio de Economía y Competitividad (Spain). Project co-financed with FEDER funds.

References

[1] Akita, K., Kuga, H.: A computer method of understanding ocular fundus images. Pattern Recogn. 16, 431–443 (1982)

[2] Bayesian, Y.Y.: Machine, clustering and number of clusters. Pattern Recognition Letters 18(11-13), 1167–1178 (1997)

[3] Chaudhuri, S., Chatterjee, S., Katz, N., Nelson, M., Goldbaum, M.: Automatic detection of the optic nerve in retinal images. In: Proceedings of the IEEE International Conference on Image Processing, Singapore, vol. 1, pp. 1–5 (1989a)

[4] Chen, B., Tosha, C., Gorin, M.B., Nusinowitz, S.: Analysis of Autofluorescent retinal images and measurement of atrophic lesion growth in Stargardt disease. Experimental Eye Research 91(2), 143–152 (2010)

[5] Corchado, J.M., De Paz, J.F., Rodríguez, S., Bajo, J.: Model of experts for decision support in the diagnosis of leukemia patients. Artificial Intelligence in Medicine 46(3), 179–200 (2009)

[6] Cvijović, Z., Radenković, G., Maksimović, V., Dimčić, B.: Application of ANOVA method to precipitation behaviour studies. Materials Science and Engineering A 397(1-2), 195–203 (2005)

[7] De Paz, J.F., Rodríguez, S., Bajo, J., Corchado, J.M.: CBR System for Diagnosis of Patient. In: Proceedings of HIS 2008, pp. 807–812. IEEE Computer Society Press (2009) ISBN: 978-0-7695-3326-1

[8] De Paz, J.F., Rodríguez, S., Bajo, J., Corchado, J.: Case-based reasoning as a decision support system for cancer diagnosis: A case study. International Journal of Hybrid Intelligent Systems 6(2), 97–110 (2009)

[9] Fdez-Riverola, F., Corchado, J.M.: CBR based system for forecasting red tides. Knowledge-Based Systems 16(5), 321–328 (2003)

[10] García-Ortiz, L., Recio-Rodríguez, J.I., Parra-Sanchez, J., González Elena, L.J., Patino-Alonso, M.C., Agudo-Conde, C., Rodríguez-Sánchez, E., Gómez-Marcos, M.A.: A new tool to assess retinal vessel caliber. Reliability and validity of measures and their relationship with cardiovascular risk 30 (April 2012), `http://www.jhypertension.com`

[11] Goldbaum, M., Katz, N., Nelson, M., Haff, L.: The discrimination of similarly colored objects in computer images of the ocular fundus. Invest. Ophthalmol. Vis. Sci. 31, 617–623 (1990)

[12] Heneghan, C., Flynn, J., O'Keefe, M., Cahill, M.: Characterization of changes in blood vessel and tortuosity in retinopathy of prematurity using image analysis. Med. Image A. 6, 407–429 (2002)

[13] Hoover, A., Kouznetsoza, V., Goldbaum, M.: Locating blood vessels in retinal images by piecewise threshold probing of a matched filter response. IEEE Trans. Med. Imag. 19, 203–210 (2000)

[14] Hunter, A., Lowell, J., Steel, D., Basu, A., Ryder, R.: Non-linear filtering for vascular segmentation and detection of venous beading. University of Durham (2002)

[15] Roger, J., Arnold, J., Blauth, C., Smith, P.L.C., Taylor, K.M., Wootton, R.: Measurement of capillary dropout in retinal angiograms by computerised image analysis. Pattern Recognition Letters 13(2), 143–151 (1992)

[16] Kalviainen, H., Hirvonen, P., Xu, L., Oja, E.: Probabilistic and non-probabilistic Hough transforms. Image Vision Comput. 13, 239–252 (1995)

[17] Kenney, J.F., Keeping, E.S.: Mathematics of Statistics, Pt. 2, 2nd edn. Van Nostrand, Princeton (1951)
[18] Lee, S., Wang, Y., Lee, E.: A computer algorithm for automated detection and quantification of microaneurysms and haemorrhages in color retinal images. In: SPIE Conference on Image Perception and Performance, vol. 3663, pp. 61–71 (1999)
[19] Li, H., Chutatape, O.: Automated feature extraction in color retinal images by a model based approach. IEEE Trans. Biomed. Eng. 51, 246–254 (2004)
[20] Lowell, J., Hunter, A., Steel, D., Basu, A., Ryder, R., Kennedy, L.: Measurement of retinal vessel widths from fundus images based on 2-D modeling. IEEE Trans. Biomed. Eng. 23, 1196–1204 (2004)
[21] Patton, N., Aslam, T.M., MacGillivray, T., Deary, I.J., Dhillon, B., Eikelboom, R.H., Yogesan, K., Constable, I.J.: Retinal image analysis: Concepts, applications and potential. Progress in Retinal and Eye Research 25(1), 99–127 (2006)
[22] Quinlan, J.R.: C4.5: Programs For Machine Learning. Morgan Kaufmann Publishers Inc. (1993)
[23] Rodríguez, S., De Paz, J.F., Bajo, J., Corchado, J.M.: Applying CBR Sytems to Micro-Array Data Classification. In: Corchado, J.M., De Paz, J.F., Rocha, M.P., Riverola, F.F. (eds.) Proceedings of IWPACBB 2008. ASC, vol. 49, pp. 102–111. Springer, Heidelberg (2010)
[24] Ruxton, G.D., Beauchamp, G.: Some suggestions about appropriate use of the Kruskal–Wallis test. Animal Behaviour 76(3), 1083–1087 (2008)
[25] Sánchez, C., Hornero, R., López, M.I., Aboy, M., Poza, J., Abásolo, D.: A novel automatic image processing algorithm for detection of hard exudates based on retinal image analysis. Medical Engineering & Physics 30(3), 350–357 (2008)
[26] Sánchez-Pi, N., Fuentes, V., Carbó, J., Molina, J.M.: Knowledge-based system to define context in commercial applications. In: 8th ACIS International Conference on Software Engineering, Artificial Intelligence, Networking, and Parallel/Distributed Computing (SNPD 2007) and 3rd ACIS International Workshop on Self-Assembling Wireless Networks (SAWN 2007), Qingdao, Tsingtao, China, Julio 27-29, pp. 694–699 (2007)
[27] Serrano, E., Gómez-Sanz, J.J., Botía, J.A., Pavón, J.: Intelligent data analysis applied to debug complex software systems. Neurocomputing 72(13), 2785–2795 (2009)
[28] Tamura, S., Okamoto, Y., Yanashima, K.: Zero-crossing interval correction in tracing eye-fundus blood vessels. Pattern Recogn. 21, 227–233 (1988)
[29] Tanabe, Y., Kawasaki, R., Wang, J.J., Wong, T.Y., Mitchell, P., Daimon, M., et al.: Retinal arteriolar narrowing predicts 5-year risk of hypertension in Japanese people: the Funagata study. Microcirculation 17, 94–102 (2010)
[30] Verde, G., García-Ortiz, L., Rodríguez, S., Recio-Rodríguez, J.I., De Paz, J.F., Gómez-Marcos, M.A., Merchán, M.A., Corchado, J.M.: Altair: Automatic Image Analyzer to Assess Retinal Vessel Caliber. In: Burduk, R., Jackowski, K., Kurzynski, M., Wozniak, M., Zolnierek, A. (eds.) CORES 2013. AISC, vol. 226, pp. 429–438. Springer, Heidelberg (2013)
[31] Wong, T.Y., Duncan, B.B., Golden, S.H., Klein, R., Couper, D.J., Klein, B.E., et al.: Associations between the metabolic syndrome and retinal microvascular signs: the Atherosclerosis Risk In Communities study. Invest Ophthalmol. Vis. Sci. 45, 2949–2954 (2004)
[32] Yang, X., Huang, Y., Crowson, M., Li, J., Maitland, M.L., Lussier, Y.A.: Kinase inhibition-related adverse events predicted from in vitro kinome and clinical trial data. Journal of Biomedical Informatics 43(3), 376–384 (2010)
[33] Yatsuya, H., Folsom, A.R., Wong, T.Y., Klein, R., Klein, B.E., Sharrett, A.: Retinal microvascular abnormalities and risk of lacunar stroke: Atherosclerosis Risk in Communities Study. Stroke 41, 1349–1355 (2010)

Improving miRNA Classification Using an Exhaustive Set of Features

Sherin M. ElGokhy[1,2], Tetsuo Shibuya[2], and Amin Shoukry[1,3]

[1] Department of Computer Science and Engineering, Egypt-Japan University of Science and Technology (E-JUST), New Borg El-Arab City, Postal Code 21934, Alexandria, Egypt
[2] Human Genome Center, Institute of Medical Science, University of Tokyo, 4-6-1 Shirokanedai, Minato-ku, Tokyo, 108-8639, Japan
[3] Computer and Systems Engineering Department, Alexandria University, Alexandria, Egypt

{sherin.elgokhy,amin.shoukry}@ejust.edu.eg, tshibuya@hgc.jp

Abstract. MicroRNAs (miRNAs) are short (∼22 nucleotides), endogenously-initiated non-coding RNAs that control gene expression post transcriptionally, either by the degradation of target miRNAs or by the inhibition of protein translation. The prediction of miRNA genes is a challenging problem towards the understanding of post transcriptional gene regulation. The present paper focuses on developing a computational method for the identification of miRNA precursors.

We propose a machine learning algorithm based on Random Forests (RF) for miRNA prediction. The prediction algorithm relies on a set of features; compiled from known features as well as others introduced for the first time; that results in a performance that is better than most well known miRNA classifiers. The method achieves 91.3% accuracy, 86% f-measure, 97.2% specificity, 93.4% precision and 79.6% sensitivity, when tested on real data. Our method succeeds in getting better results than MiPred (the best currently known RF algorithm in literature), Triplet-SVM and Virgo and EumiR.

The obtained results indicate that Random Forests is a better alternative to Support Vector Machines (SVM) for miRNA prediction, especially from the point of view of accuracy and f-measure metrics.

Keywords: MicroRNA, Support Vector Machine, Random Forests.

1 Introduction

MicroRNAs (miRNAs) are endogenous ∼22 nt RNAs that are recognized in many species as effective regulators of gene expressions. Experimental recognition of miRNAs is still slow since miRNAs are difficult to separate by cloning due to their low expression, low stability, tissue specificity and the high cost of the cloning process. Thus, computational recognition of miRNAs from genomic sequences supplies a useful complement to cloning. Different approaches

J. Sáez-Rodríguez et al. (eds.), *8th International Conference on Practical Appl. of Comput. Biol. & Bioinform. (PACBB 2014)*, Advances in Intelligent Systems and Computing 294,
DOI: 10.1007/978-3-319-07581-5_4,

for recognition of miRNAs have been proposed based on homology, thermodynamic features, and cross-species comparisons.

MiRNA recognition problem is defined over precursor miRNAs (pre-miRNAs) rather than mature miRNAs. The features that are used in the recognition process are extracted from the hairpin stem loop secondary structure of pre-miRNA sequences. However, many sequences in a genome have a similar stem-loop hairpin structure, in spite of not being genuine pre-miRNAs. So, the basic challenge; in miRNA recognition; is to distinguish real pre-miRNAs from other hairpin sequences with similar stem-loop hairpin structure (pseudo pre-miRNAs).

Homology and machine learning methods are the two major computational strategies considered for pre-miRNA prediction. Most miRNA prediction methods have been developed to find out homologous miRNA in closely related species. 'Blastn' adopts the homology principle in miRNA prediction [1]. This strategy is unable to recognize new miRNAs for which there are no known close homologies. Therefore, the attitude turned towards focusing on machine learning methods to distinguish real pre-miRNAs from pseudo pre-miRNA[2]. The early machine learning methods used to discriminate real versus pseudo pre-miRNAs are miRScan [3], miRseeker [4], miRfinder [5], miRCheck [6] and miPred [7].

Support vector machine systems have been built, aiming to obtain better results in predicting miRNAs such as miR-abela [8], Triplet-SVM [9], MiPred [2], miREncoding [10], microPred [11] and yasMiR[12].

Virgo [13] and EumiR [13] are efficient prediction classifiers that distinguish true pre-miRNAs from pseudo pre-miRNAs. They have been developed based on sequence structural features. A sequence is folded using RNA-fold and the structural context of overlapping triplets is determined. A triplet nucleotide can have 64 possibilities and each nucleotide in the triplet can have two states; '1' if it is bound and '0' if it is unbound. Thus, a feature (eg AUG001, AUG010, ... etc) can have a total of 512 possibilities. Virgo and EumiR perform better than reported pre-miRNA machine learning prediction methods for predicting non-conserved eukaryotic miRNAs.

Triplet-SVM uses a set of features that combines the local contiguous structures with sequence information to characterize the hairpin structure of real versus pseudo pre-miRNAs. These features are fed to a support vector machine classifier to differentiate between real and pseudo pre-miRNAs. RNAfold program from the RNA Vienna package has been used to predict the secondary structure of the query sequences [14]. In the predicted secondary structure, each nucleotide is paired or unpaired, represented by brackets ("("or")") and dots ("."), respectively. There are 8 possible structural combinations: "(((", "((.", "(..", "(.(", ".((", ".(.", "..(" , and"...", that lead to 32 possible structure-sequence combinations, which are denoted as "U(((", "A((.", ... etc. This defines the triplet elements. The triplet elements are used to represent the local structure-sequence features of the hairpin. The occurrence of all triplet elements are counted along a hairpin segment, developing a 32-dimensional feature vector, which is normalized and presented as an input vector to the SVM classifier[9].

The SVM classifier is trained depending on the triplet element features of a set of real human pre-miRNAs from the miRNA Registry database [15] and a set of pseudo pre-miRNAs from the NCBI RefSeq database [16]. The training set consists of 163 human pre-miRNAs (positive samples) and 168 pseudo pre-miRNAs (negative samples) randomly chosen. A 90% accuracy in distinguishing real from pseudo pre-miRNA hairpins in the human genome and up to 90% precision in identifying pre-miRNAs from other 11 species have been achieved.

MiPred is a Random Forests based classifier which differentiates the real pre-miRNAs from the pseudo pre-miRNAs using hybrid features. The features consist of the local structural sequence features of the hairpin with two thermodynamically added features (Minimum Free Energy (MFE) of the secondary structure that is predicted using the Vienna RNA software package and the P-value that is determined using the Monte Carlo randomization test [17]).

MiPred is one of the refinements of Triplet-SVM in which SVM is replaced by a Random Forests. The Random Forests prediction model has been trained on the same training data set used by the triplet-SVM-classifier. It achieved nearly 10% greater overall accuracy compared to Triplet-SVM on a new test dataset.

However, most of the above algorithms utilize a few types of features, though there are many other miRNA features. Thus we try to consider an exhaustive set of features in order to improve the performance of our classifier.

In this paper, we propose a computational method for the identification of miRNA precursors. The miRNA features are extracted from the pre-miRNA sequences, the secondary hairpin stem loop structures of miRNAs as well as thermo-dynamical and Shannon entropy features. Thus, an exhaustive set of discriminative features has been constructed and used to decide whether a given sequence is a true or pseudo miRNA using a Random Forests (RF) classifier.

Our classifier succeeds in getting better results than MiPred, Triplet-SVM, Virgo and EumiR. Also, the present work gives evidence that RF is better than SVM for miRNA prediction.

This paper is organized as follows. Section 1 gives an overview of the miRNA prediction techniques. Section 2 presents the proposed methodology. Section 3 analyses the prediction results. Section 4 concludes the paper.

2 Methodology

This section discusses the strategy of our method.

2.1 Features Used in the Proposed RF-Based Classifier

In an attempt to improve miRNA prediction, the proposed algorithm uses a set of features that have never been collectively used in previous works.

A set of discriminative features is utilized in our classifier. Scatter diagrams of some of the selected features are shown in Fig. 1. These scatter diagrams show how good the chosen features are. The selected package of features includes:

1. The local sequence structure features described in [9] correspond to the number of contiguous nucleotide triplets. These 32 features have been defined in our method as the frequency of each sequence structure triplet.
2. The minimum free energy (MFE) of the secondary structure has been predicted using the RNA Vienna package [14]. Five normalized features of MFE have been considered. MFE adjusted by the hairpin length, MFE adjusted by the stem length, MFE corrected for GC-content, MFE adjusted by the hairpin length and corrected for GC-content and MFE adjusted by the stem length and corrected for GC-content.
3. P-value is the fraction of sequences in a set of dinucleotide shuffled sequences having MFE lower than that of the start sequence [18].
4. Z-value is the number of standard deviations by which sequence's2 MFE deviates from the mean MFE of the set of dinucleotide shuffled sequences [18].
5. The partition function has been determined using RNAfold from the RNA Vienna package [14].
6. Normalized Shannon entropy of the hairpin sequence and normalized Shannon entropy of the hairpin structure. The Shannon entropy is defined as

$$H = -\sum_{i=1}^{n} (p_i * \log_2 p_i) \tag{1}$$

   where p_i is the probability of each character in the given sequence or structure [19]. This Shannon entropy is computed once for the hairpin sequence and adjusted by the hairpin sequence length, and once for the hairpin structure and adjusted by the hairpin structure length.
7. The bulge ratio is the ratio of asymmetrical bulges relative to the stem alignment length [20].
8. The adjusted base pairing propensity (dP) counts the number of base pairs in the RNA secondary structure divided by the hairpin length [21].
9. The match ratio in the hairpin stem is calculated as the number of matched positions in the stem alignment string divided by its length [20].
10. The max match count is the maximum number of matches in 24 positions in the stem alignment string of the hairpin structure [20].

Table 1 presents the list of the selected features that are defined in our approach. The table also shows the new features used for miRNA prediction. We utilize a set of 47 features, ten of which have never been used in any other previous miRNA prediction algorithm.

2.2 Classifier

Random Forests (RF) base classifier uses an ensemble of decision trees [22]. It has a higher performance than support vector machines. Although it has not been widely used in miRNA prediction, it has several advantages that make it suitable for miRNA prediction. RF has excellent predictive performance even when

Table 1. Selected set of features

● indicates that the feature has been utilized.
■ indicates that the feature has not been utilized.

Feature Explanation	Our classifier	Triplet-SVM	MiPred	other familiar classifiers
A.((, A(((,...etc the local contiguous sequence-structure triplets	●	●	●	●
The minimal free energy of the folding	●	■	●	●
Minimal free energy corrected for GC content	●	■	■	■
Minimal free energy adjusted for Hairpin length	●	■	■	●
Minimal free energy adjusted for Hairpin length corrected for GC content	●	■	■	■
Minimal free energy adjusted for stem length	●	■	■	■
Minimal free energy adjusted for stem length corrected for GC content	●	■	■	■
P value of MFE of randomized sequences	●	■	●	●
Z value of MFE of randomized sequences	●	■	■	●
The partition function	●	■	■	■
The normalized shannon entropy of the hairpin sequence	●	■	■	■
The normalized shannon entropy of the hairpin structure	●	■	■	■
The bulge ratio is ratio of asymmetrical bulges versus the stem alignment length	●	■	■	■
Adjusted base pairing propensity measures the total number of base pairs presented in the secondary structure divided by the hairpin length	●	■	■	●
the matech ratio equal the number of matches divided by the length of the stem alignment string considering bulges	●	■	■	■
the max match count is the highest number of matches in 24 positions in the stem alignment string	●	■	■	■

most data samples are noisy. RF avoids over-fitting. It also guarantees computational scalability by increasing the number of trees. There are high quality and free implementations of RF, the Waikato Environment for Knowledge Analysis (WEKA) includes one of the most popular implementations of RF [23]. In the present work we use WEKA as the platform of our RF classifier. The default values of the parameters are considered.

3 Results and Discussions

This section discusses the data sets that have been used for the training and testing of our classifier and also the obtained prediction results.

3.1 Data Sets

The training dataset consists of 300 known human pre-miRNAs; retrieved from miRBase19 [24]; and 700 pseudo hairpins; extracted from human RefSeq genes [16]. A testing data set consisting of 500 known human pre-miRNAs and 1000 pseudo hairpins - different than those used in training - retrieved from miRBase19 [24] and human RefSeq genes [16]; respectively; have been used for testing the performance of the already trained classifier.

3.2 Comparison with the Most Familiar miRNA Prediction Methods

For the problem under investigation in this paper, four measures are estimated: true positive (TP), true negative (TN), false positive (FP), and false negative (FN). These values are used to assess the performance of the proposed classifier.

$$Accuracy = \frac{TP + TN}{TP + TN + FP + FN} * 100\% \tag{2}$$

$$Sensitivity = \frac{TP}{TP + FN} * 100\% \quad (3)$$

$$Specificity = \frac{TN}{TN + FP} * 100\% \quad (4)$$

$$Precision = \frac{TP}{TP + FP} * 100\% \quad (5)$$

$$F - measure = 2 * \frac{Precision * Sensitivity}{Precision + Sensitivity} * 100\% \quad (6)$$

Table 2 compares the performance of our classifier with Triplet-SVM, MiPred, Virgo and EumiR. The proposed classifier outperforms the other classifiers for all the five adopted performance indices.

Table 2. Performance of our classifier versus Triplet-SVM, MiPred, Virgo and EumiR

Classifier	Accuracy	F-measure	Specificity	Precision	Sensitivity	AUC
Triplet-SVM	78.5%	63.8%	89.2%	72.59%	57%	0.91
Mipred	89%	81.7%	96.7%	91.7%	73.6%	0.89
Virgo	73.7%	66.4%	73.4%	58.9%	76.2%	0.72
EumiR	74.1%	66.36%	72.2%	58.3%	77%	0.73
Our RF-based classifier	91.3%	86%	97.2%	93.4%	79.6%	0.92

The receiver operating characteristic curve (ROC), is a plot which evaluates the predictive ability of a binary classifier. ROC curves show how the number of correctly classified positive examples varies relative to the number of incorrectly classified negative examples. Fig. 2 shows the ROC curve of our RF-based classifier versus the four adopted classifiers. Our classifier gives a significant performance progress comparing to Triplet-SVM, Virgo and EumiR. Our RF-based classifier, in general, is consistently better than MiPred classifier. It performs better in the more conservative region of the graph, i.e. it is better at identifying likely positives miRNAs than at identifying likely negatives miRNAs. The area under the ROC curve (AUC) is a very widely used measure of classifier performance. Table 2 displays the AUC of each adopted classifier versus our RF-based classifier which indicates that our RF-based classifier is consistently better.

3.3 Comparison with SVM-Based Classifier

The SVM classifier has been trained on the same training data set and tested on the same testing data set using the same set of features. It has been implemented by libSVM 3.17 library [25]. Table 3 compares the performance of our classifier with SVM-based classifier. The results confirm that Random Forests perform

Table 3. Performance of our RF-based classifier versus SVM-based classifier

Classifier	Accuracy	F-measure	Specificity	Precision	Sensitivity	AUC
Our RF-based classifier	91.3%	86%	97.2%	93.4%	79.6%	0.92
SVM-based classifier	88.33%	79%	98.7%	96.29%	67%	0.91

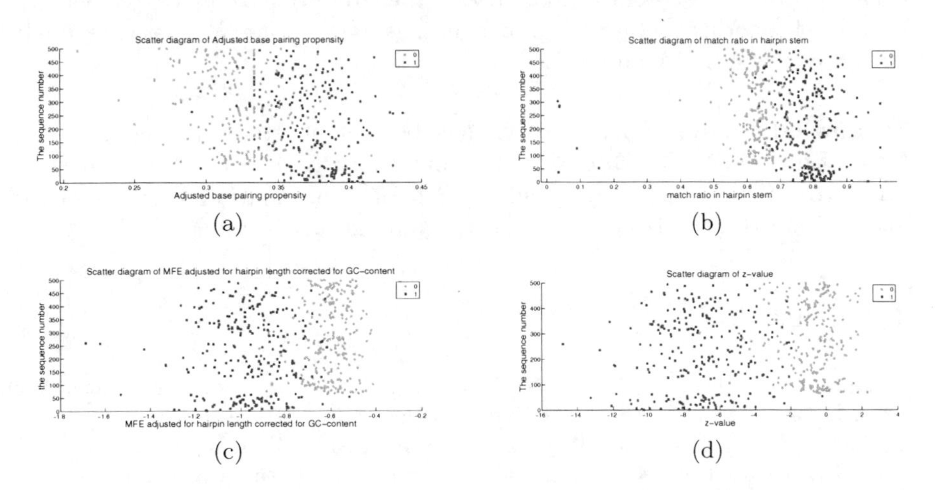

Fig. 1. The selected features scatter diagrams. (a) Scatter diagram of adjusted base pairing propensity. (b) Scatter diagram of match ratio in hairpin stem. (c) Scatter diagram of MFE adjusted for the hairpin length corrected for GC-content. (d) Scatter diagram of z-value.

• indicates feature values for pseudo miRNA samples
■ indicates feature values for real miRNA samples.

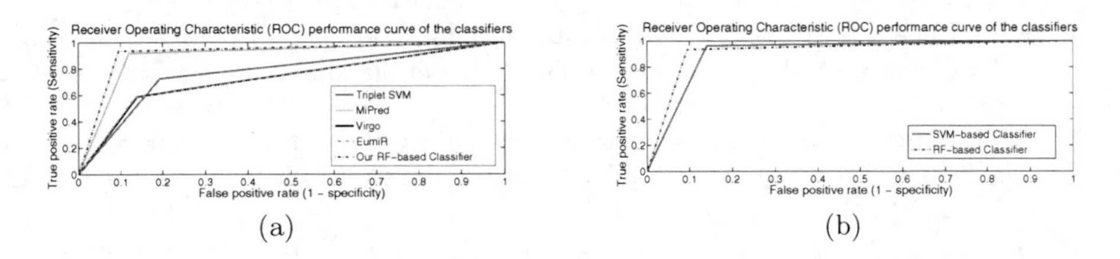

Fig. 2. Receiver Operating Characteristic performance curve. (a) RF-based classifier versus the adopted classifiers. (b) RF-based classifier versus SVM-based classifier.

well relative to SVMs [2]. Fig. 2 shows the ROC curve of our RF-based classifier versus the SVM-based classifier.

SVM-based classifier is better in specificity and precision metrics due to the margin maximization principle adopted by SVM, but this comes at the expense of accuracy and f-measure metrics that measure how well a binary classification test correctly identifies the true data.

4 Conclusion

We have developed a computational tool for miRNA prediction. The adopted technique succeeds in getting better results than Triplet-SVM, MiPred, Virgo and EumiR. Despite the scarceness of experimentation on the use of RF classifiers in miRNA prediction compared to SVM, the present work gives evidence that RF; as well as other meta-models such as ensemble methods; constitute a better alternative for miRNA mining.

Acknowledgments. This research has been supported by the Ministry of Higher Education (MoHE) of Egypt through an PH.D. fellowship. Our sincere thanks to Shibuya laboratory (Graduate School of Information Science and Technology, University of Tokyo) for hosting, guidance and support.

References

1. Altschul, S., Gish, W., Miller, W., Myers, E., Lipman, D., et al.: Basic local alignment search tool. Journal of Molecular Biology 215, 403–410 (1990)
2. Jiang, P., Wu, H., Wang, W., Ma, W., Sun, X., Lu, Z.: Mipred: classification of real and pseudo microrna precursors using random forest prediction model with combined features. Nucleic Acids Research 35, W339–W344 (2007)
3. Lim, L., Lau, N., Weinstein, E., Abdelhakim, A., Yekta, S., Rhoades, M., Burge, C., Bartel, D.: The micrornas of caenorhabditis elegans. Genes & Development 17, 991 (2003)
4. Lai, E., Tomancak, P., Williams, R., Rubin, G.: Computational identication of drosophila microrna genes. Genome Biology 4 (2003)
5. Bonnet, E., Wuyts, J., Rouz, P., Van de Peer, Y.: Detection of 91 potential conserved plant micrornas in arabidopsis thaliana and oryza sativa identies important target genes. Proc. Natl. Acad. Sci. USA 101, 11511–11516 (2004)
6. Jones-Rhoades, M., Bartel, D.: Computational identification of plant micrornas and their targets, including a stress-induced mirna. Molecular Cell 14, 787–799 (2004)
7. Ng, K., Mishra, S.: De novo svm classification of precursor micrornas from genomic pseudo hairpins using global and intrinsic folding measures. Bioinformatics 23, 1321–1330 (2007)
8. Sewer, A., Paul, N., Landgraf, P., Aravin, A., Pfeffer, S., Brownstein, M., Tuschl, T., van Nimwegen, E., Zavolan, M.: Identication of clustered micrornas using an ab initio prediction method. BMC Bioinformatics 6 (2005)
9. Xue, C., Li, F., He, T., Liu, G., Li, Y., Zhang, X.: Classification of real and pseudo microrna precursors using local structure-sequence features and support vector machine. BMC Bioinformatics 6 (2005)
10. Zheng, Y., Hsu, W., Li Lee, M., Wong, L.: Exploring essential attributes for detecting microRNA precursors from background sequences. In: Dalkilic, M.M., Kim, S., Yang, J. (eds.) VDMB 2006. LNCS (LNBI), vol. 4316, pp. 131–145. Springer, Heidelberg (2006)
11. Batuwita, R., Palade, V.: Micropred: effective classification of pre-mirnas for human mirna gene prediction. Bioinformatics 25, 989–995 (2009)

12. Pasaila, D., Mohorianu, I., Sucila, A., Pantiru, S., Ciortuz, L.: Yet another svm for mirna recognition: yasmir. Technical report, Citeseer (2010)
13. Shiva, K., Faraz, A., Vinod, S.: Prediction of viral microrna precursors based on human microrna precursor sequence and structural features. Virology Journal 6 (2009)
14. Hofacker, I., Fontana, W., Stadler, P., Bonhoeffer, L., Tacker, M., Schuster, P.: Fast folding and comparison of rna secondary structures. Monatshefte für Chemie/Chemical Monthly 125, 167–188 (1994)
15. Griffiths-Jones, S.: The microrna registry. Nucleic Acids Research 32, D109–D111 (2004)
16. Pruitt, K., Maglott, D.: Refseq and locuslink: Ncbi gene-centered resources. Nucleic Acids Research 29, 137–140 (2001)
17. Bonnet, E., Wuyts, J., Rouzé, P., Van de Peer, Y.: Evidence that microrna precursors, unlike other non-coding rnas, have lower folding free energies than random sequences. Bioinformatics 20, 2911–2917 (2004)
18. Freyhult, E., Gardner, P.P., Moulton, V.: A comparison of rna folding measures. BMC Bioinformatics 6, 241 (2005)
19. Shannon, C.E.: A mathematical theory of communication. ACM SIGMOBILE Mobile Computing and Communications Review 5, 3–55 (2001)
20. van der Burgt, A., Fiers, M.W., Nap, J.P., van Ham, R.C.: In silico mirna prediction in metazoan genomes: balancing between sensitivity and specificity. BMC Genomics 10, 204 (2009)
21. Loong, S.N.K., Mishra, S.K.: Unique folding of precursor micrornas: Quantitative evidence and implications for de novo identification. Rna 13, 170–187 (2007)
22. Breiman, L.: Random forests. Machine Learning 45, 5–32 (2001)
23. Hall, M., Frank, E., Holmes, G., Pfahringer, B., Reutemann, P., Witten, I.H.: The weka data mining software: an update. ACM SIGKDD Explorations Newsletter 11, 10–18 (2009)
24. Griffiths-Jones, S., Grocock, R., Van Dongen, S., Bateman, A., Enright, A.: mirbase: microrna sequences, targets and gene nomenclature. Nucleic Acids Research 34, D140–D144 (2006)
25. Chang, C.C., Lin, C.J.: LIBSVM: A library for support vector machines. ACM Transactions on Intelligent Systems and Technology 2, 27:1–27:27 (2011), Software available at http://www.csie.ntu.edu.tw/~cjlin/libsvm

Designing an Ontology Tool for the Unification of Biofilms Data

Ana Margarida Sousa[1], Maria Olívia Pereira[1],
Nuno F. Azevedo[2], and Anália Lourenço[1,3]

[1] CEB - Centre of Biological Engineering, University of Minho,
Campus de Gualtar, 4710-057 Braga, Portugal
{anamargaridasousa,mopereira}@deb.uminho.pt
[2] LEPABE – Dep. of Chemical Engineering, Faculty of Engineering, University of Porto,
Rua Dr. Roberto Frias, s/n, 4200-465 Porto, Portugal
nazevedo@fe.up.pt
[3] ESEI - Escuela Superior de Ingeniería Informática, Edificio Politécnico, Campus
Universitario As Lagoas s/n, Universidad de Vigo, 32004 Ourense, Spain
analia@ceb.uminho.pt, analia@uvigo.es

Abstract. The description of biofilm features presents a conceptual and practical challenge. Biofilm studies often encompass multidisciplinary approaches from Biology, Chemistry, Medicine, Material Science and Engineering, among other fields. Standardising biofilm data is essential to be able to accomplish large-scale collaborative and complementary analysis. To define a common standard format to exchange the heterogeneous biofilm data, it is first necessary to define a set of minimum information for the documentation of biofilm experiments. Then, data should be organised and semantically integrated. This paper describes the first ontology designed to share structured vocabulary for the annotation of the general biofilm experimental workflow – the Biofilm Ontology (BO). This ontology is intended for the broad research community, including bench microbiologists, clinical researchers, clinicians, curators and bioinformaticians.

Keywords: Biofilms, minimum information about a biofilm experiment, ontology, functional modelling.

1 Introduction

Biofilms are organised communities of microorganisms attached to each other and/or to a surface, and involved in a self-produced polymeric matrix [1]. Biofilms are ubiquitous to natural, clinical and industrial environments and thus, their ecological impact is transversal to many economic and social areas [2].

The study of biofilms is a multidisciplinary knowledge field, at the crossroads of Biology, Chemistry, Medicine, Material Science and Engineering, among others. Biofilms are extremely complex environments and their function resembles that of a multicellular organism [1]. Depending on the biofilm ecological niche, and the particular relationships established among species, microorganisms have different

J. Sáez-Rodríguez et al. (eds.), *8th International Conference on Practical Appl. of Comput. Biol. & Bioinform. (PACBB 2014)*, Advances in Intelligent Systems and Computing 294,
DOI: 10.1007/978-3-319-07581-5_5, © Springer International Publishing Switzerland 2014

metabolic and genetic profiles [3]. Notably, each biofilm presents a unique biological signature. Therefore, biofilm studies often encompass multidisciplinary approaches to the characterisation of the structure and activity of the sessile community. For example, these studies may include mass spectrometry, microscopy, flow cytometry, antimicrobial susceptibility, respiratory activity, metabolomic, proteomic, transcriptomic and genomic techniques. Consequently, their outputs tend to vary greatly in type and semantics, including numeric, image, and spectra data (Table 1).

Table 1. Characteristic types of biofilm data and techniques used

Type of biofilm data	Technique
Numeric	Spectrophotometric methods (CV, XTT, ATP detection, Lowry protein assay, Dubois assay and Alamar blue), CFU, antimicrobial susceptibility
Images	Microscopy techniques (SEM, TEM, CLSM, FISH), colony morphology characterisation, gram-staining, proteomic techniques (SDS-PAGE, 2D electrophoresis)
Spectra	Mass spectrometry, MALDI-TOF, chromatography

Legend: CV – crystal violet, XTT - 2,3-Bis-(2-Methoxy-4-Nitro-5-Sulfophenyl)-2H-Tetrazolium-5-Carboxanilide; CFU – colony-forming units; SEM – scanning electron microscopy; TEM – transmission electron microscopy, CLSM – confocal microscopy, FISH - fluorescence *in situ* hybridization, SDS-PAGE - sodium dodecyl sulfate polyacrylamide gel electrophoresis; MALDI-TOF - Matrix Assisted Laser Desorption/Ionization Time of Flight.

Currently, the Minimum Information About a Biofilm Experiment (MIABiE) international consortium (http://miabie.org) is working on the definition of guidelines to document biofilm experiments and the standardisation of the nomenclature in use. Specifically, this consortium has established the minimum set of information be recorded and published about an experiment in order for the procedure and results to be unambiguously and comprehensive interpreted [4]. These issues were discussed with field experts - at the Eurobiofilms 2013 meeting in Ghent, Belgium (9-12 September 2013) - as means to accommodate for as much of this complexity and variability as possible, and anticipate new requirements resulting from the increasing use of high-throughput methods. The creation of controlled vocabulary in support of the description of biofilm studies is inherent to this standardisation effort. Controlled vocabulary based on well-engineered ontologies may support powerful querying and computational analysis tools. Therefore, this paper brings forward the development of the first ontology on Biofilms, identifying the main areas of terminology to be accounted for as well as discussing its integration with other upper and domain ontologies, such as ontology for scientific experiments (EXPO), Chemical Entities of Biological Interest (ChEBI), and Functional Genomics Ontology (FuGO). The adequateness and extensibility of the new ontology, named Biofilm Ontology (BO), has been discussed with MIABiE members and other domain experts, and validated against different studies in the BiofOmics, the centralised database on biofilms experiments [5]. Here, we introduce BO and exemplify some of its main descriptive abilities.

2 The Organising Principles of the Biofilm Ontology

There is no standardised methodology for building ontologies. However, Open Biological and Biomedical Ontologies (OBO) foundry has introduced some useful guidelines and principles regarding the different stages of the ontology development life-cycle that helped the construction of the BO [6]. In particular, the organisation of the BO was based on the following main criteria:

- BO is restricted to the biofilm knowledge domain and, therefore, it contains just model concepts and relations that are relevant to the representation of biofilm data;
- BO should be used for annotating data in databases and for textual documentation as such, it should be understandable to people and unambiguously interpreted by software;
- BO development should be pragmatic, that is as new devices, techniques or applications arise, it should be possible to integrate new branches without affecting the existent ontology structure;
- any biofilm experiment should be comprehensively described by a combination of BO instances;
- whenever possible, terms should have a synonyms list to avoid misinterpretations and to enable consistent data curation and repositories searching;

Then, BO development followed the typical ontology life-cycle previously described [7]. The purpose and scope of the ontology were well identified and centred on the Biofilm domain, following MIABiE directives. Knowledge acquisition relied on varied sources of information, namely: discussion of the BO with other Biofilm experts, metadata associated to the experiments in BiofOmics database [5], research papers on biofilms, and some learning from other bio-ontologies. The construction of the ontology undertook the conceptualisation and integration of the biofilms concepts identified, formally defining their properties, meaning and relationships with other concepts.

Top-down and bottom-up approaches were combined. First, the top-down approach led to the insertion of the BO into a generic upper ontology. Generic ontologies describe general and domain-independent knowledge, aiming to avoid the duplication of terms related to template structures typical of scientific experiments of any research field. As such, BO would focus only on the specificities of biofilms studies, delegating general experimental characterisation to upper ontologies. Then, the bottom-up approach was implemented to gather and organise the biofilms-specific concepts into ontological instances and establish relations among them. Since Biofilms are a multidisciplinary knowledge field, many concepts will be cross-linked to ontologies in related domains, avoiding term duplication and enforcing data interoperation across platforms and resources, in benefit of broad research community.

BO was represented in formal language and comprehensively documented to promote its use in databases, bioinformatics tools and other resources created in support of Biofilm research. Its consistency, completeness and conciseness was evaluated by practical exercise, i.e. applying its terms to the description of published studies, and by discussing the ontology with field experts at international conferences, such as the Eurobiofilms 2013 meeting in Ghent, Belgium (9-12 September 2013).

3 The Structure of the Biofilm Ontology

BO development follows the principles of the Open Biological and Biomedical Ontologies (OBO) foundry [6]. Terms are organised in a hierarchical structure with four main branches: biofilm model, biofilm growth conditions, biofilm removal and biofilm characterisation (Fig 1). These branches represent the main steps/components of a typical biofilms experiment. Next, we will provide a brief overview of each of these branches.

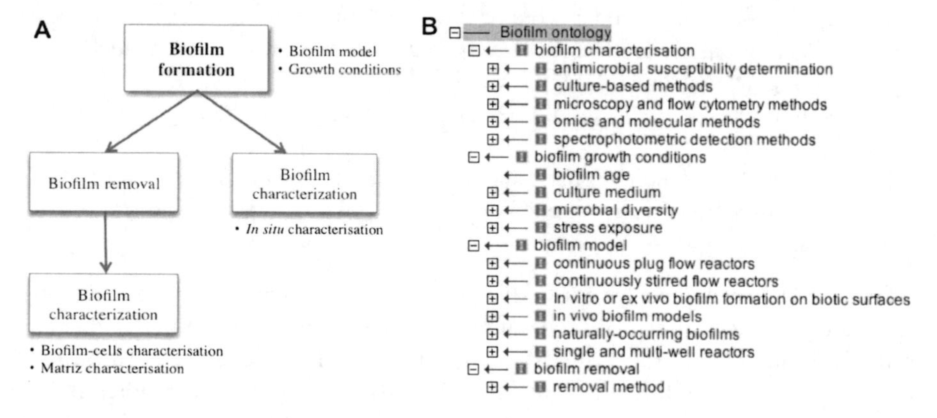

Fig. 1. (A) Typical biofilm experiment workflow; (B) Hierarchical structure of the BO. The four main branches, which compose BO, 'biofilm characterisation', 'biofilm growth conditions', 'biofilm model', and 'biofilm removal' represents the typical steps of a biofilm experiment. Each branch is composed by multiple child-terms that, by its turns, have also their descendent terms as denoted by the signal '+'. All the BO terms are related among them through the "is_a" relationship type.

Biofilm Model

This BO branch intends to encompass all systems in which biofilms can be formed, analysed or retrieved, including the models of naturally-occurring biofilms, single and multi-well reactors, continuously stirred flow reactors, continuous plug flow reactors, *in vivo* biofilms, and *in vitro* or *ex vivo* cultivation of biofilms on biotic surfaces [8].

Biofilm Growth Conditions

Besides identifying the model used to form the biofilms, the description of the experiment must also specify the operational conditions. These conditions include temperature, culture medium and supplements, hydrodynamic conditions (static or dynamic conditions), pH, oxygen availability (aerobic, anaerobic or microaerophilic conditions), time of biofilm growth and maturation, and stress conditions.

Biofilm Removal

Typically, the characterisation of biofilm cells requires the removal of the biofilms from surfaces where they were formed and further cell detachment. This is a critical step that may alter the physiological state of cells and bias the results of the experiment. This module of the BO collects the methods or techniques commonly used, such as sonication or scrapping, and the operation conditions, such as the device used, period of time of biofilm removal (for example, 5 cycles of 30 s of sonication) and solvent used to collect biofilm-cells (for example, water, phosphate buffer or other buffer solution).

Biofilm Characterisation

This branch describes the methods according to the type of analysis conducted, as follows: culture-based methods, such as CFU counting and colony morphology characterisation; non culture-based methods, such DAPI, CV and FISH; microscopy (SEM, TEM and CLSM) and flow cytometry methods; spectrophotometric methods, such as XTT, ATP detection and Lowry protein assay; "Omics" and molecular methods, such proteomics, genomics; and antimicrobial susceptibility testing, such determination of the minimum inhibitory and bactericidal concentrations and the minimum biofilm eradication concentration. Techniques that are common to other research areas are included, but BO delegates on their definition into well-established domain-specific ontologies and controlled vocabularies (Table 2).

Table 2. Ontologies or controlled vocabularies (CV) relevant in biofilm field and cross-reference by BO

Ontology/ controlled vocabulary	Source
Chemical entities of biological Interest (CHEBI)	http://obo.cvs.sourceforge.net/obo/obo/ontology/chemical/chebi.obo
Colony morphology characterization (CMO)	http://mibbi.sourceforge.net/projects/MIABiE.shtml
Gene ontology (GO)	http://obo.cvs.sourceforge.net/obo/obo/ontology/genomic-proteomic/gene_ontology.obo
Functional Genomics Investigation Ontology (FuGO)	http://sourceforge.net/projects/fugo/
MALDI imaging ontology (IMS)	http://www.maldi-msi.org/download/imzml/imagingMS.obo
PSI-Mass Spectrometry CV (MS)	http://psidev.cvs.sourceforge.net/viewvc/psidev/psi/psi-ms/mzML/controlledVocabulary/psi-ms.obo
PRIDE CV	http://code.google.com/p/ebi-pride/source/browse/trunk/pride-core/schema/pride_cv.obo
Protein ontology (PRO)	http://obo.cvs.sourceforge.net/obo/obo/ontology/genomic-proteomic/pro.obo
PSI-Sample Processing and Separations (SEP)	https://psidev.svn.sourceforge.net/svnroot/psidev/psi/sepcv/trunk/sep.obo

4 Cooperation and Integration with Other Bio-Ontologies

There are several upper ontologies that may assist BO in describing generic experimental setup and procedures. None of these ontologies is an ideal representation of general knowledge and, therefore, there must be a compromise between the "imperfection" degree of the upper ontology and the practical needs of BO, as practical domain ontology. Here, the Suggested Upper Merged Ontology (SUMO) [available at http://www.ontologyportal.org/], proposed by the IEEE Standard Upper Ontology Working Group, was selected to formalise concepts that are seen as meta, generic or abstract to a broad range of domain areas (e.g. medical, financial, engineering).

BO was also integrated with the ontology for scientific experiments (EXPO), which includes the fundamental concepts about experiment design, methodology, and the representation of results that are domain independent [9]. For example, any experiment has a goal ('EXPO:ExperimentalGoal') and it aims to test a hypothesis ('EXPO:ExperimentalHypothesis') that results may confirm it ('EXPO:FactSupport') or reject it ('EXPO:RejectSupport').

BO integration and cooperation with other ontologies is not limited to generic or top-level ontologies. BO also cooperates with other domain ontologies. BO is focused on biofilm-specific data issues and, therefore, data coming from other knowledge domains, such as flow cytometry, proteomic techniques or microarrays, should be annotated according to the data standards of the respective consortia (Table 2).

5 Applications of BO

The availability of BO allows consistent documentation of biofilm experiments, and facilitates large-scale and computer-aided data processing and analysis. Ultimately, the BO aims to support the comparison of inter-laboratory experiments and the generation of new experimental hypotheses.

BiofOmics [5], the first ever public repository dedicated to biofilm experiments, is a biological resource that takes advantage of this ontology. At first, BO will assist in intelligent experiment screening and comparison, recognising experiments with similar setup and goals of analysis. By using the BO, researchers may compare their own data against other datasets in the repository. Specifically, by means of ontology-based meta-analysis it is possible to compute and score correlations between datasets and a given ontology term, e.g. "biofilm susceptibility", "matrix composition", "viable cells", or "biofilm mass". This sort of analysis across published biofilm data may provide insights into important research questions, such as: 1) the identification of consistently expressed genes by microorganisms forming biofilms; 2) the description of the mechanisms of resistance and persistence of biofilms in numerous circumstances, e.g. resistance to antibiotic treatments and to the host immune defences, persistence to environmental stress, including pH, microbial predation, or starvation; 3) the identification of microbial biomarkers that may guide the development of new drug therapeutics. In addition, the BO represents a valuable resource to data curation. With its support, curators are able to interpret and validate richer and more detailed experimental setups and results, assisting data submitters in the comprehensive description of their data. The utility of BO is better demonstrated with some examples of descriptions of biofilm experiments extracted from literature (Fig. 2). Typically, studies including biofilm experiments describe the conditions in which biofilms are formed and the subsequent analysis performed to characterise the community under study.

Example #1

P. aeruginosa ATCC 10145 and *S. aureus* ATCC 25293 were allowed to form 24-h-old biofilms as described previously. (...) Afterwards, each bacterial suspension was transferred to 6-well polystyrene plate, where biofilms were developed aerobically on a horizontal shaker (120 rpm) at 37 °C for 24 h. Then, biofilms were scrapped into sterile water, homogenized, being the biofilm-cells serially diluted with sterile ultrapure water and spread on solid media plates." Extracted from Sousa et al. (2013). *Improvements on colony morphology identification towards bacterial profiling*. J Microbiol Methods

BO main branches
Biofilm model
Biofilm growth conditions
Biofilm removal
Biofilm characterisation

Example #2

"Biofilm samples were analyzed by DGGE with a D-Code universal mutation detection system (Bio-Rad, Hemel Hempstead, United Kingdom). Polyacrylamide (8%) gels (16 by 16-cm, 1 mm deep) were run with 1 TAE buffer diluted from 50 TAE buffer (40 mM Tris base, 20 mM glacial acetic acid, 1 mM EDTA)." Extrated from McBain et al. (2004). *Effects of Quaternary-Ammonium-Based Formulations on Bacterial Community Dynamics and Antimicrobial Susceptibility*. Applied and Environmental Microbiology

Example #3

"Briefly, biofilms were grown for 24 h on the pegs of the CBD and subsequently stained with acridine orange. The pegs were examined using a Leica DM IRE2 spectral confocal and multiphoton microscope with a Leica TCS SP2 acoustic optica beam splitter (AOBS) (Leica Microsystems). (...) Image stacks were processed using Imaris 6.3.1 (Bitplane) to generate images for publication." Extracted from Workentine et al. (2010). *Phenotypic and metabolic profiling of colony morphology variants evolved from Pseudomonas fluorescens biofilms*. Environmental Microbiology

Fig. 2. Examples of the use of BO terms in the description of biofilms studies

6 Conclusions

Biofilms remains a major challenge to several scientific fields. The global context of biofilms makes data integration essential for sharing and retrieval the scientific information that semantic annotation using to ontologies have a crucial role. Despite there are several ontologies of specific knowledge domains, it was verified a lack of semantic resources for biofilm-specific terms. This paper describes the development of the Biofilm Ontology (BO), which aims to address the issue of representing biofilm information, achieving interoperability between scientific areas, research groups and laboratories.

In this paper, it was presented the current state of BO and how it is related to other ontologies, upper and domain ontologies, relevant for the biofilm domain. The actual version covers all steps of a typical biofilm experiment workflow, including biofilm formation (composed by the branches 'biofilm model' and 'biofilm growth conditions'), 'biofilm removal' and 'biofilm characterisation'. However, BO is prepared for future alterations. BO design contemplates existing procedures and dependencies, but it is flexible to account for future extensions. Through active dissemination and group discussion, the biofilm community is invited to collaborate in the population and update of BO.

Acknowledgements. The authors thank the projects PTDC/SAU-ESA/646091/2006/ FCOMP-01-0124-FEDER-007480FCT; PEst-OE/EQB/LA0023/2013; "BioHealth - Biotechnology and Bioengineering approaches to improve health quality", NORTE-07-0124-FEDER-000027, co-funded by the Programa Operacional Regional do Norte (ON.2 – O Novo Norte), QREN, FEDER; RECI/BBB-EBI/0179/2012 - Consolidating Research Expertise and Resources on Cellular and Molecular Biotechnology at CEB/IBB, FCOMP-01-0124-FEDER-027462, FEDER; and the Agrupamento INBIOMED from DXPCTSUG-FEDER unha maneira de facer Europa (2012/273). The research leading to these results has received funding from the European Union's Seventh Framework Programme FP7/REGPOT-2012-2013.1 under grant agreement n° 316265, BIOCAPS. This document reflects only the author's views and the European Union is not liable for any use that may be made of the information contained herein. The authors also acknowledge PhD Grant of Ana Margarida Sousa SFRH/BD/72551/2010.

References

1. Lopez, D., Vlamakis, H., Kolter, R.: Biofilms. Cold Spring Harb. Perspect Biol. 2(7), a000398 (2010)
2. Yang, L., et al.: Current understanding of multi-species biofilms. Int. J. Oral Sci. 3(2), 74–81 (2011)
3. Stewart, P.S., Franklin, M.J.: Physiological heterogeneity in biofilms. Nature Reviews Microbiology 6(3), 199–210 (2008)
4. Lourenço, A., et al.: Minimum information about a biofilm experiment (MIABiE): standards for reporting experiments and data on sessile microbial communities living at interfaces. Pathogens and Disease (in press, 2014)
5. Lourenço, A., et al.: BiofOmics: a Web platform for the systematic and standardized collection of high-throughput biofilm data. PLoS One 7(6), e39960 (2012)
6. Smith, B., et al.: The OBO Foundry: coordinated evolution of ontologies to support biomedical data integration. Nat. Biotechnol. 25(11), 1251–1255 (2007)
7. Stevens, R., Goble, C.A., Bechhofer, S.: Ontology-based knowledge representation for bioinformatics. Brief Bioinform. 1(4), 398–414 (2000)
8. Buckingham-Meyer, K., Goeres, D.M., Hamilton, M.A.: Comparative evaluation of biofilm disinfectant efficacy tests. Microbiol. Methods 70(2), 236–244 (2007)
9. Soldatova, L.N., King, R.D.: An ontology of scientific experiments. J. R Soc. Interface 3(11), 795–803 (2006)

BEW: Bioinformatics Workbench for Analysis of Biofilms Experimental Data

Gael Pérez Rodríguez[1], Daniel Glez-Peña[1], Nuno F. Azevedo[2], Maria Olívia Pereira[3], Florentino Fdez-Riverola[1], and Anália Lourenço[1,2]

[1] ESEI - Escuela Superior de Ingeniería Informática, Edificio Politécnico, Campus Universitario As Lagoas s/n, Universidad de Vigo, 32004 Ourense, Spain
[2] LEPABE – Dep. of Chemical Engineering, Faculty of Engineering, University of Porto, Rua Dr. Roberto Frias, s/n, 4200-465 Porto, Portugal
[3] IBB - Institute for Biotechnology and Bioengineering, Centre of Biological Engineering, University of Minho, Campus de Gualtar, 4710-057 Braga, Portugal
gprodriguez2@esei.uvigo.es, {dgpena,riverola,analia}@uvigo.es, nazevedo@fe.up.pt, mopereira@deb.uminho.pt, analia@ceb.uminho.pt

Abstract. Biofilms research has evolved considerably in the last decade and is now generating large volumes of heterogeneous data. MIABiE, the international initiative on Biofilms, is devising guidelines for data interchange, and some databases provide access to biofilms experiments. However, the field is lacking appropriate bioinformatics tools in support of increasing operational and analytical needs. This paper presents a flexible and extensible open-source workbench for the operation and analysis of biofilms experiments, as follows: (i) the creation of customised experiments, (ii) the collection of various analytical results, following community standardisation guidelines and (iii) on-demand reporting and statistical evaluation.

Keywords: Biofilms, standard operating procedures, data interchange, data analysis.

1 Introduction

Biofilms are surface-attached cellular agglomerates that are widespread in Nature and exhibit great abilities to adapt to environmental changes [1, 2]. Their actions can be seen as beneficial or detrimental to humans, depending on their ecological impact and our ability to act upon them [3, 4]. Notably, the resistance and resilience of biofilms attracts much attention from the biomedical research community due to the continuous emergence of multi-resistant strains in clinical settings, which are rendering conventional antibiotics ineffective [5, 6].

Biofilms research has witnessed a considerable development in the last decade. Conventional microbiological experimentation is giving place to large-scale multidisciplinary experimentation [7]. Cell viability, biomass formation, respiratory activity, morphological characterisation, and transcriptome and proteome profiling are just some examples of the methods of analysis now in use. Moreover, and due to the

J. Sáez-Rodríguez et al. (eds.), *8th International Conference on Practical Appl. of Comput. Biol. & Bioinform. (PACBB 2014)*, Advances in Intelligent Systems and Computing 294,
DOI: 10.1007/978-3-319-07581-5_6, © Springer International Publishing Switzerland 2014

diversity of settings where biofilms can be found, the environmental scenarios recreated in the lab can be quite different and sometimes challenging to implement. Repeatability, ruggedness and reproducibility tests are therefore conducted to ensure the quality of the acquired data and, in particular, the ability to compare results between laboratories [8, 9].

The MIABiE initiative (http://miabie.org), encompassing an international body of Biofilms experts, is working on the definition of guidelines to document biofilms experiments and the standardisation of the nomenclature in use. Biofilms databases such as BiofOmics (http://biofomics.org) [10] and MorphoCol (http://morphocol.org) [11] are endorsing these guidelines and making experimental data freely available. However, the community is missing tools to take the best advantage of these resources, and it is notable the inexistence of bioinformatics tools dedicated to biofilms data operation and analysis. In particular, unstructured data operation, using Excel or similar tools, compromises data standardisation and thus, any form of computer-aided processing and analysis.

This paper presents the first software tool dedicated to Biofilms: the Biofilms Experiment Workbench (BEW). The primary aim is to cover for primary intra-laboratory data collection and analysis necessities. Previous work on data standardisation and computerised data structuring [11], now complemented by MIABiE guidelines, established the starting point of development. A specialised markup language is defined now as grounds to document experiments, and to effectively promote data interchange across resources and software tools. Moreover, the application is developed with AIBench, an open-source Java desktop application framework [14], to ensure a flexible, cross-platform and interoperable development suitable to sustain future interactions with other Biofilms-related tools.

The next sections detail the markup language as well as the design and main functionalities of BEW.

2 Biofilms Markup Language

The current inability to exchange experiments between laboratories has its roots in the lack of a common format for describing Biofilms experiments. The modelling of Biofilm experiments is challenged by the complexity and variability of the studies. Studies may vary widely in aspects as important as: the conditions tested (related to the goals of analysis), the microorganisms studied (with implicit growth and other biological specifics), the methods of analysis used (data is only comparable for similar methods), specific data pre-processing (e.g. the calculation of dilution rates or log reductions), and the number of replicates and reproductions performed to ensure the validity of the study.

These issues were discussed with field experts - at the Eurobiofilms 2013 meeting in Ghent, Belgium (9-12 September 2013) - as means to accommodate for as much of this complexity and variability as possible, and anticipate new requirements resulting from the increasing use of high-throughput methods. Then, XML, the eXtensible Markup Language [12], was elected to formalise the data structure because of its portability and widespread acceptance as a standard data language for Bioinformatics [13].

At the end, we came by a very generic skeleton for the description of a biofilm experiment that considers four major conceptual elements:

- identification and authorship, including the name of the institution and the authors of the experiment, the title and a short description of the experiment, and any associated publications;
- method of analysis, which comprises a set of test conditions and a number of resulting data series;
- condition, i.e. a experimental condition tested by a given method of analysis (e.g. a value of temperature or a dose of an antimicrobial agent);
- data series that describe the data obtained by a method of analysis for a given set of conditions.

The definition of an experiment in XML simply consists of lists of one or more of these various components (Fig. 1).

```
<?xml version="1.0" encoding="UTF-8" standalone="no"?>
<ns0:bml xmlns:ns0="" xmlns:xsi="" xsi:schemaLocation="">
  <experiment contact="" date="" experimentName="" organization="" publication="">
    <authors><![CDATA[]]></authors>
    <notes><![CDATA[]]></notes>
    <methods>
      <method methodName="" methodUnits="">
        <conditions>
          <condition conditionName="" conditionUnits="">...</condition>
          ...
        </conditions>
        <dataSeriesSet>
          <dataSerie>
            <conditionValues>...#...</conditionValues>
            <measurements>...#...</measurements>
          </dataSerie>
          ...
        </dataSeriesSet>
      </method>
      ...
    </methods>
    <constantConditions>
      <constantCondition condition="" conditionUnits="" conditionValue=""/>
      ...
    </constantConditions>
  </experiment>
</ns0:bml>
```

Fig. 1. Skeleton of the definition of a biofilm experiment in XML format, showing all possible top-level elements

3 BEW: The Biofilms Workbench

BEW is a desktop-based application dedicated to the management and analysis of the data resulting from Biofilms experiments. The application was developed with AIBench, an open-source Java desktop application framework for scientific software development in the domain of translational biomedicine [14]. As illustrated in Fig. 2, BEW incorporates a plug-in for the R statistical computing tool (http://www.r-project.org/) and the libraries JFreeChart for data plotting (http://www.jfree.org/jfreechart/), JXL to read and write Excel sheets (http://jexcelapi.sourceforge.net/), and JSoup for working with

HTML documents (http://jsoup.org/). These third-parties support main data processing and data analysis operations whilst enable future adaptation or extension (notably regarding statistical analysis).

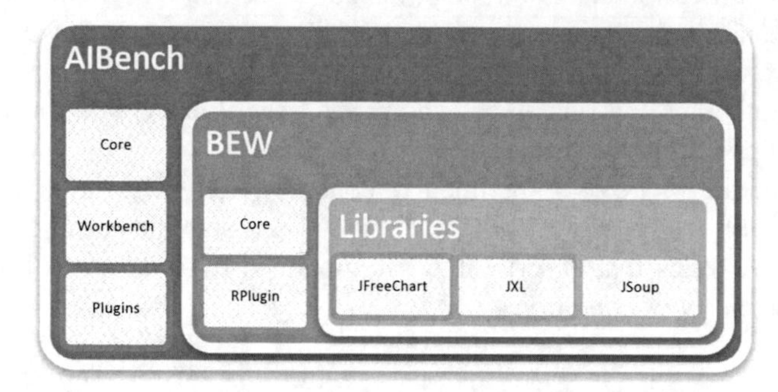

Fig. 2. General architecture of the software components in BEW

The next subsections describe the core management and analysis functionalities currently provided by the software.

3.1 Experiment Management

In accordance to standardisation guidelines and to cope with data legacy, BEW supports both .xml and .xls data representation formats. The XML is the native format (as described in section 2) and wrapping functionalities are provided to import/export experiments from/to the standardised and hierarchy-based Excel format [11].

The experiment should be comprehensively described in order to guarantee its community-wide unambiguous interpretation. First, the experiment is "identified" in terms of authorship (author and institution), author's summary and any associated publications. Then, the experimental setup is detailed and, more specifically, the overall settings or constant conditions, and the conditions tested by each of the methods of analysis. Given the wide range of analyses now conducted for biofilms, user interface was made as flexible and intuitive as possible to enable the construction of variable nature and extent data "worksheets" (Fig. 3). In particular, metadata standardisation and systematic data structuring were encapsulated in logical selections to detach the user from unnecessary computational details.

3.2 Statistical Analysis

BEW is equipped with a powerful analytical component that supports on-demand construction of data plots as well as statistical data testing.

Plots and statistical tests usually present in Biofilms publications served as references of expected functionalities, but the component was developed such to enable the use of a wider range of plots and tests. Two free third-party tools support BEW analytical component: the JFreeChart library provides for plotting abilities while the R plug-in grants access to broad statistical analysis.

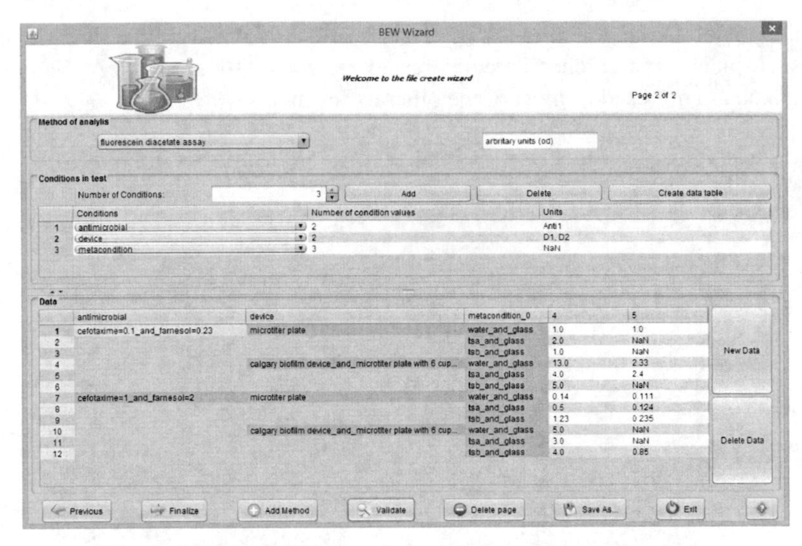

Fig. 3. Snapshot of the description of an experimental analysis

Data plotting was made as flexible as possible in order to accommodate the visualisation of the results produced by virtually any combination of test conditions (Fig. 4).

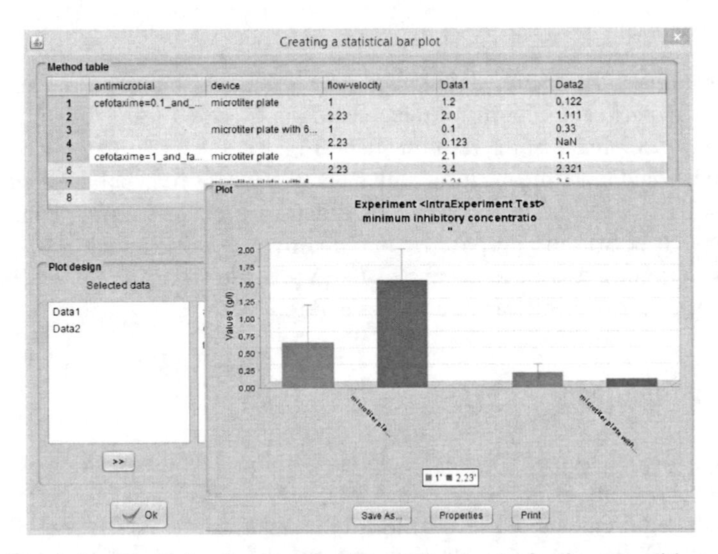

Fig. 4. Snapshot of a plot representing the test results for three conditions

Likewise, BEW has no pre-established pipeline for statistical analysis, i.e. the user decides which tests should be executed for his data. This is crucial given the heterogeneity of existing experimentation. For example, antimicrobial susceptibility testing usually aims to analyse the variance of results obtained under different sets of therapeutic conditions (Fig. 5). In turn, repeatability and reproducibility experiments are quite unique in the goal of analysis – standard operation – and the interpretation of results. Typically, researchers are interested in conducting ANOVA Gauge R&R tests, which are not used in most of the other experiments.

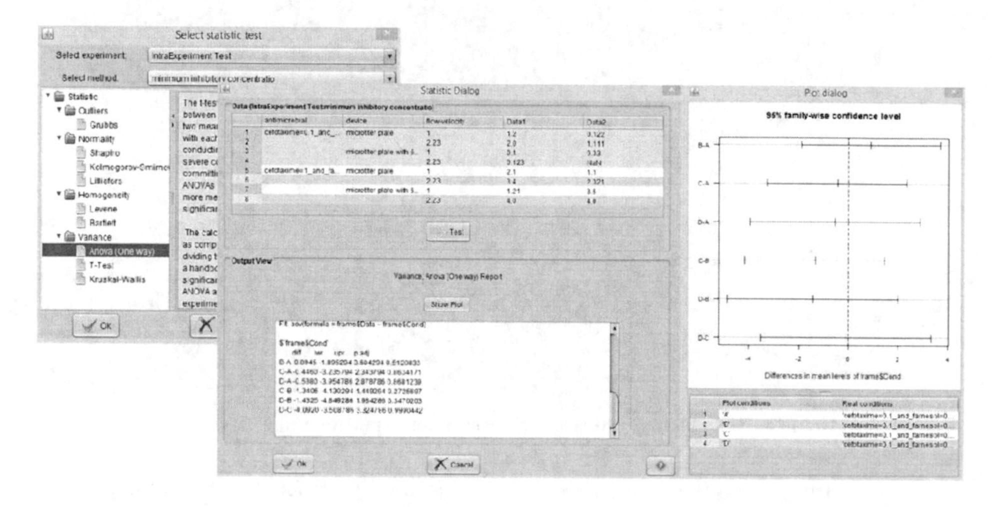

Fig. 5. Snapshot of the execution of an analysis of variance

3.3 Reports

Report capabilities are useful to automatically document the experiment, notably to describe the experiment in terms of methods of analysis and tested conditions, as well as to provide a summary of the resulting data (Fig. 6).

This functionality is meant to complement the XML file, being of practical use when depositing an experiment in a public database or submitting a manuscript to peer-review. It serves the purpose of introducing the work developed, prior to the actual inspection of results by means of the XML file. Also, it can be used to pre-visualise the experiments in Web resources and tools.

4 Conclusions

In this paper we present BEW, a desktop application devoted to the management of Biofilms experiments. It is the very first tool developed to meet the data processing and analysis requirements of the Biofilms domain. As such, its first contribution is a XML-based markup language for the general description of biofilm experiments.

Widespread use of Biofilms markup language in resources and software packages will benefit users as well as developers. With greater interaction between tools, and a common format for publications and databases, researchers will be able to perform systematic experiment comparison and data interchange.

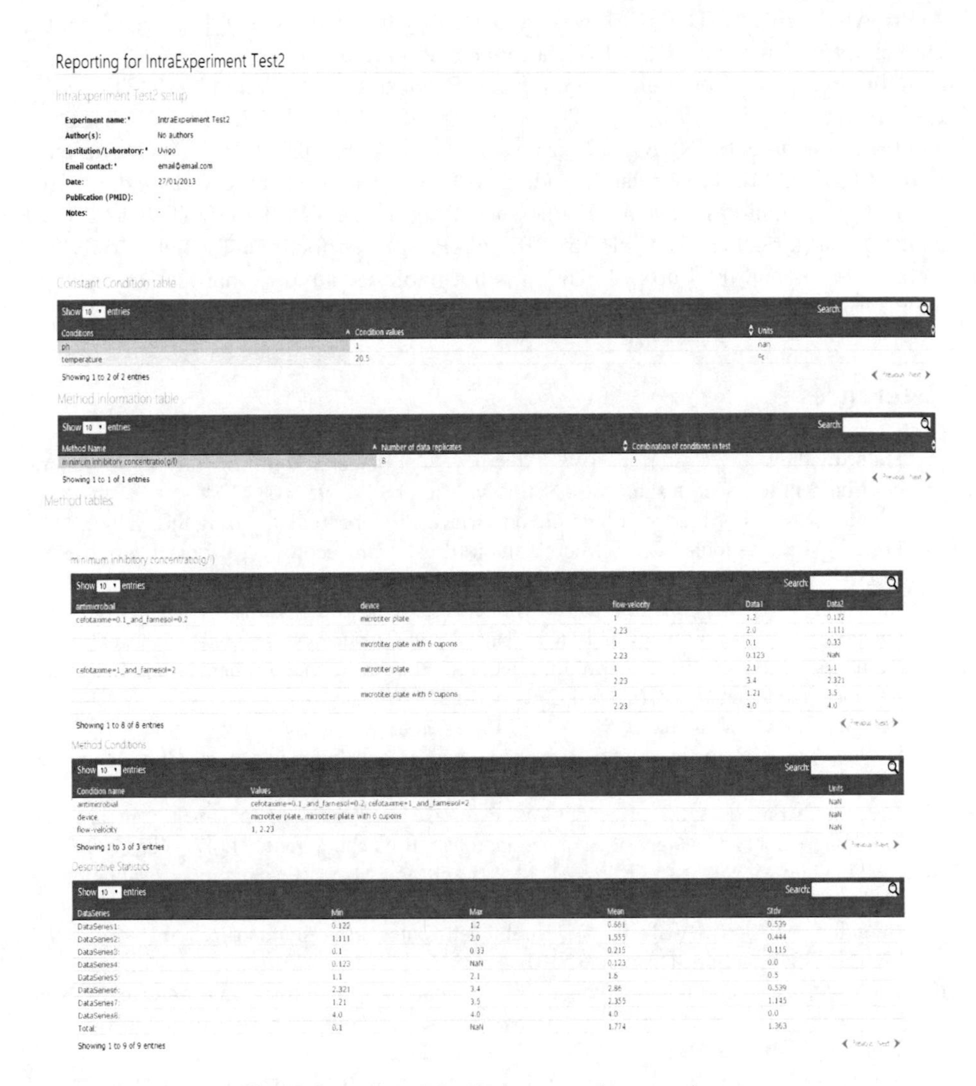

Reporting for IntraExperiment Test2

IntraExperiment Test2 setup

Experiment name:* IntraExperiment Test2
Author(s): No authors
Institution/Laboratory:* Uvigo
Email contact:* email@email.com
Date: 27/01/2013
Publication (PMID): -
Notes: -

Constant Condition table

Show 10 entries Search:

Conditions	Condition values	Units
ph	1	nan
temperature	20.5	ºc

Showing 1 to 2 of 2 entries

Method information table

Show 10 entries Search:

Method Name	Number of data replicates	Combination of conditions in test
minimum inhibitory concentratio(g/l)	8	5

Showing 1 to 1 of 1 entries

Method tables

minimum inhibitory concentratio(g/l)

Show 10 entries Search:

antimicrobial	device	flow-velocity	Data1	Data2
cefotaxime=0.1_and_farnesol=0.2	microtiter plate	1	1.2	0.122
		2.23	2.0	1.111
	microtiter plate with 6 cupons	1	0.1	0.33
		2.23	0.123	NaN
cefotaxime=1_and_farnesol=2	microtiter plate	1	2.1	1.1
		2.23	3.4	2.321
	microtiter plate with 6 cupons	1	1.21	3.5
		2.23	4.0	4.0

Showing 1 to 8 of 8 entries

Method Conditions

Show 10 entries Search:

Condition name	Values	Units
antimicrobial	cefotaxime=0.1_and_farnesol=0.2, cefotaxime=1_and_farnesol=2	NaN
device	microtiter plate, microtiter plate with 6 cupons	NaN
flow-velocity	1, 2.23	NaN

Showing 1 to 3 of 3 entries

Descriptive Statistics

Show 10 entries Search:

DataSeries	Min	Max	Mean	Stdv
DataSeries1:	0.122	1.2	0.661	0.539
DataSeries2:	1.111	2.0	1.555	0.444
DataSeries3:	0.1	0.33	0.215	0.115
DataSeries4:	0.123	NaN	0.123	0.0
DataSeries5:	1.1	2.1	1.6	0.5
DataSeries6:	2.321	3.4	2.86	0.539
DataSeries7:	1.21	3.5	2.355	1.145
DataSeries8:	4.0	4.0	4.0	0.0
Total:	0.1	NaN	1.774	1.363

Showing 1 to 9 of 9 entries

Fig. 6. Snapshot of the Web report of an experiment

Moreover, BEW architecture is designed to be flexible and easily extensible, in anticipation of the various requirements emerging from the growing multidisciplinary of these experiments and the diverse nature of studies. At first, BEW aims to support researcher's daily operation, both in dealing with the generated data and executing the necessary statistical discussion. In a near future, BEW is expected to go beyond intra-laboratory frontiers, notably by enabling the automatic deposition of standardised experimental data in community databases, such as BiofOmics, and enabling inter-laboratory experiment comparisons.

Acknowledgements. This work was supported by the [INOU13-07] project from the University of Vigo, the IBB-CEB, the Fundação para a Ciência e Tecnologia (FCT) and the European Community fund FEDER, through Program COMPETE [FCT Project number PTDC/SAU-SAP/113196/2009/FCOMP-01-0124-FEDER-016012], and the Agrupamento INBIOMED from DXPCTSUG-FEDER unha maneira de facer Europa (2012/273). The research leading to these results has received funding from the European Union's Seventh Framework Programme FP7/REGPOT-2012-2013.1 under grant agreement n° 316265, BIOCAPS. This document reflects only the author's views and the European Union is not liable for any use that may be made of the information contained herein.

References

1. Hall-Stoodley, L., Costerton, J.W., Stoodley, P.: Bacterial biofilms: from the natural environment to infectious diseases. Nat. Rev. Microbiol. 2, 95–108 (2004)
2. Donlan, R.M.: Biofilms: microbial life on surfaces. Emerg. Infect. Dis. 8, 881–890 (2002)
3. Davey, M.E., O'toole, G.A.: Microbial biofilms: from ecology to molecular genetics. Microbiol. Mol. Biol. Rev. 64, 847–867 (2000)
4. Jain, A., Gupta, Y., Agrawal, R., Khare, P., Jain, S.K.: Biofilms–a microbial life perspective: a critical review. Crit. Rev. Ther. Drug Carrier Syst. 24, 393–443 (2007)
5. Römling, U., Balsalobre, C.: Biofilm infections, their resilience to therapy and innovative treatment strategies. J. Intern. Med. 272, 541–561 (2012)
6. Peters, B.M., Jabra-Rizk, M.A., O'May, G.A., Costerton, J.W., Shirtliff, M.E.: Polymicrobial interactions: impact on pathogenesis and human disease. Clin. Microbiol. Rev. 25, 193–213 (2012)
7. Azevedo, N.F., Lopes, S.P., Keevil, C.W., Pereira, M.O., Vieira, M.J.: Time to "go large" on biofilm research: advantages of an omics approach. Biotechnol. Lett. 31, 477–485 (2009)
8. Buckingham-Meyer, K., Goeres, D.M., Hamilton, M.A.: Comparative evaluation of biofilm disinfectant efficacy tests. J. Microbiol. Methods 70, 236–244 (2007)
9. Hamilton, M.A.: KSA-SM-03 - Desirable attributes of a standardized method. Stand. METHODS Test. Surf. Disinfect (2010)
10. Lourenço, A., Ferreira, A., Veiga, N., Machado, I., Pereira, M.O., Azevedo, N.F.: BiofOmics: A Web Platform for the Systematic and Standardized Collection of High-Throughput Biofilm Data. PLoS One 7, e39960 (2012)
11. Sousa, A.M., Ferreira, A., Azevedo, N.F., Pereira, M.O., Lourenço, A.: Computational approaches to standard-compliant biofilm data for reliable analysis and integration. J. Integr. Bioinform. 9, 203 (2012)
12. Bray, T., Paoli, J., Sperberg-McQueen, C.M.: Extensible markup language, XML (1998), http://www.w3.org/TR/1998/REC-xml-19980210
13. Sreenivasaiah, P.K., Kim, D.H.: Current trends and new challenges of databases and web applications for systems driven biological research. Front. Physiol. 1, 147 (2010)
14. Glez-Peña, D., Reboiro-Jato, M., Maia, P., Rocha, M., Díaz, F., Fdez-Riverola, F.: AIBench: a rapid application development framework for translational research in biomedicine. Comput. Methods Programs Biomed. 98, 191–203 (2010)

Discrimination of Brazilian Cassava Genotypes (*Manihot esculenta* Crantz) According to Their Physicochemical Traits and Functional Properties through Bioinformatics Tools

Rodolfo Moresco[1,*], Virgílio G. Uarrota[1], Eduardo da C. Nunes[1], Bianca Coelho[1], Edna Regina Amante[2], Vanessa Maria Gervin[2], Carlos Eduardo M. Campos[3], Miguel Rocha[4], and Marcelo Maraschin[1]

[1] Plant Morphogenesis and Biochemistry Laboratory, Core of Natural Products, Federal University of Santa Catarina, Florianopolis, Brazil
[2] Laboratory of Fruits and Vegetables, Department of Food Science and Technology, Federal University of Santa Catarina, Florianopolis, Brazil
[3] Laboratory of X-ray Diffraction, Department of Physics, Federal University of Santa Catarina, Florianopolis, Brazil
[4] Department of Informatics, School of Engineering, University of Minho, Braga, Portugal
Rodolfo.moresco@posgrad.ufsc.br

Abstract. *Manihot esculenta* currently ranks as the third most important species source of calories in the world. The most important part of the plant is the root, rich in starch. The starch fraction is basically composed of amylose and amylopectin, and different ratios of contents of these two polysaccharides determine the physicochemical traits and functional properties peculiars to genotypes. In this study, principal component analysis (PCA) and clusters analysis were applied to a set of physicochemical and functional variables of ten starch samples of *M. esculenta* genotypes. Moreover, a further chemometric approach was used to a FTIR spectral data set. The analytical techniques employed, associated with chemometric analyzes, allowed distinguishing and/or grouping the genotypes according to their physicochemical traits and functional peculiarities. It was also observed a good relationship between the descriptive models built using the physicochemical dataset and the FTIR dataset from the carbohydrate fingerprint region, allowing a more detailed and robust understanding of possible differences and/or similarities of the studied genotypes.

Keywords: Cassava genotypes, *Manihot esculenta* Crantz, Chemometrics, FTIR.

1 Introduction

Cassava (Manihot esculenta Crantz) currently ranks as the third most important species source of calories in the world among the group of basic food crops, including

[*] Corresponding author.

J. Sáez-Rodríguez et al. (eds.), *8th International Conference on Practical Appl. of Comput. Biol. & Bioinform. (PACBB 2014)*, Advances in Intelligent Systems and Computing 294,
DOI: 10.1007/978-3-319-07581-5_7,

rice and corn [1]. The most important part of the plant is the root, rich in starch, used in food and feed or as raw material for various industries. The starch fraction is composed of two polysaccharides, i.e., amylose and amylopectin. The ratio of contents of these two macromolecules is variable according to the genetic source, which will determine specific starch paste characteristics [2]. It is inferred to exit a genetic diversity associated to a chemodiversity among the M. esculenta genotypes, i.e., starch types with distinct chemical structures and consequently physicochemical and functional properties. In this study, principal component analysis (PCA) and clusters analysis were applied to a set of physicochemical variables of ten starch samples of M. esculenta genotypes, aiming at to build descriptive and classification models to aid in the understanding of eventual association of those variables with the functional starch properties. Moreover, a further chemometric approach was used by applying PCA and clusters analysis to a FTIR spectral data set. In such an analytical approach, the rapid and effective extraction of relevant and not redundant information from a set of complex data enables a more detailed and robust understanding of possible differences and / or similarities in the studied samples, leading to a better discrimination thereof.

2 Material and Methods

2.1 Selection of Genotypes

Roots of ten *M. esculenta* genotypes were produced in the 2010/2011 season in the germplasm bank of the cassava breeding program of Santa Catarina Agricultural Research and Rural Extension Company. The ten genotypes used in this study have been traditionally and regionally named: Apronta mesa, Pioneira, Oriental, Amarela, Catarina, IAC 576-70, Salézio, Estação, Crioulo de Videira, and Rosada. Their selection was based on an extensive cultivation by small farmers, representing economic, social, and environmental importance. Starch samples were extracted from each genotype and analyzed physicochemical and functionally.

2.2 Physicochemical and Functional Properties

The amylose content of starch samples was determined using the colorimetric procedure [3]. The granule test for density determination was performed with the aid of a helium gas pycnometer (Quantachrome®) [4]. The lipid and water adsorption capacity were determined from the method described by Beuchat (1977) [5]. The swelling power of starch of the genotypes in study was determined by the method of Leach, McCowen, & Schoch (1959), with modifications [6, 7]. The grain size was measured from scanning electron micrographs (2000x magnification) with the aid of ImageJ® software (release 1.45k). The relative crystallinity index was determined through X-ray diffractometry [8]. Variations of the consistency of starch during pasting, under the effect of temperature and agitation (temperature of pasting, maximum viscosity, temperature at maximum peak viscosity, time to achieve the maximum peak viscosity, minimal viscosity, final viscosity at cooling cycle, and setback) were measured in a viscometer Rapid Visco Analyzer (RVA).

2.3 Fourier Transform Infrared Vibrational Spectroscopy (FTIR)

FTIR spectra of starches from the genotypes in study were collected on a Bruker IFS-55 (Model Opus. v 5.0, Bruker Biospin, Germany). The processing of the spectra used the Essential FTIR (v.1.50.282) software and considered the definition of the spectral window of interest (3000-600cm^{-}), the baseline correction, normalization, and the optimization of the signal/noise ratio (smoothing).

2.4 Statistical Analysis

The FTIR spectral data and the physicochemical variables measured were subjected to multivariate statistical analysis, using the methods of principal components (PCA) and clusters with the use of R language (v.2.15.2) and the tools from the Chemospec [9] and HyperSpec packages [10].

3 Results and Discussion

In the principal component analysis applied to the data of the physicochemical and functional variables, the first three components counted for 30.02, 29.63, and 15.00% of the variation, respectively, expressing 74.66% of the total variance (**Fig. 1**). The variables swelling power, granule density, and crystallinity index calculated by XRD showed to be positively correlated in PC1+. In their turn, the rheological data, breakdown and peak viscosity, and lipid adsorption represented the most significant variables in PC1-. The second component mainly correlated to the amylose content and setback (retrogradation tendency) in PC2+ and to crystallinity index calculated from the FTIR and XRD data set in PC2-. The amylose content (PC2+) showed a strong negative correlation with the crystallinity index calculated by IR ($r = -0.702$) and XRD ($r = -0.709$), density ($r = -0.398$), granule size ($r = -0.407$), and lipid adsorption ($r = -0.214$) (PC2-). Contrarily, a positive correlation with water adsorption ($r = 0.071$) and with the rheological data, e.g., retrogradation tendency (setback, $r = 0.708$) pasting temperature ($r = 0.297$), maximum viscosity ($r = 0.462$), and swelling power ($r = 0.108$) (PC2+) was observed to exist. The construction of the descriptive model based on the calculation of the principal components allowed some inferences about the behavior of the functional properties of the starches studied derived from their physicochemical characteristics. These results are consistent with the information that correlates the crystalline structure of the granules to the double helices formed by the branches of amylopectin [11]. The region where amylopectin is found is more dense or crystalline. Because it is more compact, this region is less hydrated and more resistant to hydrolysis process [3]. The amorphous regions of the starch granules (mainly formed by amylose) are less dense, absorbing more water at temperatures below the gelatinization temperature. Additionally, the amount of amorphous regions has been traditionally associated to a greater retrogradation tendency of starch granules, because of its strong tendency to rebind via hydrogen bonds with other adjacent amylose molecule [12].

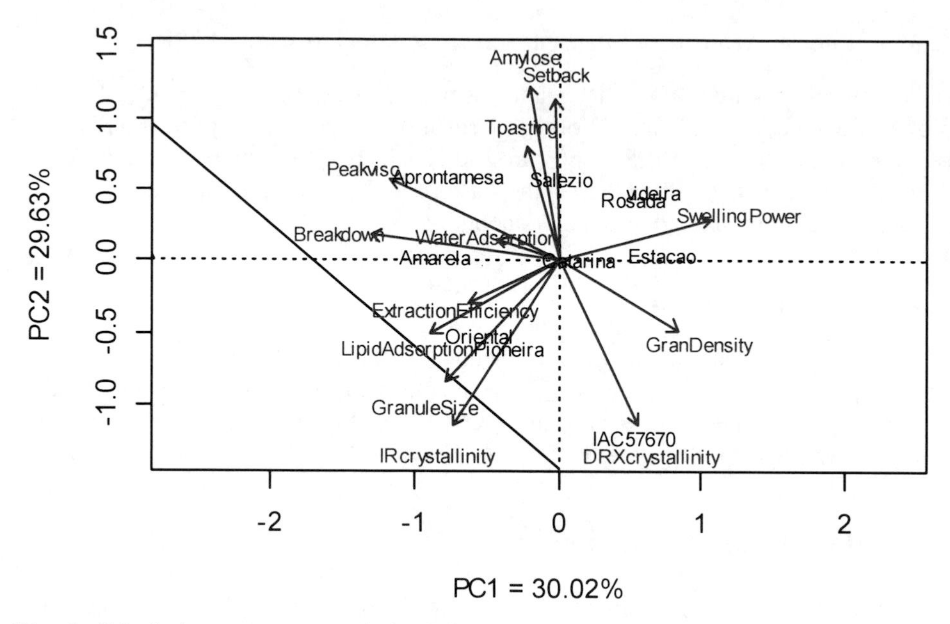

Fig. 1. Principal components analysis (PCAs) scores scatter plot of the physicochemical variables of starches of Brazilian cassava genotypes

In a second series of experiments, a cluster analysis was applied to the physicochemical and functional data set. The similarities were defined based on the Euclidean distance between two samples using the arithmetic average (UPGMA). Genotypes with the highest similarity in their physicochemical and functional characteristics are represented by cluster hierarchical analysis in Fig. 2. The cophenetic correlation was 65.71%.

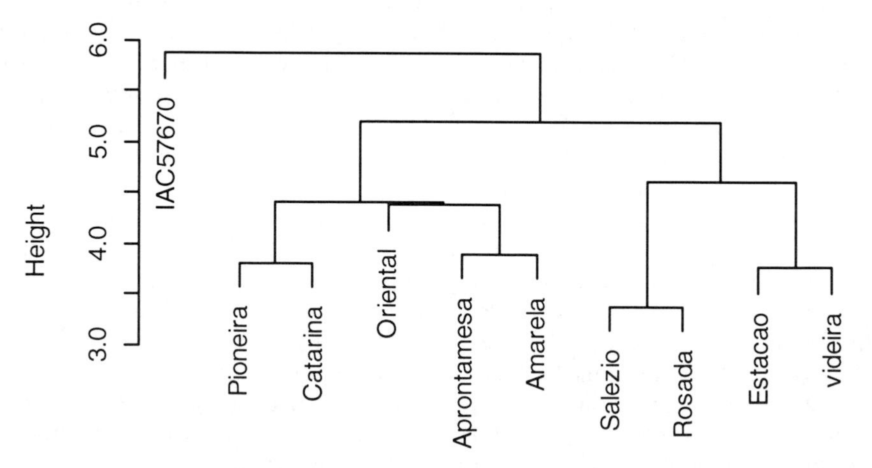

Fig. 2. Similarity of starches of cassava genotypes in respect to their physicochemical and functional variables. Hierarchical cluster dendogram analysis (UPGMA method) with 65.71% of cophenetic correlation.

The visual analysis of the FTIR spectral profiles of the starch samples (Fig 3) extracted from the cassava roots revealed the presence of functional groups associated to chemical constituents in the 2900-700cm^{-1} spectral window. More properly of interest, carbohydrates signals from the amylose and amylopectin components of the study samples presented strong absorbance intensities in the 1200-950cm^{-1} spectral range, considered a fingerprint region of this class of primary metabolites [13].

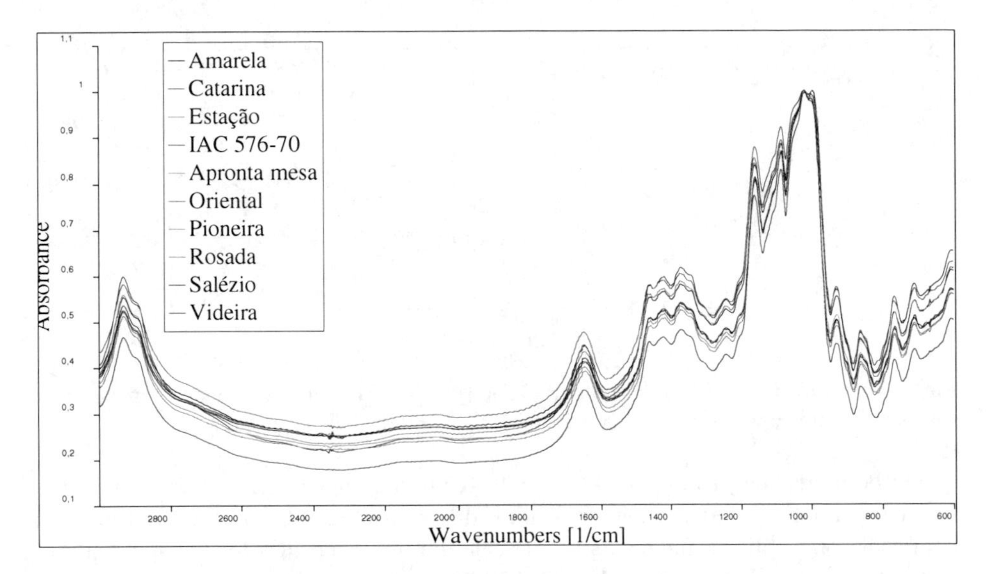

Fig. 3. FTIR spectra of starches of ten cassava genotypes in the 3000–600 cm^{-} wavenumber region

When the principal components were calculated from the full FTIR spectra (3000-600cm^{-1}) data matrix, PC1 and PC2 contributed to explain in 98.46 % the data set variance. However, a clear discrimination of the samples was not found, rather samples occurred spread out by all the sections of the factorial distribution graph. Such findings prompted us to build a second classification model by applying PCA to the carbohydrate fingerprint region of the FTIR dataset, in connection with the focus of this study on the starch fraction of the samples. Fig. 4 shows that PC1 and PC2 accounted for 99.2% of the existing variance, clearly revealing the existence of two genotype groups according to their similarities, i.e., the genotypes Rosada, Crioulo de Videira, Salézio, and Apronta mesa in PC2+, as the genotypes Pioneira, Oriental, and Amarela grouped in PC2-, spreading out along the PC1 axis. As previously indicated by hierarchical analysis, the IAC 576-70 genotype seems to have a typical metabolic profile, occurring away from all the other samples (Fig 4).

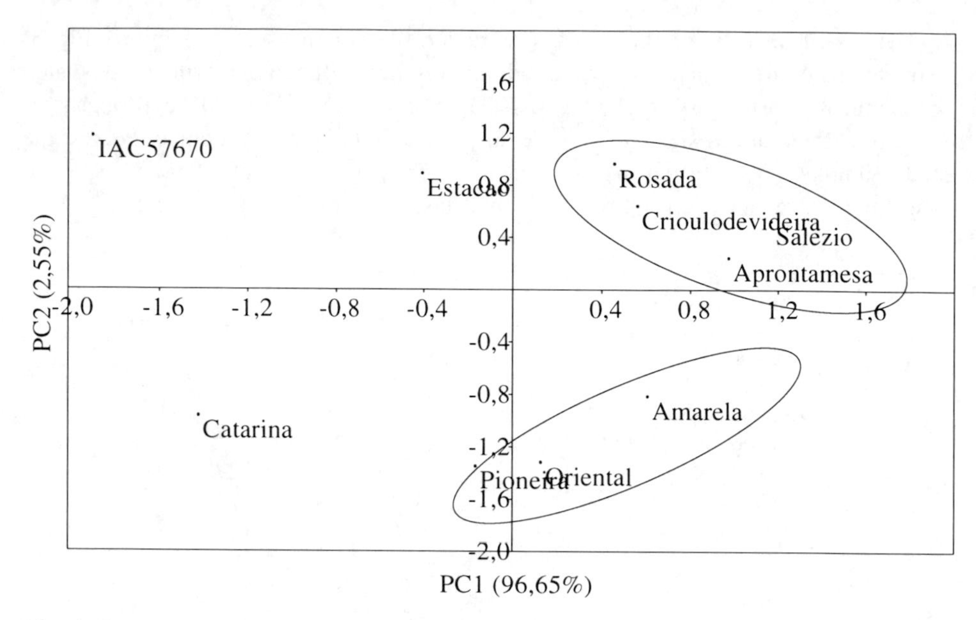

Fig. 4. Factorial distribution of PC1 and PC2 for the ATR-FTIR spectral data set typical of the carbohydrate fingerprint region (1200-950cm^{-1}) for the studied cassava starches

Furthermore, the genotypes IAC 576-70, Catarina, and Estação were detected to occur separately from the others, revealing different metabolic profiles and peculiar chemical compositions. Interestingly, the calculation of the PCs for the spectral data of the fingerprint regions of lipids (3000-2800cm^{-1}) [14] and proteins (1650-1500cm^{-1}) [15] also allowed a clear discrimination among those samples (data not shown), especially for the genotypes IAC 576-70 and Catarina.

4 Conclusions

The metabolomic approach used in this study allowed us to identify and characterize the chemical variability of the cassava genotypes, supporting the working hypothesis. Moreover, the analytical techniques employed and the rheological tests performed, associated with chemometric analyzes (PCAs and clusters), allowed to distinguish and/or group the genotypes according to their physicochemical and functional peculiarities. In this sense, the spectral profiles of IAC 576-70, Catarina, and Estação genotypes showed to differ from the others according to their metabolic profiles, suggesting a peculiar chemical composition.

It is assumed, therefore, that the concomitant use of the technologies employed in this work is of great interest to investigate the structural characteristics of the starch fraction of *M. esculenta* germplasm. Taken as one, the results can be used as a support tool for cassava biochemically assisted genetic breeding programs, optimizing the selection and the maintenance process of the genetic diversity in germplasm banks, for instance.

References

1. FAO (Food and Agriculture Organization of the United Nations), Rome, Italy, http://www.fao.org/ag/agp/agpc/gcds/ (access in: April 13, 2012)
2. Cereda, M.P., Franco, C.M.L., Daiuto, E.R., Demiate, I.M., Carvalho, L.J.C.B., Leonel, M.: Propriedades gerais do amido. culturas de tuberosas amiláceas latino-americanas, vol. 1. Fundação Cargill, São Paulo (2001)
3. Williams, P.C., Kuzina, F.D., Hlynka, I.: A rapid colorimetric procedure for estimating the amylose content of starches and flours. Cereal Chemistry 7, 412–420 (1970)
4. Chang, C.S.: Measuring density and porosity of grain kernels using a gas pycnometer. Cereal Chemistry 65, 13–15 (1988)
5. Beuchat, L.R.: Functional and electrophoretic characteristics of succinylated peanut flour protein. Journal of Agricultural and Food Chemistry 25, 258–261 (1977)
6. Leach, H.W., MCcowen, L.D., Schoch, T.J.: Structure of the starch granule. Swelling and solubility patterns of various starches. Cereal Chemistry 36, 534–544 (1959)
7. Aryee, F.N.A., Oduro, I., Ellis, W.O., Afuakwa, J.J.: The physicochemical properties of flour samples from the roots of 31 varieties of cassava. Food Control 17, 916–922 (2006)
8. Hayakawa, K.: Quality characteristics of hexaploid wheat (Triticumaestivum L.): properties of starch gelatinization and retrogradation. Cereal Chemistry 74, 576–580 (1997)
9. Hanson, A.B.: ChemoSpec: An R Package for Chemometric Analysis of Spectroscopic Data and Chromatograms (Package Version 1.51-0) (2012)
10. Beleites, C.: Import and Export of Spectra Files. Vignette for the R package hyperSpec. (2011)
11. Hoover, R.: Composition, molecular structure, and physicochemical properties of tuber and root starches: a review. Carbohydrate Polymers 45, 253–267 (2001)
12. Singh, N., Singh, J., Kaur, L., Sodhi, N.S., Gill, B.S.: Morphological, termal and rheological properties of starches from different botanical sources. Review Food Chemistry 81, 219–231 (2003)
13. Cerná, M., Barros, A.S., Nunes, A., Rocha, S.M., Delgadillo, I., Copíková, J., Coimbra, M.A.: Use of FTIR spectroscopy as a tool for the analysis of polysaccharide food additives. Carbohydrate Polymers 51, 383–389 (2003)
14. Lambert, J.B., Shurvell, H.F., Lightner, D.A., Cooks, R.G.: Organic Structural Spectroscopy. Prentice Hall, Upper Saddle River (2001)
15. Silverstein, R.M.: Identificação Espectrométrica de Compostos Orgânicos, 5ª edn. Guanabara Koogan, Rio de Janeiro (1994)

Prediction of Active Residues of β-galactosidase from *Bacteroides thetaiotaomicron*

Vladimir Vukić[1], Dajana Hrnjez[2], Spasenija Milanović[2], Mirela Iličić[2], Katarina Kanurić[2], and Edward Petri[3]

[1] Institute of Field and Vegetable Crops, Maksima Gorkog 30, 21000 Novi Sad, Serbia
[2] University of Novi Sad, Faculty of Technology, Bulevar Cara Lazara 1, 21000 Novi Sad, Serbia
[3] University of Novi Sad, Faculty of Sciences, Department of Biology and Ecology, Trg Dositeja Obradovica 2, 21000 Novi Sad, Serbia
vladvukic@gmail.com, dajana@tf.uns.ac.rs,
{senadm,panim,stay}@uns.ac.rs

Abstract. *Bacteroides thetaiotaomicron*, a Gram-negative anaerobe and symbiotic commensal microbe, dominates the human intestinal tract; where it provides a range of beneficial metabolic tasks not encoded in the human genome. *B. thetaiotaomicron* uses various polysaccharides as its carbon and energy source, providing valuable monosaccharides for its host. Regarding dairy technology, the most important characteristic of *B. thetaiotaomicron* is its ability to degrade lactose.β-galactosidase from *B. thetaiotaomicron* belongs to thesubfamily GH-35. There is a lack of structural information about *B. thetaiotaomicron* β-galactosidase, including the active site and residues involved in lactose degradation. The aim of this research was to predict the residues of *B. thetaiotaomicron* β-galactosidase involved in substrate catalysis, to construct a model of its active site, and to predict residues involved in substrate binding.Amino acid sequences were retrieved from UNIPROT database. Sequence clustering and alignments were performed using UGENE 1.11.3.Docking studies were performed using Surflex-Dock. Our results indicate that proton donor and nucleophillic residues could be GLU182 and GLU123, respectively.These active residues of *B. thetaiotaomicron* β-galactosidase have not been reported previously.

Keywords: *Bacteroides thetaiotaomicron,* β-galactosidase, active site, active residue, docking.

1 Introduction

Bacteroides thetaiotaomicron, which dominates the human intestinal tract is a Gram-negative anaerobic microbe and symbiotic commensal microbe,. It comprises 6% of all bacteria and 12% of all *Bacteroides* in the human intestine; where it provides a range of beneficial metabolic tasks not encoded in the human genome [1, 2,]. In particular, it has the ability to use energy from carbohydrates and other nutrients

J. Sáez-Rodríguez et al. (eds.), *8th International Conference on Practical Appl. of Comput. Biol. & Bioinform. (PACBB 2014)*, Advances in Intelligent Systems and Computing 294,
DOI: 10.1007/978-3-319-07581-5_8, © Springer International Publishing Switzerland 2014

present in human intestine [3]. However, *B. thetaiotaomicron* can also be an opportunistic pathogen, especially dangerous in post-operative infections of the peritoneal cavity and bacteraemia [4]. *B. thetaiotaomicron* has a 6.26 Mb genome with 4779 protein coding genes [5]. Most investigated genes and proteins are glucosidases, due to their high importance in gluco-oligosaccharide metabolism [6, 7]. In addition togluco-oligosaccharides, *B. thetaiotaomicron* uses various polysaccharides as a carbon and energy source: including amylose, amylopectin, and pullulan, as well as malto-oligosaccharides [8].Other host benefits provided contribute to postnatal gut development, metabolic capabilities and host physiology [9].Regarding dairy technology, the most important characteristic of *B. thetaiotaomicron* is its ability to degrade lactose, improving lactose digestion in the human intestinal tract.

β-galactosidase (EC 3.2.1.23) catalyzes the hydrolysis of β(1-3) and β(1-4) galactosyl bonds in oligo- and disaccharides. All known β-galactosidases belong to the GH-A superfamily of glycoside hydrolases and to the 1, 2, 35 and 42 subfamilies. β-galactosidase from *B. thetaiotaomicron* belongs to the subfamily GH-35 [10]. Databases researchhas revealed a lack of structural information about *B. thetaiotaomicron* β-galactosidase, including its active site and residues involved in lactose degradation. Therefore, the aim of this research was to predict the residues of *B. thetaiotaomicron* β-galactosidase involved in substrate catalysis, to construct a model of its active site, and to predict residues involved in substrate binding.

2 Material and Methods

Sequences. Sequences of analyzed β-galactosidases were retrieved from UNIPROT: *B. thetaiotaomicron* (UniProt: Q8AB22), *Aspergilus oryzae* Q2UCU3, *Penicillium* sp. (UniProt: Q700S9), *Trichoderma reesei,* (UniProt: Q70SY0), *Homo sapiens* (UniProt: P16278) [11]. Structural coordinates for *B. thetaiotaomicron* β-galactosidase were retrieved from the Protein Data Bank (PDB ID: 3OBA:A) [12]. NCBI BlastP was performed against the PDB database for identification of highly similar sequences [13].

Clustering and Docking. Sequence similarity based clustering was performed using the program UGENE 1.11.3 [14].The neighbor joining method with a filter of 30% similarity was used for sequence clustering. After clustering, multiple sequence alignments for all clusters and for all sequences were performed using the program MUSCULE [15].Highly conserved regions and potential active site residues were identified by sequence alignment. Protein and ligand structures were energy minimized using standard MMFF94 force fields [16, 17, 18, 19, 20, 21]. Partial atomic charges were calculated using the MMFF94 method. The Powell method, distance dependent dielectric constant and convergence gradient method with a convergence criterion of 0.005 kcal/mol were used. Protein structures were prepared for docking simulations using default parameters, and polar hydrogen atoms were added. Docking studies were performed using the program Surflex-Dock with flexible H atoms [22].

Statistics. Duncan's multiple range test ("Statistica9") was performed to evaluate statistically significant differences between the studied parameters. Differences were considered statistically significant if $p \leq 0.05$.

3 Results and Discussion

Whole sequence analysis and alignment revealed that *B. thetaiotaomicron* β-galactosidase has more than 70% sequence similarity with *Pencillium sp, T. ressei, A. oryzae* and less than 60% similarity with *H. sapiens* β-galactosidase. All of these β-galactosidase variants belong to the GH-35 subfamily.

The closest relative of the GH-35 subfamily with experimentally determined active site residues and 3D structure is β-galactosidase from *Thermus thermophillus*, which belongs to the GH-42 subfamilly [23]. Sequence alignment of β-galactosidase from GH-35 with β-galactosidase from *Thermus thermophillus* revealed overlapping residues (results not shown). In *B. thetaiotaomicron*, this residue is GLU182. Interestingly, in *B. thetaiotaomicron* this residue could be proton donor, as it is suggested by sequence similarity with *Homo sapiens* which also belong to GH-35 subfamily and have experimentally determined active residues (Fig.1) [24]. Therefore, by similarity it could be suggested that GLU182 is the proton donor active site residue.

Analysis of conserved domains of potential proton donor active site residues indicated 25% conserved residues in all 5 analyzed sequences, 33.3% in 4 out of 5 and 25 % in 3 out of 5 analyzed sequences. According to UNIPROT, nucleophillicactive residues in *Homo sapiens* and *A. orizae* are GLU268 and GLU298, respectively. Detailed analysis revealed only 15.3 % conserved residues in all 5 sequences, 10.2% in 4 out of 5 sequence and 35% in 3 out of 5 analyzed sequences. On the other hand, the conserved domain in Figure 1B shows significantly higher conservation ($p<0.05$). It has 38.9% conserved residues in all 5 analyzed sequence, 10.25% in 4 out of 5 and 43.6% in 3 out of 5 sequences. On the basis of these data, these conserved domains might interact with substrate molecules and therefore be the nucleophillic residue in the enzyme active site. These active residues of *B. thetaiotaomicron* β-galactosidase are not reported in previous reports or databases.

In order to predict the identity of the nucleophillic in the active site, docking analysis was performed for *B. thetaiotaomicron* β-galactosidase. The active site was set using all three potential active residues with additional spheres of 5 Å, in order to include all possible binding residues and to predict the most probable active site conformation and ligand docking. Lactose was docked into the active site (Fig 2). Detailed analysis of docking revealed the preferred lactose conformation as well as active and binding residues (Fig. 3). Docking results show that the closest residue to the lactose cleavage site is GLU123 (4.39 Å). GLU259 which is suggested in databases to be the nucleophillic residue is located far from the lactose cleavage site (5.26Å) and according to our results cannot make H bonds with lactose. Therefore, according to our docking model and sequence alignment, contrary to previously suggested residues in databases, the nucleophillic residue could be GLU123. This residue makes hydrogen bonds with lactose and is positioned near the covalent bound between galactose and glucose moieties in lactose (Fig 3). Other substrate binding residues are TYR77, ASN181, GLU259 and TYR325. So, in the case of lactose, two additional binding residues could be located in the enzyme active site pocket. First is GLU123 as a nucleophillic and second is GLU182 as the proton donor residue. It is noteworthy that significant similarity in the pattern of binding site locations was observed in the case of these five examples of β-galactosidase although they originate from distantly related species.

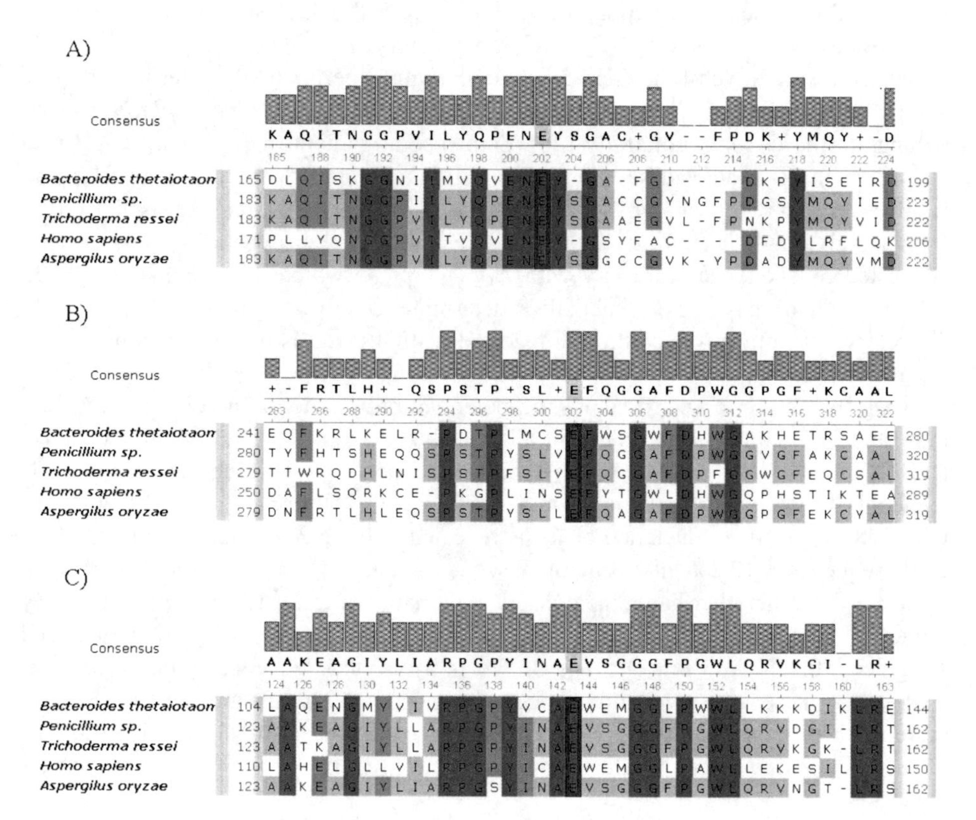

Fig. 1. Proposed and proton donor and nucleophillic residues in active site of *B. thetaiotaomicron* β-galactosidase; A) proton donor residues; B) nucleophilic active residues proposed by similarity with Human β-galactosidase; C) nucleophilic active residues proposed by conserved regions and docking analysis

In future research, based on our findings, it would be interesting to experimentally determine the active resides of *B. thetaiotaomicron*. Furthermore, it would be interesting to examine β-galactosidases from all GH-35 subfamily by docking analysis and to compare them with *B. thetaiotaomicron*.

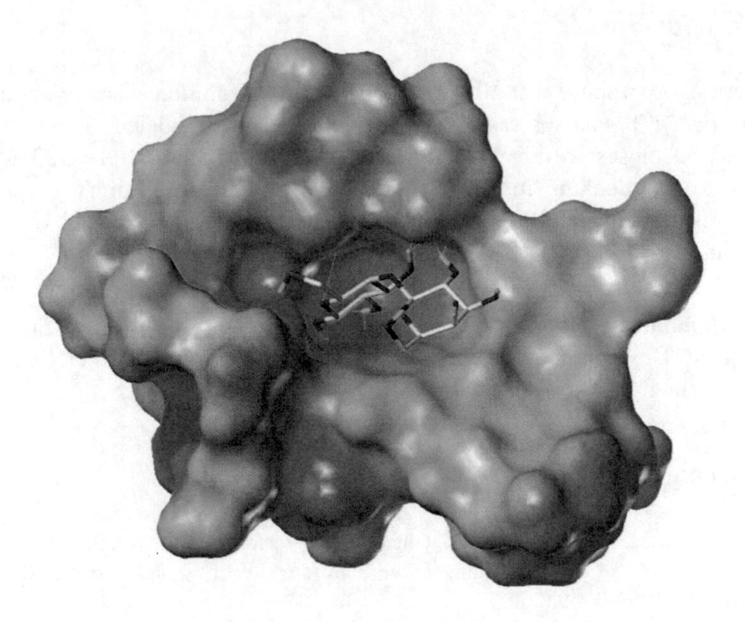

Fig. 2. Lactose in the active site of β-galactosidase *B. thetaiotaomicron*

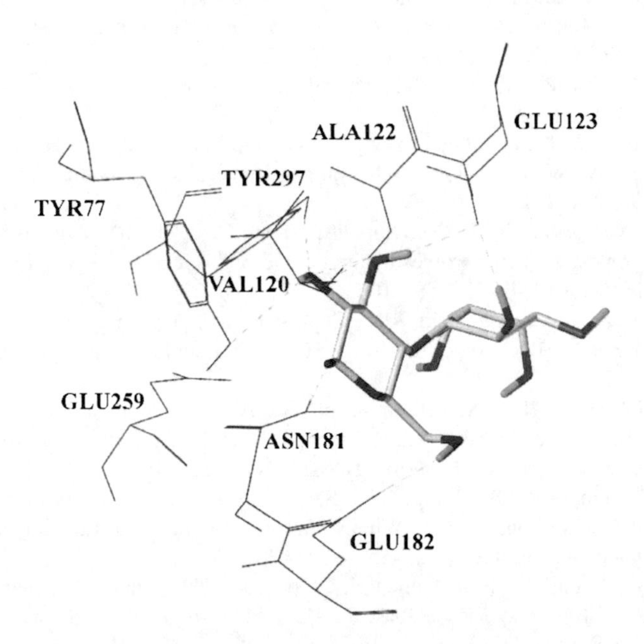

Fig. 3. Docking analysis of analysed β-galactosidase *B. thetaiotaomicron*

4 Conclusions

By similarity, we propose that GLU182 acts as the proton donor active residue in *B. thetaiotaomicron* β-galactosidase. The active site was modeled. Docking results reveal that the closest residue to the lactose cleavage site is GLU123 (4.39 Å). According to our docking model and sequence alignment, contrary to previously suggested residues in databases, the nucleophilic residue could be GLU123. Other substrate binding residues are TYR77, ASN181, GLU259 and TYR325.

Acknowledgments. Authors want to thank Ministry of Education, Science and Technological Development of Republic of Serbia for the financial support of research presented in this article, Project No. III 46009.

References

1. Sonnenburg, J.L., Xu, J., Leip, D.D., Chen, C.H., Westover, B.P., Weatherford, J., Buhler, J.D., Gordon, J.I.: Glycan foraging in vivo by an intestine-adapted bacterial symbiont. Science 307, 1955–1959 (2005)
2. Gill, S.R., Pop, M., De Boy, R.T., Eckburg, P.B., Turnbaugh, P.J., Samuel, B.S., Gordon, J.I., Relman, D.A., Fraser-Liggett, C.M., Nelson, K.E.: Metagenomic analysis of the human distal gut microbiome. Science 312, 1355–1359 (2006)
3. Henrissat, B., Bairoch, A.: Updating the sequence-based classification of glycosyl hydrolases. Biochem. J. 316, 695–696 (1996)
4. Wexler, H.M.: Bacteroides: the good, the bad, and the nittygritty. Clin. Microbiol. Rev. 20, 593–621 (2007)
5. Xu, J., Bjursell, M.K., Himrod, J., Deng, S., Carmichael, L., Chiang, H.C., Hooper, L.V., Gordon, J.I.: A genomic view of the human-Bacteroides thetaiotaomicron symbiosis. Science 299, 2074–2076 (2003)
6. Jinhu, W., Xiang, S., Yi, Z., Yongjun, L., Chengbu, L.: QM/MM investigation on the catalytic mechanism of Bacteroides thetaiotaomicron α-glucosidase BtGH97a. Biochimica et Biophysica Acta 1824, 750–758 (2012)
7. Marcia, M.C., Jenny, L.A., David, R.R.: Expression and purification of two Family GH31 α-glucosidases from Bacteroides thetaiotaomicron. Protein Expression and Purification 86, 135–141 (2012)
8. Shipman, J.A., Cho, H.K., Siegel, H.A., Salyers, A.A.: Physiological Characterization of SusG, an Outer Membrane Protein Essential for Starch Utilization by *Bacteroides thetaiotaomicron*. Journal of Bacteriology 181, 7206–7211 (1999)
9. Wang, J., Shoemaker, N.B., Wang, G.R., Salyers, A.A.: Characterization of a Bacteroides Mobilizable Transposon, NBU2, Which Carries a Functional Lincomycin Resistance Gene. Journal of Bacteriology 182, 3559–3571 (2000)
10. Cantarel, B.L., Coutinho, P.M., Rancurel, C., Bernard, T., Lombart, V., Henrissat, B.: The carbohydrate active enzymes database (CAZy): an expert resource for glycogenomics. Nucleic Acids Res. 37, D233–D238 (2009)
11. UniProt, `http://www.uniprot.org`
12. Protein Data Bank, `http://www.rcsb.org`
13. National Center for Biotechnology Information, `http://www.ncbi.nlm.nih.gov`
14. UniproUgene, `http://ugene.unipro.ru/`

15. Edgar, R.C.: MUSCLE: multiple sequence alignment with high accuracy and high throughput. Nucleic Acids Res. 32, 1792–1797 (2004)
16. Halgren, A.T.: Merck molecular force field. I. Basis, form, scope, parameterization, and performance of MMFF94. J. Comput. Chem. 17, 490–519 (1996)
17. Halgren, A.T.: Merck molecular force field. II. MMFF94 van der Waals and electrostatic parameters for intermolecular interactions. J. Comput. Chem. 17, 520–552 (1996)
18. Halgren, A.T.: Merck molecular force field. III. Molecular geometries and vibrational frequencies for MMFF94. J. Comput. Chem. 17, 553–586 (1996)
19. Halgren, A.T., Nachbar, B.R.: Merck molecular force field. IV. Conformational energies and geometries for MMFF94. J. Comput. Chem. 17, 587–615 (1996)
20. Halgren, A.T.: Merck molecular force field. V. Extension of MMFF94 using experimental data, additional computational data, and empirical rules. J. Comput. Chem. 17, 616–641 (1996)
21. Halgren, A.T.: MMFF: VI. MMFF94s option for energy minimization studies. J. Comput. Chem. 20, 720–729 (1999)
22. Jain, A.N.: Surflex: fully automatic flexible molecular docking using a molecular similarity-based search engine. J. Med. Chem. 46, 499–511 (2003)
23. Maksimainen, M., Paavilainen, S., Hakulinen, N., Rouvinen, J.: Structural analysis, enzymatic characterization, and catalytic mechanisms of *β*-galactosidase from *Bacillus circulans*ssp. *alkalophilus*. FEBS J. 279, 1788–1798 (2012)
24. McCarter, J.D., Burgoyne, D.L., Miao, S., Zhang, S., Callahan, J.W., Withers, S.G.: Identification of Glu-268 as the Catalytic Nucleophile of Human Lysosomal b-Galactosidase Precursor by Mass Spectrometry. J. Biol. Chem. 272, 396–400 (1997)

Detection of Intramolecular Tunnels Connecting Sequence of Sites in Protein Structures

Ondrej Strnad, Barbora Kozlikova, and Jiri Sochor

Faculty of Informatics, Masaryk University,
Botanicka 68a, 602 00, Brno, Czech Republic
{xstrnad2,xkozlik,sochor}@fi.muni.cz

Abstract. Proteins are essential for functioning of all living organisms and studying their inner structure and functions has been of a high importance. Many studies concentrated on detection of various inner structures inside macromolecules (e.g. tunnels, channels, pores) which play an essential role in the functioning of a large number of proteins. Here we present a novel approach to a detection of intramolecular tunnels. These pathways may facilitate the transport of reaction intermediates among buried active sites. The results obtained by the proposed algorithm were compared to intramolecular tunnels whose presence in given structures is already known. The algorithm is able to also identify other inner structures, such as channels or pores.

Keywords: protein structure, tunnel, intramolecular tunnel, active site.

1 Introduction

Proteins are very complex structures containing various clefts, protrusions and empty space. These inner structures can have specific functions - for example, the inner cavities can contain active sites, or the tunnels and channels can serve as transport pathways for small molecules, ions and water molecules. These pathways play an essential role in the function of many proteins. Their role can vary:

- **Tunnels** – serve as transport paths for a ligand accessing the deeply buried active site [1],[4].
- **Channels** or **pores** – can facilitate the transport of ions or molecules across biological membranes [7].
- **Intramolecular tunnels** – connect distinct active sites and serve as transport pathways for reaction intermediates. For instance, bifunctional enzymes contain two active sites connected by an intramolecular tunnel [6],[10].

This study presents a novel approach to the definition of the latter group of structures, the intramolecular channels. Their presence can be crucial for the catalytic functions of proteins where the enzymatic reaction requires different environments (for example polar and non-polar). These environments can be located in spatially separated sites inside the protein structure and a substrate is supposed to be transported between them.

J. Sáez-Rodríguez et al. (eds.), *8th International Conference on Practical Appl. of Comput. Biol. & Bioinform. (PACBB 2014)*, Advances in Intelligent Systems and Computing 294,
DOI: 10.1007/978-3-319-07581-5_9,

The whole process of migration of intermediates resembles the working process where each worker works on one part of the final product. The material comes to the first worker, who modifies it according to instructions and passes the intermediate to the second worker. In each phase the intermediate has to satisfy given requirements. After passing through all workers, the product is finalized and can leave the assembly line. The same situation appears in the protein structure - after passing through all predefined sites the product is finalized and leaves the protein structure.

Our approach is based on the CAVER 3.0 algorithm [3]. We extend this algorithm in order to compute the pathways between arbitrarily located sites (points of interest) in the protein.

2 Related Work

Commonly available methods for detection of tunnels or channels are designed mainly for the computation of pathways between one active site and the bulk solvent (tunnels) or pathways leading throughout the protein structure (channels or pores). With a specific set of input parameters, most of them are also able to detect inner pathways connecting several active sites. However, the identification of the proper combination of input parameters is complicated and it requires detailed knowledge of these methods and their principles. To the best of our knowledge, we are not aware of any available technique for computation of intramolecular tunnels connecting more deeply buried active sites.

Recently, several methods for identification and analysis of pathways in protein structures have been developed. HOLE 2.2 [12] is applicable to the analysis of a single channel or pore traversing the protein. Yaffe et al. introduced MolAxis 1.4 [14] which is able to identify pathways between two sites in a static snapshot as well as in molecular dynamics simulations. MOLE 2.0 [11] is able to detect channels and pores in seven basic steps. The starting and ending points can be defined by the user or automatically positioned into each detected cavity. The PoreWalker [8] tool enables an automatic detection of pores. However, none of the existing tools is designed for the detection of pathways between more than two sites.

Our approach is based on CAVER 3.0 [3], that was originally designed for the computation of tunnels leading from the bulk solvent to a buried active site of the protein. It incorporates the Voronoi diagram computation [2] for geometric division of the space enclosing the molecule. The resulting tunnels lead along the Voronoi edges.

To cope with atoms of different radii, CAVER 3.0 uses an approximation, which gives more precise and chemically-relevant tunnels. Each atom is approximated by a set of spheres of the same size where the basic Voronoi diagram approach can be utilized. CAVER 3.0 searches for all possible tunnels starting from the user specified site. The detected tunnels are annotated by their priority, which is derived from the combination of the bottleneck width of the tunnel, its length, curvature and other parameters. This information helps our extended algorithm to compute all connections between the given points of interest and filter out tunnels with the lowest priority.

3 Algorithm for Detection of Tunnels among Several Active Sites

The proposed algorithm is designed for computing pathways connecting arbitrary sites inside the macromolecule controlled by a set of input parameters. Thus in the further description of the algorithm we generalize the notion of the active site to a term **point of interest** (POI).

The algorithm is able to find all pathways according to the user's needs for large macromolecular structures consisting of more than hundreds of thousands of atoms. As mentioned above, the algorithm itself requires as the input a set of points of interest enclosed inside a molecular structure. These points serve as the starting and ending points of the searched pathways. Algorithm 1 summarizes the major stages of the computation.

Algorithm 1. Detection of Tunnels Among More Points of Interest

Input: atom positions, set of points of interest $P_0, ..., P_n$
Output: t_r connecting $P_0, ..., P_n$

1. **foreach** pair P_i, P_{i+1}
2. compute center C of the line $[P_i\ P_{i+1}]$
3. compute planes $p(C)$, $p(P_i)$, $p(P_{i+1})$ perpendicular to line $[P_i\ P_{i+1}]$
4. create a set of atoms $A(P_i)$ geometrically between $p(P_i)$ and $p(C)$
5. create a set of atoms $A(P_{i+1})$ geometrically between $p(P_{i+1})$ and $p(C)$
6. compute a set of tunnels T_i, T_{i+1} for $A(P_i)$, $A(P_{i+1})$
7. remove from T_i, T_{i+1} all tunnels that do not intersect $p(C)$
8. create concatenated set of tunnels $\{t(T_i,\ T_{i+1})\}$
9. t_i = select *the best* tunnel from $\{t(T_i,\ T_{i+1})\}$
10. t_r = concat $(t_i, ..., t_n)$

return t_r

Step 1 – Input Definition: In the first step, the user has to define the position of points of interest. The only limitation is that these points have to be encapsulated by the convex hull of the protein structure. This can be reached by three possible options:

- Points of interest are selected from a set of active sites listed in one of the databases containing definitions of active sites for many structures (Catalytic Site Atlas [9] or UniProt [13]).
- The user can select the positions of points of interest manually – by giving the coordinates in three-dimensional space or by defining the set of residues or atoms surrounding the desired site.
- The user can also take advantage of cavities present in proteins. Active sites are located in inner cavities and thus their detection can also help to reveal the position of active sites.

Users can specify other input parameters, which control many features of detected pathways, such as their minimal width. The complete list of parameters can be found in [3]. Our approach uses two most important parameters – the *probe size* and the *shell radius*. The *probe size* determines the minimal width of the bottleneck of searched tunnels whereas the *shell radius* determines the surface of the molecule. The computation of a tunnel is terminated when the tunnel intersects the surface of the protein.

The computation is driven by an ordered sequence of points of interest in which the pathways should be searched for. The algorithm then detects two types of pathways: (1) pathways connecting points of interest in a given order, (2) tunnels from the outside environment to the first point of interest and tunnels connecting the last point of interest with the outer environment.

Step 2 – Selecting Relevant Subsets of Atoms: Firstly, the center point C of the line connecting points (P_0, P_1) is determined and planes $p(P_0), p(P_1), p(C)$ perpendicular to line (P_0, P_1) are detected. According to these planes, two additional subsets of relevant atoms are constructed. The first subset $A(P_0)$ contains atoms geometrically positioned between $p(P_0)$ and $p(C)$ and the second subset $A(P_1)$ consists of all atoms between $p(C)$ and $p(P_1)$. These subsets are illustrated in Fig. 1 in the right part, where two subsets corresponding to points of interest P_0, P_1 from the left part of the figure are depicted.

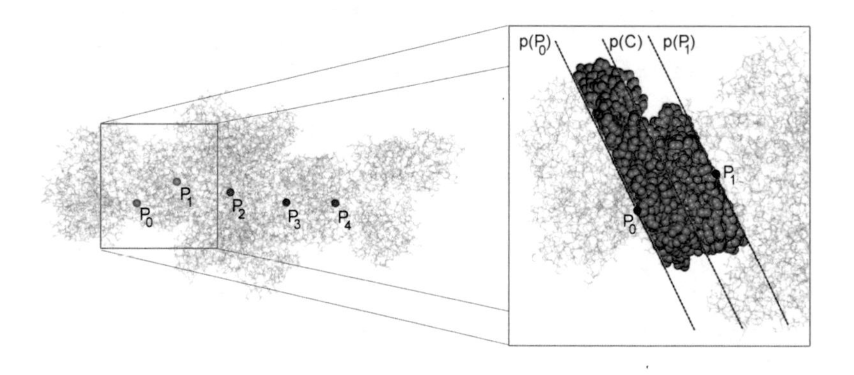

Fig. 1. Carbon Monoxide Dehydrogenase (PDB ID 1OAO). Left: five POIs defined arbitrarily. Right: illustration of two subsets of atoms corresponding to POIs pair P_0, P_1.

Step 3 – Computation of Pathways between Two POIs: The CAVER 3.0 algorithm searching for all relevant tunnels is launched for the site P_0. The tunnel searching is limited only to the set of atoms $A(P_0)$. Similarly, the same process is applied for the site P_1 and the set $A(P_1)$. As a result, the sets of detected tunnels T_0 (T_1 respectively) are obtained. The left part of Fig. 2 shows all tunnels detected between two given points of interest P_0, P_1. Since these sets T_0, T_1 contain all tunnels present in constrained sets of atoms they must be further filtered. Tunnels which do not intersect with $p(C)$ are removed; the remainder are stored for processing in the next phase. Figure 2 (right) shows a filtered subset of tunnels, which are relevant for the further steps of the algorithm.

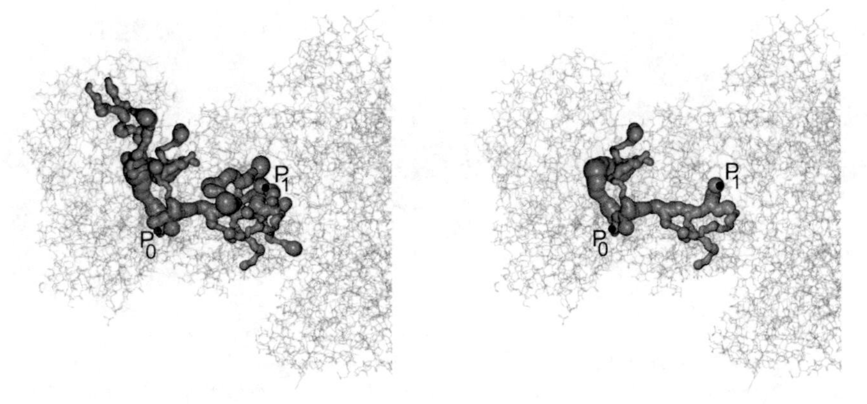

Fig. 2. Left: illustration of all tunnels detected in each subset. Right: tunnels that intersect the plane $p(C)$ are selected, the rest is filtered out.

Step 4 – Concatenation of Computed Tunnels in Constrained Sets: After the selection of relevant tunnels, the connection process is launched. All tunnels are in the form of an ordered set of spheres (see Fig. 3). In order to eliminate cases where a tunnel leads in parallel to the plane $p(C)$, the tunnels are trimmed right after the sphere that intersects $p(C)$. For every tunnel t_i from T_i and tunnel t_{i+1} from T_{i+1}, the test checking if they overlap is performed. A new tunnel $t(P_i, P_{i+1})$ is created if both last spheres from t_i and t_{i+1} overlap or the distance of their centers is lower than the user defined welding threshold. All new $t(P_i, P_{i+1})$ are then inserted into the set T. Figure 3 illustrates this step.

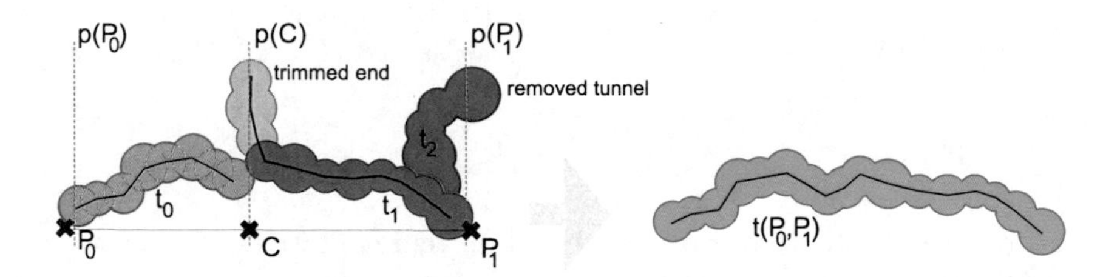

Fig. 3. Left: t_1 is trimmed, t_2 is excluded because it does not intersect $p(C)$. Right: the tunnel $t(P_0, P_1)$ concatenating t_0 with t_1.

Step 5 – Forming the Final Pathway: Finally, the set of partial tunnels T contains all tunnels connecting P_i with P_{i+1}, where $0 = i < n - 1$. For every $t(P_i, P_{i+1})$ from the set T only the path with the best evaluation is selected. This evaluation is based on several tunnel characteristics and their combination, e.g., the tunnel bottleneck, length and curvature. The resulting pathway is obtained by recursive concatenating $t(P_i, P_{i+1})$ with $t(P_{i+1}, P_{i+2})$. Figure 4 (left part) shows the resulting pathway detected between five points of interest in the 1OAO structure. Section 4 contains another examples of intramolecular tunnels detected by our algorithm.

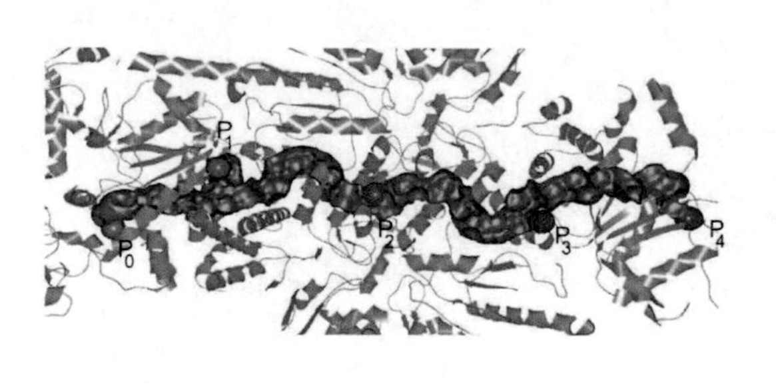

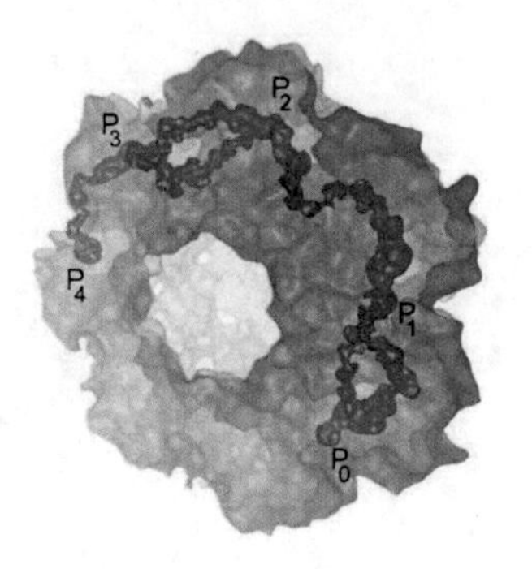

Fig. 4. Tunnels are visualized using solvent accessible surface. Left: 10A0 with the intramolecular pathway connecting five arbitrarily chosen POIs. Right: 1AON with 5 POIs controlling the trajectory of the pathway.

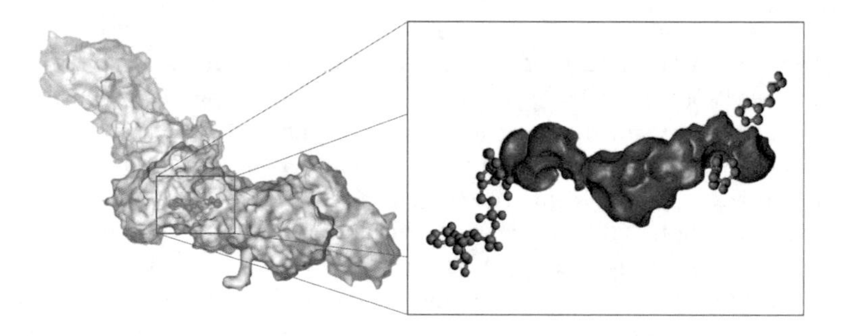

Fig. 5. Intramolecular tunnel of 4-Hydroxy-2-Ketovalerate Aldolase (1NVM) connecting the active site formed by the H21 and Y291 residues (blue) with the active site located near to NAD (green).

4 Results

The correctness of the algorithm was tested on representatives of enzymes possessing multiple active sites connected by internal tunnels for the transport of intermediates. Intramolecular tunnels are commonly present in ammonia-transferring enzymes. Many such enzymes have been already studied in some detail, including the following example structures. Carbamoyl phosphate synthetase (PDB ID of wild type (WT) 1BXR) where the first part of the intramolecular tunnel serves for ammonia transportation whereas the second part enables carbamate transportation. Glucosamine 6-phosphate synthase (WT 2J6H), glutamate synthase (WT 1OFD), imidazole glycerol phosphate synthase (WT 1KA9), cytidine triphosphate synthetase (WT 1VCM) have tunnels for ammonia transportation. Among others, Tryptophan synthase (WT 3CEP) tunnel enables indole transportation and 4-Hydroxy-2-Ketovalerate Aldolase/Acylating Acetaldehyde Dehydrogenase (WT 1NVM) contains the tunnel for acetaldehyde intermediate transport (see Fig. 5).

The algorithm was implemented in Java and integrated into the CAVER Analyst software tool. All tests were performed on a common PC/2.5Ghz in a single-threaded 32-bit environment. Testing on above listed structures and their mutants confirmed the robustness of the algorithm.

Table 1 gives results only for several studied structures. It shows results from testing on real structures (1NVM, 1BKS) and to confirm the robustness and generality of our approach, we performed testing also on several large macromolecules (here examples of 1AON, 1OAO) where 5 points of interest were selected arbitrarily. For cases when active sites inside structures were known, we compared the resulting tunnels with those biochemically relevant presented in various publications (e.g., [5], [15]).

Table 1. $\mathbf{P_0}$, $\mathbf{P_1}$ - selected POIs, $|\mathbf{A}(\mathbf{P_0})| + |\mathbf{A}(\mathbf{P_1})|$ - a reduced number of atoms used when searching for tunnels

PDB	Atoms	$\mathbf{P_0}$	$\mathbf{P_1}$	$\lvert\mathbf{A}(\mathbf{P_0})\rvert + \lvert\mathbf{A}(\mathbf{P_1})\rvert$	Time (s)	Length	Bottleneck
1NVM	19612	x=57.20 y=15.85 z=-7.11	x=66.30 y=30.99 z=10.50	6743	13	35.9Å	0.76Å
1BKS	4906	x=50.87 y=16.98 z=5.67	x=85.51 y=11.98 z=9.59	2930	15	48.44Å	0.72Å
1AON	58804	5 arbitrary sites		avg. 2373	83	484.6Å	0.71Å
1OAO	23226	5 arbitrary sites		avg. 2701	23	153.9Å	0.97Å

5 Conclusion and Future Work

We proposed the novel algorithm for computation of pathways among an arbitrary number of ordered points of interest inside macromolecular structures. Our method is applicable to the detection of intramolecular tunnels as well as other structures, such as channels and pores. In comparison with the existing methods, the algorithm enables an advanced user-driven control of the expected pathway. This is reached by defining the intermediate points of interest. Our approach is general enough to detect inner tunnels by changing only two basic parameters of the original CAVER 3.0 algorithm. It provides users with fast and intuitive tool for geometry-based analysis of various protein structures. We tested the relevance of our algorithm on various molecular structures and provided the reader with an example of the results.

References

1. Agre, P., Brown, D., Nielsen, S.: Aquaporin water channels: unanswered questions and unresolved controversies. Current Opinion in Cell Biology 7, 472–483 (1995)
2. Aurenhammer, F.: Voronoi diagrams: A survey of a fundamental geometric data structure. ACM Computing Surveys 23, 345–405 (1991)

3. Chovancova, E., Pavelka, A., Benes, P., Strnad, O., Brezovsky, J., Kozlikova, B., Gora, A., Sustr, V., Klvana, M., Medek, P., Biedermannova, L., Sochor, J., Damborsky, J.: CAVER 3.0: a Tool for the Analysis of Transport Pathways in Dynamic Protein Structures. PLoS Computational Biology 8(10), e1002708 (2012)
4. Gouaux, E., Mackinnon, R.: Principles of Selective Ion Transport in Channels and Pumps. Science 310, 1461–1465 (2005)
5. Gora, A., Brezovsky, J., Damborsky, J.: Gates of Enzymes. Chemical Reviews 113(8), 5871–5923 (2013)
6. Huang, X., Holden, H.M., Raushel, F.M.: Channeling of Substrates and Intermediates in Enzyme-Catalyzed Reactions. Annual Review of Biochemistry 70, 149–180 (2001)
7. Klvana, M., Pavlova, M., Koudelakova, T., Chaloupkova, R., Dvorak, P., Prokop, Z., Stsiapanava, A., Kuty, M., Kuta-Smatanova, I., Dohnalek, J., Kulhanek, P., Wade, R.C., Damborsky, J.: Pathways and Mechanisms for Product Release in the Engineered Haloalkane Dehalogenases Explored Using Classical and Random Acceleration Molecular Dynamics Simulations. Journal of Molecular Biology 392, 1339–1356 (2009)
8. Pellegrini-Calace, M., Maiwald, T., Thornton, J.M.: PoreWalker: A Novel Tool for the Identification and Characterization of Channels in Transmembrane Proteins from Their Three-Dimensional Structure. PLoS Computational Biology 5(7), e1000440 (2009)
9. Porter, C.T., Bartlett, G.J., Thornton, J.M.: The Catalytic Site Atlas: A Resource of Catalytic Sites and Residues Identified in Enzymes Using Structural Data. Nucleid Acids Research 32 (2004)
10. Raushel, F.M., Thoden, J.B., Holden, H.M.: Enzymes with Molecular Tunnels. Accounts of Chemical Research 36, 539–548 (2003)
11. Sehnal, D., Svobodova Varekova, R., Berka, K., Pravda, L., Navratilova, V., Banas, P., Ionescu, C.M., Otyepka, M., Koca, J.: MOLE 2.0: Advanced Approach for Analysis of Biomacromolecular Channels. Journal of Cheminformatics 5 (2013)
12. Smart, O.S., Neduvelil, J.G., Wang, X., Wallace, B.A., Sansom, M.S.: HOLE: A Program for the Analysis of the Pore Dimensions of Ion Channel Structural Models. Journal of Molecular Graphics 14(6), 354–360 (1996)
13. The UniProt Consortium: Reorganizing the Protein Space at the Universal Protein Resource (UniProt). Nucleic Acids Research (2011), doi:10.1093/nar/gkr981
14. Yaffe, E., Fishelovitch, D., Wolfson, H.J., Halperin, D., Nussinov, R.: MolAxis: A Server for Identification of Channels in Macromolecules. Nucleid Acids Research 36, W210–W215 (2008)
15. Floquet, N., Mouilleron, S., Daher, R., Maigret, B., Badet, B., Badet-Denisot, M.A.: Ammonia Channeling in Bacterial Glucosamine-6-Phosphate Synthase (Glms): Molecular Dynamics Simulations and Kinetic Studies of Protein Mutants. FEBS Letters 581(16), 2981–2987 (2007)

Improving Positive Unlabeled Learning Algorithms for Protein Interaction Prediction

Doruk Pancaroglu and Mehmet Tan

Department of Computer Engineering, TOBB University
of Economics and Technology, Ankara, Turkey
{dpancaroglu,mtan}@etu.edu.tr

Abstract. In binary classification, it is sometimes difficult to label two training samples as negative. The aforementioned difficulty in obtaining true negative samples created a need for learning algorithms which does not use negative samples. This study aims to improve upon two PU learning algorithms, AGPS[2] and Roc-SVM[3] for protein interaction prediction. Two extensions to these algorithms is proposed; the first one is to use Random Forests as the classifier instead of support vector machines and the second is to combine the results of AGPS and Roc-SVM using a voting system. After these two approaches are implemented, their results was compared to the original algorithms as well as two well-known learning algorithms, ARACNE [9] and CLR [10]. In the comparisons, both the Random Forest (called AGPS-RF and Roc-RF) and the Hybrid algorithm performed well against the original SVM-classified ones. The improved algorithms also performed well against ARACNE and CLR.

Keywords: Protein Interaction Networks, Binary Classification, Positive Unlabeled Learning, Random Forests, Support Vector Machines.

1 Introduction

Supervised machine learning algorithms generally utilize both positive and negative examples for training purposes. However, in many domains such as medicine, procuring negative examples is harder and in many cases, not possible. For some cases, the negative examples are simply not that 'interesting' to keep in a database for further analysis, such as the non-interaction of a protein and a ligand. This problem created a need for algorithms that are using positive and unlabeled examples. Thus, Positive Unlabeled (PU) learning algorithms are proposed for this purpose: learning without negative examples.

While PU learning algorithms are used in many different fields (such as text categorization), the scope of this paper is the protein-protein interaction networks (PPI Networks). A PPI Network is a graph, where a node in the graph is a protein, and an edge represents the presence of interaction between two proteins. It is accepted that given a set of proteins, PPI network is sparse; the number of interacting proteins are almost always much smaller than the number of non-interacting proteins.

Interactions between two proteins can be tested in a laboratory environment. A positive interaction result in the laboratory can be used as is, (but does not necessarily mean that they actually do some work together due to localizations of the proteins in

J. Sáez-Rodríguez et al. (eds.), *8th International Conference on Practical Appl. of Comput. Biol. & Bioinform. (PACBB 2014)*, Advances in Intelligent Systems and Computing 294,
DOI: 10.1007/978-3-319-07581-5_10, © Springer International Publishing Switzerland 2014

the body), but the lack of interaction does not imply that two proteins will never interact. In theory, an interaction may not occur because of environmental conditions, or simply, time constraints. This is one of the reasons of the difficulty of obtaining negative examples; it is not easy to prove that two proteins will never interact.

Previously, in our group, we performed a comprehensive assessment of some PU Learning methods [1]. The methods analyzed in the study are AGPS [2], Roc-SVM [3], PSoL [4] (including $PSoL_m$ and $PSoL_o$), Carter [5], PosOnly [6], Bagging SVM [7] and S-EM [8]. Two algorithms that stand out in the comparison are AGPS and Roc-SVM. Both are named as two step algorithms (first extract some reliable negative examples, then perform classical binary classification) and both algorithms produced comparable results in Precision, Recall, F-measure and Matthew Correlation Coefficient (MCC) values. Both algorithms use support vector machines for classification.

In this paper, we aimed to improve the performance of AGPS and Roc-SVM algorithms by different means, due to the promising results in the mentioned study. First, we overviewed the two algorithms and changed the classification method to Random Forest (RF) from SVM. Then we created a method to merge the algorithms' results by voting and proposed a hybrid approach. We compared the results of the separate algorithms and the hybrid one. Finally, we compared the results of the proposed methods with two well-known biological network inference algorithms: ARACNE [9] and CLR [10].

The organization of the paper is as follows: In section 2, we overview the AGPS and Roc-SVM algorithms. In section 3, we describe the use of Random Forest in the algorithms and the hybrid approach. In section 4, we report the performance of the proposed algorithms and compare them to the originals. In section 5, we summarize our work and conclude the paper.

2 Background

2.1 AGPS

AGPS (Annotating Genes With Positive Samples), proposed by Zhao X-M et al. [2] is a two-step algorithm for gene function prediction by PU learning. AGPS consists of three main steps; generating the initial negative set, expanding the negative set and classification. For further details about AGPS please refer to [2] or our previous work [1].

2.2 Roc-SVM

Roc-SVM (Rocchio Technique and SVM), created by Li X. and Liu B. [3] is a two-step algorithm that combines the Rocchio method used to select strong negatives and SVM to create classifiers. Again, the libSVM library created by C. Chang and C. Lin [11] is used for Roc-SVM. For further details about Roc-SVM please refer to [3] or our previous work [1].

3 Proposed Algorithms

3.1 Random Forest Approach

The random forest approach aims to implement the random forest classification method instead of the SVM classification method used in AGPS and Roc-SVM. The algorithms that utilize random forest are named AGPS-RF and Roc-RF.

Random forests method was developed by Leo Breiman [12]. It can be called as an ensemble learning method, which basically generates many decision trees and combine (by using aggregation or voting) their respective results. Random forests use this voting approach, which is called bagging (bootstrap aggregation). Classifiers constructed in bagging are independent from each other. The difference that random forests bring is the additional randomness in the bagging phase.

Purpose & Aims of Using Random Forest. The motivation behind using random forests classifier instead of SVM is to use random forests' ensemble approach, i.e. using multiple classifiers and combining the results. The current state of AGPS and Roc-SVM use a single, final classifier. Manually storing multiple classifiers during runtime and creating an ensemble method failed, as the classifiers were too large to create an accurate prediction.

In comparisons such as the one made by J. Nappi, D. Regge and H. Yoshida [13], the one made by Y. Tang, S. Krasser, Y. He, W. Yang and D. Alperovitch [14], and the one made by G. Rios and H. Zha [15], random forests are found to perform well (yet the margin varies in comparisons) against SVM while working with microarray data. As our data is two-class, both random forest and SVM can work in optimal conditions. The E.coli gene expression data set we used in our test is rather large, and random forest is found to perform well in large data sets as well.

Implementing Random Forest. The implementation requires all the training and prediction done by SVM to be replaced by random forest classifiers. The algorithms can be found below in Algorithm 1 and 2. (Q/U = Unlabeled examples, RN = Strong Negatives, P = Positive Examples, P1/2 = Parts of P, N = Negative Examples)

1. Learning Step

— U_{new} = Ku + P2;

(a) Initial Negative Set Generation

- Construct the classifier f(1) using P1 and U_{new} (one-class SVM is used because random forest does not support one class)
- Classify Unew using f(1). The negative set N(1) will be used in Stage B as the initial negative set
- $U_{new} = U_{new}$ - N(1).

(b) Negative Set Expansion (Iterative)

- Random forest instance I(i) is trained using P(1) and N(1)
- The training instance is evaluated and an evaluation model is created
- A classifier tree T(i) is built using the trained instance I(i)
- U_{new} is classified using T(i), while N(2) is the predicted negative set, where $|N(2)| \leq k|P1|$
- $U_{new} = U_{new}$ - N(2)
- Continue until $| U_{new} | < k | P1 |$

(c) Classifier and Negative Set Selection

- Classify U_{new} using the classifiers produced in Stage B
- Find the best classifier (by comparing their prediction accuracies) and take the best classifiers' negative set as TNS
- Return negative set TNS

2. Classification

— Classify U using P and TNS, where P = P1 + P2

Algorithm 1. AGPS-RF algorithm

1. Step

- Initial negative set generation is done, producing RN
- RN is subtracted from Q to produce Unew.
- Q is divided by 10 (10-fold cross validation) to produce U_{evo}.
- U_{evo} will be the set of unlabeled examples that will be classified at the end.

2. Step

- Train the classifier f(i) using P and RN, where i is the iteration number and increases by one at each iteration.
- Store the first classifier in the iteration, f(1), for later use.

3. Step

- U_{new} is classified using f(i).
- The examples that are negative in the classification result of Unew are named N(i).

4. Step

- If there are no negatives, i.e. the N(i) is empty, stop and go to Step 5.
- If there are negatives in N(i), subtract N(i) from U_{new}
- After the subtraction, go to Step 2.

5. Step

- The last classifier produced in the iteration is named f(last).
- P is classified using f(last).

6. Step

- If the results produced from the classification of P using f(i) in Step 5 have more than 5% of the total number of examples classified as negative, designate f(1) (the first classifier built in the algorithm) as the final classifier.
- If less than 5% of the total number of examples are classified as negative, f(last) is retained as the final classifier.

7. Step

- Classify U_{evo} (produced in Step 1) using the final classifier, which is chosen in Step 6.

Algorithm 2. Roc-RF algorithm

3.2 Hybrid Approach

Purpose & Aims of Using Hybrid Approach. The hybrid approach aims to provide a union of both algorithms via comparing the results. It is different than an ensemble classifier because it does not produce multiple classifiers and vote their respective results. The motivation for using hybrid approach is to bring on a voting scheme that can be used on the AGPS and Roc-SVM algorithms. As the Random Forests classification method mentioned in the previous chapter also utilizes a voting scheme inherently, we aimed to implement a voting aspect on the original AGPS and Roc-SVM algorithms.

Implementation. The voting process compares the results of the two algorithms for the same example and sets the result of a given example if two algorithms' results are same. For example, if AGPS and Roc-SVM both produce positive results for the presence of a relation between two proteins, then the hybrid result is also positive. However, if the results of AGPS and Roc-SVM differ on a given example, the probability values that show whether the classification is correct or not is taken into account (The values are produced at runtime). The result whose probability is more certain is selected as the stronger one and that algorithm's result is taken into account.

4 Experimental Results

4.1 Setting

The dataset used in the evaluation is the Escherichia coli (E. coli) gene expression data set by Faith et al. [16]. The data set contains 4345 genes and each gene has 445 samples. IntAct protein-protein interaction database [17] is used as the set of known protein interactions. This set constitutes the positive set in our classification. This is the same data set we used in our previous comparison study [1], enabling us to compare our results accurately. (For further details of the dataset, please refer to our previous work [1]).

To implement a random forest classifier in our algorithms coded in java, we used WEKA's [18] random forest libraries. As WEKA is also written in java, it provided greater operability and ease of use.

F-Measure and Matthews Correlation Coefficient (MCC) are used to compare and evaluate the performance of the algorithms. These values are calculated using the number of True/False Positive and True/False Negative examples classified by algorithms.

Table 1. Average performances of AGPS-RF, Roc-RF and Hybrid algorithms compared to the original AGPS and Roc-SVM

Measure/Algorithm	AGPS-RF	AGPS	Roc-RF	Roc-SVM	Hybrid
F-measure	0.214	0.212	0.593	0.228	0.515
MCC	0.130	0.133	0.567	0.261	0.294

4.2 Results for AGPS-RF, Roc-RF and Hybrid Algorithms

Though the results from AGPS-RF and Roc-RF have both improved over their original counterparts, AGPS-RF had a very slight increase while Roc-RF had a very large one. This can be attributed to the Rocchio method and the wholly iterative nature of the Roc-RF algorithm. In AGPS, the average F-measure was 0.21246, while in AGPS- RF, the average F-measure was 0.21420. This points to a slight (insignificant) increase in F-measure. The improvement is much more significant in Roc-RF. In Roc-SVM, the average F-measure was 0.2284, while in Roc-RF, the average F-measure was 0.59222. The increase is 159.2% compared to Roc-SVM.

Compared to AGPS's and Roc-SVM's values of average F-measures (which were 0.21246 and 0.22840 respectively), the hybrid approach yielded an average F-measure of 0.515, which is respectively a 140% and a 125% improvement over both AGPS and Roc-SVM. The results for AGPS-RF, Roc-RF and the Hybrid Algorithm can be seen in Table 1.

4.3 Comparison to other Network Inference Algorithms

In the previous sections, we reported the comparison results for AGPS-RF, Roc-RF and hybrid algorithms with the original AGPS and Roc-SVM. As the proposed algorithms actually derive a biological network, the next step in evaluation would be to compare our algorithms to well known network inference algorithms that use microarray data. This section reports the results of the experiments that compare our methods to two biological network derivation algorithms, CLR and ARACNE. Although these algorithms are not supervised (i.e. they do not exploit previously known interactions as input) this evaluation can show the relative performances of the proposed algorithms.

CLR (Context Likelihood of Relatedness) is an algorithm developed by J. Faith et al [10]. It is defined as a novel extension to the relevance network algorithms. CLR is essentially a relevance network algorithm, but it improves this by introducing a background correction step for eliminating falsely identified relations. CLR makes predictions using the statistical likelihood of each information value within the network. The pairs with the highest probabilities are assumed to be interacting.

ARACNE (Algorithm for the Reconstruction of Accurate Cellular Networks) is a novel algorithm created by Margolin et al. in 2006 [9]. It uses the same concept of using graph nodes to represent examples and edges to represent the presence of an interaction between two examples. ARACNE classifies the examples by inferring statistical dependencies from pairwise marginals. Similar to the relevance network base of the CLR algorithm, the probabilities that fail to pass a certain threshold are eliminated. The second step continues by eliminating the indirect interactions between the remaining mutual interaction pairs to reduce false positives.

The evaluation results of CLR and ARACNE run using our datasets are shown in Table 2. The table also has the F-measure values of AGPS-RF, Roc-RF and the Hybrid algorithm for easy comparison.

Table 2. Comparison to CLR and ARACNE

Measure/Algorithm	ARACNE	CLR	AGPS-RF	Roc-RF	Hybrid
F-measure	0.315	0.401	0.214	0.592	0.515
MCC	0.262	0.376	0.130	0.567	0.294

Our E.Coli data run with CLR produced an average F-measure of 0,40. ARACNE performed with an average F-measure of 0,315.

AGPS-RF yielded an average F-measure of 0,214 in our tests. CLR provides a 46% better average F-measure than AGPS-RF's average F-measure. ARACNE's average F-measure of 0,315 is 32% better than AGPS-RF's average F-measure.

Roc-RF produced an average F-measure value of 0,592 in our tests. This value is 32,4% better than CLR's average F-measure and 46,7% better than ARACNE's average F-measure.

Our hybrid algorithm produced an average F-measure of 0,515 in our tests. This value is 22% better than CLR's average F-measure and 38,8% better than ARACNE's average F-measure.

In these comparison tests with ARACNE and CLR, Roc-RF performed better than ARACNE and CLR in terms of average F-measure. This is consistent with the respective improvements of AGPS-RF and Roc-RF too, wherein AGPS-RF had a very small increase in average F-measure while Roc-RF had a large increase in average F-measure. The Hybrid Algorithm also performed well against ARACNE and CLR. On the other hand, AGPS-RF did not perform better in terms of average F-measure.

5 Conclusion

In this paper, we proposed two improvements over AGPS and Roc-SVM algorithms previously proposed for positive unlabeled learning. In our evaluations, the random forest approach, which is applied separately in AGPS-RF and Roc-RF had better results than their original SVM counterparts. A hybrid algorithm for AGPS and Roc-SVM is also implemented and compared to the newly proposed algorithms. Finally, we also compared all the proposed algorithms to CLR and ARACNE, two well-known methods for biological network inference from microarray data.

Compared to other well-known algorithms such as CLR and ARACNE, random forest approach produced different results. While the AGPS-RF algorithm produced an average F-measure that is lower than CLR's or ARACNE's, the Roc-RF algorithm produced an average F-measure value that is higher than both CLR's and ARACNE's. It can be concluded that the Roc-RF is worth further investigations for improvement as Roc-RF is superior to all algorithms in terms of all measures.

The Hybrid algorithm, on the other hand, proved a better average F-measure value compared to single AGPS or Roc-SVM algorithms. Compared to CLR and ARACNE, the Hybrid algorithm did well too, but Roc-RF performed significantly better.

For future extensions to this work, we plan to apply Roc-RF to other biological network inference problems such as drug-target or miRNA-mRNA interaction prediction. These problems include heterogeneous networks, Roc-RF algorithm can be applied to these domains easily with slight modifications.

References

1. Kilic, C., Tan, M.: Positive unlabelled learning for deriving protein interaction networks. Netw. Modeling Anal. in Health Inform. and Bioinform. 1(3), 87–102 (2012)
2. Zhao, X.-M., Wang, Y., Chen, L., Aihara, K.: Gene function prediction using labeled and unlabeled data. BMC Bioinformatics 9, 57 (2008)
3. Li, X., Liu, B.: Learning to classify texts using positive and unlabeled data. In: IJCAI 2003: Proceedings of the 18th International Joint Conference on Artificial Intelligence, pp. 587–592 (2003)
4. Wang, C., Ding, C., Meraz, R.F., Holbrook, S.R.: PSoL: a positive sample only learning algorithm for finding non-coding RNA genes. Bioinformatics 22(21), 2590–2596 (2006)
5. Carter, R.J., Dubchak, I., Holbrook, S.R.: A computational approach to identify genes for functional RNAs in genomic sequences. Nucleic Acids Res. 29(19), 3928–3938 (2001)
6. Elkan, C., Noto, K.: Learning classifiers from only positive and unlabeled data. In: KDD 2008: Proceeding of the 14th ACM SIGKDD International Conference on Knowledge Discovery and Data Mining, pp. 213–220. ACM, New York (2008)
7. Mordelet, F., Vert, J.-P.: A bagging SVM to learn from positive and unlabeled examples (2010)
8. Liu, B., Lee, W.S., Yu, P.S., Li, X.: Partially supervised classification of text documents. In: Proceedings of the Nineteenth International Conference on Machine Learning, ICML (2002)

9. Margolin, A.A., Nemenman, I., Basso, K., Wiggins, C., Stolovitzky, G., Dalla Favera, R., Califano, A.: ARACNE: an algorithm for the reconstruction of gene regulatory networks in a mammalian cellular context. BMC Bioinformatics 7(suppl. 1), S7 (2006)
10. Faith, J.J., Hayete, B., Thaden, J.T., Mogno, I., Wierzbowski, J., et al.: Large-Scale Mapping and Validation of Escherichia coli Transcriptional Regulation from a Compendium of Expression Profiles. PLoS Biol. 5(1), e8 (2007), doi:10.1371/journal.pbio.0050008
11. Chang, C.-C., Lin, C.-J.: LIBSVM: a library for support vector machines. ACM Trans. Intell. Syst. Technol. 2, 27:1–27:27 (2011)
12. Statistics, L.B., Breiman, L.: Random Forests. Machine Learning, 5–32 (2001)
13. Näppi, J.J., Regge, D., Yoshida, H.: Comparative Performance of Random Forest and Support Vector Machine Classifiers for Detection of Colorectal Lesions in CT Colonography. In: Yoshida, H., Sakas, G., Linguraru, M.G. (eds.) Abdominal Imaging. LNCS, vol. 7029, pp. 27–34. Springer, Heidelberg (2012)
14. Tang, Y., Krasser, S., He, Y., Yang, W., Alperovitch, D.: Support Vector Machines and Random Forests Modeling for Spam Senders Behavior Analysis. In: Proceedings of IEEE Global Communications Conference (IEEE GLOBECOM 2008), Computer and Communications Network Security Symposium, New Orleans, LA (2008)
15. Rios, G., Zha, H.: Exploring support vector machines and random forests for spam detection. In: Proceedings of the First Conference on Email and Anti-Spam, Mountain View, CA, USA (2004)
16. Faith, et al.: Many microbe microarrays database: uniformly normalized affymetrix compendia with structured experimental metadata. Nucleic Acids Res. 36(Database issue), D866–D870 (2008), doi:10.1093/nar/gkr1088
17. Kerrien, S., Aranda, B., Breuza, L., Bridge, A., Broackes-Carter, F., Chen, C., Duesbury, M., Dumousseau, M., Feuermann, M., Hinz, U., Jandrasits, C., Jimenez, R.C., Khadake, J., Mahadevan, U., Masson, P., Pedruzzi, I., Pfeiffenberger, E., Porras, P., Raghunath, A., Roechert, B., Orchard1, S., Hermjakob, H.: The IntAct molecular interaction database in 2012. Nucleic Acids Res. 40(1), D841–D846 (2011), doi:10.1093/nar/gkr1088
18. Witten, H., Frank, E.: Data Mining: Practical Machine Learning Tools and Techniques with Java Implementations. Morgan Kaufmann (October 1999), http://www.cs.waikato.ac.nz/ml/weka/

Finding Class C GPCR Subtype-Discriminating N-grams through Feature Selection

Caroline König[1], René Alquézar[1,2], Alfredo Vellido[1,3], and Jesús Giraldo[4]

[1] Departament de Llenguatges i Sistemes Informàtics, Univ. Politècnica de Catalunya, BarcelonaTech, 08034, Barcelona, Spain
[2] Institut de Robòtica i Informàtica Industrial, CSIC-UPC, 08034, Barcelona, Spain
[3] Centro de Investigación Biomédica en Red en Bioingeniería, Biomateriales y Nanomedicina (CIBER-BBN), 08193, Cerdanyola del Vallès, Spain
[4] Institut de Neurociències - Unitat de Bioestadìstica, Univ. Autònoma de Barcelona, 08193, Cerdanyola del Vallès, Spain
{ckonig,alquezar,avellido}@lsi.upc.edu, jesus.giraldo@uab.es

Abstract. G protein-coupled receptors (GPCRs) are a large and heterogeneous superfamily of receptors that are key cell players for their role as extracellular signal transmitters. Class C GPCRs, in particular, are of great interest in pharmacology. The lack of knowledge about their full 3-D structure prompts the use of their primary amino acid sequences for the construction of robust classifiers, capable of discriminating their different subtypes. In this paper, we describe the use of feature selection techniques to build Support Vector Machine (SVM)-based classification models from selected receptor subsequences described as n-grams. We show that this approach to classification is useful for finding class C GPCR subtype-specific motifs.

Keywords: G-Protein coupled receptors, pharmaco-proteomics, feature selection, n-grams, support vector machines.

1 Introduction

G protein-coupled receptors (GPCRs) are cell membrane proteins with a key role in regulating the function of cells due to their transmembrane location. This is the result of their ability to transmit extracellular signals, activating intra-cellular signal transduction pathways, ability that makes them particularly attractive for pharmacological research.

The functionality of a protein depends at large on its structural configuration in 3-D, which determines its ability for a given ligand binding. Despite active research, the 3-D structure is currently only determined in full for approximately a 12% of the human GPCR superfamily [6]. As a result, GPCR classes that lack a known 3-D structure require alternatives such as the analysis of their primary amino acid sequence, which is well-known and reported in many open curated databases.

J. Sáez-Rodríguez et al. (eds.), *8th International Conference on Practical Appl. of Comput. Biol. & Bioinform. (PACBB 2014)*, Advances in Intelligent Systems and Computing 294,
DOI: 10.1007/978-3-319-07581-5_11,

This paper specifically focuses on the class C subset of a publicly available GPCR database. These data were analyzed in a previous study [8] using a supervised, multi-class classification approach that yielded relatively high accuracies in the discrimination of the seven constituting subtypes of the class. This previous work used several transformations based on the physicochemical properties of the sequence amino acids. In the current study, we go one step further and apply feature selection prior to classification with SVMs from n-gram subsequence features. A relevant objective of this work is the analysis of the constructed classifiers in order to find subfamily-specific motifs that might reveal information about ligand binding processes. A further motivation for this study is the fact that no major motifs are currently known for class C GPCRs [11].

2 Materials

GPCRs are cell membrane proteins that transmit signals from the extracellular to the intracellular domain, prompting cellular response. This makes them of great relevance in pharmacology. The GPCRDB [12], a popular curated database of GPCRs, divides the superfamily into five major classes (namely, A to E) based on ligand types, functions, and sequence similarities. As stated in the introduction, this study concerns class C, which has of late become an increasingly important target for new therapies, particularly in areas such as pain, anxiety, neurodegenerative disorders and as antispasmodics.

The investigated data (from version 11.3.4 as of March 2011) comprises of 1,510 class C GPCR sequences, belonging to seven subfamilies: 351 metabotropic glutamate (mG), 48 calcium sensing (CS), 208 GABA-B (GB), 344 vomeronasal (VN), 392 pheromone (Ph), 102 odorant (Od) and 65 taste (Ta).

3 Methods

In this work, SVMs were used for the supervised classification of the alignment-free amino acid sequences into the seven subclasses of class C GPCRs. Given the multi-class problem setting, the svmLib implementation [2] was used. The amino acid sequences of varying lengths were first transformed into fixed-size feature representations. For this, we used in previous work transformations based on the physicochemical properties of the sequences [8]. Instead, in this work we use short protein subsequences in the form of n-gram features. The n-grams were created from three different existing alphabets that have previously been used for the classification of GPCR sequences [4]. Different feature selection methods are also used to reduce the dimensionality of the data with the objective of finding the parsimonious set of n-grams that might best discriminate the class C subtypes.

3.1 Amino Acid Alphabets

According to [5], many amino acids have similar phisicochemical properties, which makes them equivalent at a functional level. An appropriate grouping of amino acids reduces the size of the alphabet and may decrease noise. In this work, besides the basic 20-amino acid alphabet, we used two alternative amino acid groupings (See Table 1): the Sezerman (SEZ) alphabet, which includes 11 groups, and the Davies Random (DAV), including 9 groups. They have both been evaluated [4] in the classification of GPCRs into their 5 major classes.

Table 1. Amino acid grouping schemes

GROUPING	1	2	3	4	5	6	7	8	9	0	X
SEZ	IVLM	RKH	DE	QN	ST	A	GT	W	C	YF	P
DAV	SG	DVIA	RQN	KP	WHY	C	LE	MF	T		

3.2 N-grams

The concept of n-grams has widely been used in protein analysis ([1],[9]). A successful application of text classification methods for the classification of class A GPCRs was presented in [3]. While a discretization of the n-gram features was used in that study, we instead used the relative frequencies of the n-grams, which are non-discrete variables. Therefore, the n-gram feature representation corresponds here to the measurement of the relative frequency of each n-gram in a sequence. Due to the exponential growth of the size of n-grams, we limit the reported research to n-grams of size 1,2 and 3.

3.3 Feature Selection

Many irrelevant features are likely to exist in the different n-gram frequency representations of the data. To ameliorate the classification process by minimizing the negative impact of irrelevant features, we used two different feature selection approaches in this study: sequential forward feature selection with an SVM-classifier and a filter method computing two-sample t-tests among the C GPCR subtypes.

A sequential forward selection algorithm [7] was used to find the reduced set of features that best discriminated the data subtypes. This kind of algorithm is a so called wrapper method, where the classification model search is performed within the subset feature search [10].

This algorithm starts from an empty candidate feature set and adds, in each iteration, the feature which most improves the accuracy (i.e., that which minimizes the misclassification rate). The algorithm uses an SVM classifier in which the accuracy is evaluated using a 5-CV to test the candidate feature set. The algorithm stops when the addition of a further feature does not increase the accuracy over a threshold set at $1e^{-6}$.

A two-sample t-test was used to evaluate the discriminating power of each feature as a filtering approach. This univariate statistical test analyzes whether there are foundations to consider two independent samples as coming from populations (normal distributions) with unequal means by analyzing the values of the given feature. In our case, we used t-tests with 0.01 confidence. If the t-test suggested that this hypothesis was true (i.e. the null hypothesis was rejected), the feature was considered to significantly distinguish between the two different subtypes of class C GPCRs. As we face a multi-class classification problem, the t-test results were examined for the 21 feasible two-class combinations of the 7 class C subtypes. We decided to calculate the two-sample t-test values at this detail because the multi-class svmLib implementation internally performs a comparison of the data between each class (one-vs-one implementation). Therefore, the t-test exactly evaluates the data considered in each binary classifier, making the ranking of the features possible according to their overall significance (i.e., in how many binary classifiers a feature is significant).

4 Experiments

4.1 N-gram Representation

First, we built classification models with n-grams for each of the three alphabets (AA, SEZ, DAV). Table 2 shows the classification results obtained and the size of the feature set for each alphabet. We observe that the size of the n-gram feature set decreases significantly with the size of the alphabet, but that the best classification results are obtained for the AA alphabet, which is the largest. Nevertheless, the construction of an SVM model with 3-grams for all three alphabets was unsuccessful, probably due to the existence of a large set of irrelevant 3-grams. For this reason, feature selection was implemented.

Table 2. N-gram classification results, where N is the size of a feature set and ACC stands for classification accuracy (ratio of correctly classified sequences)

	AA		SEZ		DAV	
N-GRAM	N	ACC	N	ACC	N	ACC
1-gram	20	0.87	11	0.82	9	0.78
2-gram	400	0.93	121	0.926	81	0.91
1,2-gram	420	0.93	132	0.921	90	0.916

4.2 Sequential Forward Feature Selection

Table 3 shows the results of the sequential forward selection performed on each n-gram dataset. For each alphabet (AA,SEZ,DAV), this table shows a comparison between the original size of the n-grams (N) and the number of selected features found by the algorithm, as well as the corresponding classification accuracy. The experiments show that the feature selection algorithm was successful, as it was able to find, in almost all cases, a reduced subset of features providing approximately the same prediction accuracy. There were two exceptions: in the case of the 1-grams of the SEZ and DAV subsets, the algorithm was not able to reduce the number of features, probably due to the small size of the feature set. The other exception is the 1,2,3-gram feature set of the AA-alphabet: due to the large number of features the computational cost of the forward selection algorithm is too high. For this reason, we decided to apply a filtering method to reduce the candidate feature subset as a previous step to the forward selection.

Table 3. N-gram classification results using feature selection

	AA			SEZ			DAV		
N-GRAM	N	FS	ACC	N	FS	ACC	N	FS	ACC
1-gram	20	17	0.88	11	-	-	9	-	-
2-gram	400	48	0.93	121	25	0.906	81	31	0.9
1,2-gram	420	54	0.926	131	37	0.916	90	42	0.92
1,2,3-gram	8420	-	-	1331	34	0.925	818	34	0.923

4.3 t-Test Filtering

In order to handle the 1,2,3-gram feature sets, which, due to their size, were either impossible or very difficult to use in the previous methods, we decided to use the t-test filtering method to establish a ranking of the features. Table 4 shows this ranking according to the overall significance of the attributes. This means that, for each alphabet, we counted how many features were significant (column N) in at least 20,19,18, etc. two-class tests. The ACC values shown for each subset are the classification accuracies of a SVM-classifier built on each feature set.

These results provide evidence of the usefulness of this simple ranking, as we were able to find subsets that outperform the classification accuracies obtained with the previous methods. For example, the 1,2,3-gram representation of the AA alphabet achieves an accuracy of 0.943 with 585 attributes, whereas the 2-gram representation achieves a 0.93. In the case of the SEZ alphabet, an accuracy of 0.943 was obtained with this filtered 1,2,3-gram representation, as compared to 0.926 with the 2-gram representation. Using the DAV alphabet, we found a subset with 238 features that yielded a 0.933 accuracy, whereas the 1,2,3-gram representation with forward selection yielded a 0.92.

4.4 t-Test Filtering and Forward Selection

The filtering method described in the previous section found feature subsets with high classification accuracy. Nevertheless, given their high dimensionality, we decided to apply the forward selection algorithm to these subsets. Table 5 shows the results of applying forward selection starting from the n-gram subset reported in the last row of Table 4 (features relevant in at least 12 classifiers), for each alphabet. The initial number of features (FEAT), the number of selected features (N) and the corresponding classification accuracies are shown. Forward selection was quite successful at reducing the number of attributes while retaining an accuracy of approximately 0.94 in all three cases.

Table 4. t-test subset selection

	AA		SEZ		DAV	
SIGNIF	N	ACC	N	ACC	N	ACC
20	1	0.37	2	0.5	0	-
19	15	0.88	8	0.77	10	0.83
18	49	0.931	39	0.9	23	0.88
17	105	0.933	79	0.922	58	0.91
16	212	0.937	149	0.93	99	0.92
15	357	0.936	253	0.936	164	0.926
14	585	0.943	386	0.935	238	0.933
13	909	0.937	505	0.943	325	0.93
12	1284	0.942	633	0.94	429	0.927

Table 5. Forward selection on 12- t-test subsets

AA			SEZ			DAV		
FEAT	N	ACC	FEAT	N	ACC	FEAT	N	ACC
1284	49	0.939	633	59	0.939	429	60	0.94

4.5 Discussion

N-grams and Feature Selection. The experimental results have shown the interest of using feature selection: data dimensionality can be notably reduced without compromising classification quality. Forward selection has been shown to be an effective method, although is computationally too costly when the size of the feature set increases. In this situation, a fast univariate t-test filtering method becomes an appropriate solution to reduce the feature candidate set as a preprocessing step of the forward selection algorithm.

Analysis of t-Test Values. An analysis of the t-test values (hypothesis value and *p*-value) allows measuring to what degree a feature discriminates between two classes. Test values are first analyzed to detect the 3-grams with the best discrimination capabilities. We subsequently analyze if these 3-grams may be part of larger n-grams which are also discriminative.

The analysis of the test values of the reduced feature set of the AA alphabet (See Table 5: 49 features: 33 3-grams, 13 2-grams, 3 1-grams) shows that the 3-grams CSL, ITF and FSM are the most significantly discriminative. In particular, CSL is the most significant one according to the t-test values of 20 two-sample tests. This feature was found not to be significant only for the mG-Ph discrimination.

The ITF n-gram is deemed to be significant in 18 tests and an analysis of longer n-grams (results not reported) showed that the the ITFS 4-gram is specially discriminating, with a significant impact on the discrimination of 19 binary

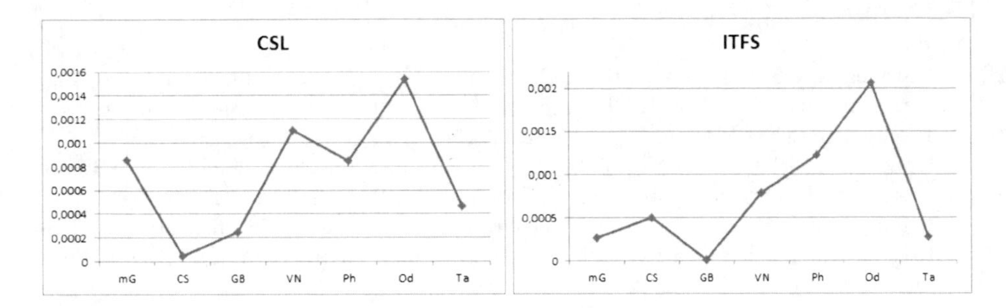

Fig. 1. Mean values of the CSL and ITFS n-gram features for the 7 class C GPCR subtypes

classifiers (i.e., all but mG-Ta and CS-Ta). Furthermore, the ITFSM 5-gram is still highly discriminative, showing significant values for 17 tests.

Another relevant 3-gram is FSM, which is significant for 18 two-class tests. An analysis of longer n-grams showed that the FSML 4-gram is highly discriminative (in 18 tests: all but mG-GB, mG-Ta and GB-Ta). The FSMLI 5-gram was also found to be significant for 15 tests. Figure 1 shows the mean values of n-gram features CSL and ITFS for the 7 class C GPCR subtypes.

5 Conclusions

Class C GPCRs, a family of receptors of great interest in pharmacology, are usually investigated from their primary sequences. This study has addressed the problem of class C GPCR subtype discrimination according to a novel methodology that transforms the sequences according to the frequency of occurrence of the low level n-grams of different amino acid alphabets. This is followed by dimensionality reduction through combination of a two-sample t-test and forward feature selection, as a preprocessing step prior to classification with SVMs. Reduced sets of n-grams that yield similar classification accuracies have been found for each of the three transformation alphabets.

The analysis of the features of the AA alphabet using the values obtained in the t-tests has provided insight about the n-grams that are best at discriminating between the GPCR subtypes. This might be considered as preliminary evidence of the existence of subtype-specific motifs that might reveal information about ligand binding processes. For this reason, the proposed method will be extended in future work to the analysis of larger n-grams. From this analysis, we expect to find larger n-grams that might actually be considered as potentially true subtype-specific motifs.

Acknowledgments. This research was partially funded by MINECO TIN2012-31377 and SAF2010-19257, as well as Fundació La Marató de TV3 110230 projects.

References

1. Caragea, C., Silvescu, A., Mitra, P.: Protein Sequence Classification Using Feature Hashing. In: 2011 IEEE International Conference on Bioinformatics and Biomedicine (BIBM), pp. 538–543. IEEE (2011)
2. Chang, C., Lin, C.: LIBSVM: A Library for Support Vector Machines. ACM Transactions on Intelligent Systems and Technology 2(3), 27:1–27:27 (2011)
3. Cheng, B., Carbonell, J., Klein-Seetharaman, J.: Protein classification based on text document classification techniques. Proteins: Structure, Function, and Bioinformatics 58(4), 955–970 (2005)
4. Can Cobanoglu, M., Saygin, Y.l., Sezerman, U.: Classification of GPCRs Using Family Specific Motifs. IEEE/ACM Transactions on Computational Biology and Bioinformatics 8(6), 1495–1508 (2011)
5. Davies, M.N., Secker, A., Freitas, A., Clark, E., Timmis, J., Flower, D.R.: Optimizing amino acid groupings for GPCR classification. Bioinformatics 24(18), 1980–1986 (2008)
6. Katritch, V., Cherezov, V., Stevens, R.C.: Structure-Function of the G Protein Coupled Receptor Superfamily. Annual Review of Pharmacology and Toxicology 53(1), 531–556 (2013)
7. Kittler, J.: Feature Set Search Algorithms. In: Chen, C.H. (ed.) Pattern Recognition and Signal Processing, pp. 41–60. Sijthoff and Noordhoff, Alphen aan den Rijn (1978)
8. König, C., Cruz-Barbosa, R., Alquézar, R., Vellido, A.: SVM-based classification of class C GPCRs from alignment-free physicochemical transformations of their sequences. In: Petrosino, A., Maddalena, L., Pala, P. (eds.) ICIAP 2013 Workshops. LNCS, vol. 8158, pp. 336–343. Springer, Heidelberg (2013)
9. Mhamdi, F., Elloumi, M., Rakotomalala, R.: Textmining, features selection and datamining for proteins classification. In: Proceedings of the 2004 International Conference on Information and Comunication Technologies: From Theory to Applications, pp. 457–458. IEEE (2004)
10. Saeys, Y., Inza, I., Larrañaga, P.: A review of feature selection techniques in bioinformatics. Bioinformatics 23(19), 2507–2517 (2007)
11. Trzaskowski, B., Latek, D., Yuan, S., Ghoshdastider, U., Debinski, A., et al.: Action of molecular switches in GPCRs– theoretical and experimental studies. Current Medicinal Chemistry 19(8), 1090–1109 (2012)
12. Vroling, B., Sanders, M., Baakman, C., Borrmann, A., Verhoeven, S., Klomp, J., Oliveira, L., de Vlieg, J., Vriend, G.: GPCRDB: information system for G protein-coupled receptors. Nucleic Acids Research 39(suppl. 1), D309–D319 (2011)

Geometric Approach to Biosequence Analysis

Boris Brimkov[1] and Valentin E. Brimkov[2]

[1] Computational & Applied Mathematics, Rice University, Houston, TX 77005, USA
boris.brimkov@rice.edu
[2] Mathematics Department, SUNY Buffalo State College, Buffalo, NY 14222, USA
brimkove@buffalostate.edu

Abstract. Tools that effectively analyze and compare sequences are of great importance in various areas of applied computational research, especially in the framework of molecular biology. In the present paper, we introduce simple geometric criteria based on the notion of *string linearity* and use them to compare DNA sequences of various organisms, as well as to distinguish them from random sequences. Our experiments reveal a significant difference between biosequences and random sequences – the former having much higher deviation from linearity than the latter – as well as a general trend of increasing deviation from linearity between primitive and biologically complex organisms. The proposed approach is potentially applicable to the construction of dendograms representing the evolutionary relationships among species.

Keywords: String linearity, deviation from linearity, biosequence comparison, discrete monotone path.

1 Introduction

The automated analysis of biosequences includes a great variety of problems; some avenues of the ongoing research are surveyed in [8,12,1]. Typically, the considered problems are approached using combinatorial techniques such as combinatorial pattern matching and combinatorics on words. In this paper we instead use—probably for the first time—a *geometric* approach in an attempt to address questions that are important for understanding biological evolution.

A number of past studies have attempted to address by quantitative means the question of what distinguishes biosequences from random sequences. While by its very nature such a goal has been found "quite elusive" [4], there is substantial evidence in support of the argument that biosequences feature properties that are typical of random sequences (for example, near-total incompressibility [6]). Thus, biosequences are regarded as "slightly edited random sequences" [13], and modern proteins are believed to be "memorized" ancestral random polypeptides which have been slightly modified by the evolutionary selection process in order to optimize their stability under specific physiological conditions [2]. Biosequences appear to be hardly distinguishable from their random permutations, although the latter are clearly incongruous with living organisms [5,9,14]. While this may seem quite obvious from a biological point of view, there have also

J. Sáez-Rodríguez et al. (eds.), *8th International Conference on Practical Appl. of Comput. Biol. & Bioinform. (PACBB 2014)*, Advances in Intelligent Systems and Computing 294,
DOI: 10.1007/978-3-319-07581-5_12,

been numerous computational arguments that support this claim. For example, in [10] Pande et al. present results of mapping some protein sequences onto so-called Brownian bridges, which revealed a certain deviation from randomness. In another study, by estimating the differential entropy and context-free grammar complexity, Weiss et al. have shown that the complexity of large sets of non-homologous proteins is lower than the complexity of the corresponding sets of random strings by approximately 1% [13]. As a first major result of the present work, we introduce simple geometric criteria by which biosequences very strongly differ from random sequences of the same length. In view of the above-mentioned 1% difference demonstrated in [13], by "very strongly" we refer to differences in the order of several hundred percent, registered for 25 biosequences compared to random sequences over the same alphabet and length.

Furthermore, provided the widely adopted postulates of the theory of evolution and in view of the available theoretical and experimental results, it is natural to conjecture that in the evolutionary process of organisms from primitive to biologically complex, their corresponding biosequences have been evolving from random or close to random toward ones that feature increasing deviation from randomness. As a second major result, our experiments based on the introduced measures confirm this expectation (although not in equally indisputable terms as for the comparison between random sequences and biosequences).

The paper is organized as follows. In the remainder of this section we introduce some technical notions and notations, including ones from the theory of words. In Section 2 we introduce the notions of string linearity and deviation from linearity, and study several related properties. In Sections 3 and 4, we present our experimental results and offer a short discussion. We conclude with final remarks and open questions in Section 5.

In the framework of this project we have obtained several other related theoretical and experimental results, which are not presented here because of the page limit. Some of these, together with a more detailed description of the experiments outlined in this article are available in a technical report [3].

1.1 Definitions and Notations

By $|X|$ we denote the cardinality of set X and by $\overline{xy}$ the straight line segment with endpoints x and y. By $d(x,y)$ we denote the Euclidean distance between points x and y, and by $d(x,Y)$ the distance between point x and set Y, i.e., $d(x,Y) = \inf_{y \in Y}\{d(x,y)\}$.

Given a list T of nonnegative real numbers $t_1 \dots t_k$ (not all of which equal 0), a *normalization* of T is obtained by multiplying each value in T by $\frac{100}{t_{\max}}$ where $t_{\max} = \max_{1 \le i \le k}\{t_i\}$.

In string $s = s_0 \dots s_m$ over an alphabet X, s_i is the ith term of s ($0 \le i \le m$), which is some element of X. The number of elements in s is called the *length* of s and denoted $|s|$. A *substring* of a string s is obtained by selecting some or all consecutive elements of s.

2 String Geometrization

Let $s = s_0 \ldots s_m$ be a string on an alphabet $X = \{x_1, \ldots, x_n\}$. We inductively construct an ordered set $L(s)$ of points $p_0, \ldots, p_m \in \mathbb{Z}^n$ corresponding to string s as follows. We set p_0 to be the origin of the Cartesian coordinate system. Let $p_i = (p_{i,1}, \ldots, p_{i,n})$ be the ith element of L for $0 \leq i < m$. If $s_{i+1} = x_j$ for some j $1 \leq j \leq n$, then we set $p_{i+1} = (p_{i,1}, \ldots, p_{i,j} + 1, \ldots, p_{i,n})$. Thus, we obtain a *monotone discrete path* $L(s)$ with $|L(s)| = |s| = m + 1$, in which the coordinates of a point are pairwise greater than or equal to the corresponding coordinates of any preceding point.[1]

Having such a discrete path constructed, one can study its geometric and combinatorial properties, which in turn can provide useful information about the original string s.

Let p_0 be the origin and $p = (p_1, \ldots, p_n)$ be a point in n-dimensional space $\mathbb{Z}^n$. Denote by $\mathbb{H}$ the set of all monotone discrete paths between p_0 and p. It is easy to see that $|\mathbb{H}| = \frac{(p_1 + \ldots + p_n)!}{p_1! \ldots p_n!}$. Each path $H \in \mathbb{H}$ consists of $1 + \sum_{i=1}^{n} p_i$ points, with initial point p_0 and terminal point p. If for every point h in H, the voxel (i.e., the unit grid cell) centered around h intersects the line segment $\overline{p_0 p}$, we call H a *linear path*. Accordingly, we call a string *linear* if its corresponding monotone path is linear. It is easy to see that the following facts hold:

Fact 1. *Given a line segment* $\overline{p_0 p}$*, there is at least one linear path from* p_0 *to* p*.*

Fact 2. *If* H *is a linear path from* p_0 *to* p*, then* $d(h, \overline{p_0 p}) \leq \frac{\sqrt{n}}{2}$ $\forall h \in H$*.*

Next we define some string characteristics that are instrumental to the experimental studies presented in the subsequent sections.

Let s be a string and $L(s) = p_0 \ldots p_m$ be its corresponding monotone path. We define the *maximum deviation of s from linearity* as $mdv(s) = \max_{i=0}^{m}\{d(p_i, \overline{p_0 p_m})\}$, and *average deviation of s from linearity* as $adv(s) = \left(\sum_{i=0}^{m} d(p_i, \overline{p_0 p_m})\right)/(m + 1)$. Note that when $n = 4$ (which is the case for biosequences), the *adv* and *mdv* of a linear string are at most 1.

3 Deviation from Linearity of Random Sequences and Biosequences: Experimental Study

3.1 General Description of Experimental Procedures

The notion of string linearity furnishes an easily implementable tool to compare the biosequences of various organisms. It is reasonable to conjecture that biologically complex organisms have highly structured DNA whose corresponding monotone path strongly deviates from a straight line, while primitive organisms have less structured DNA, whose corresponding monotone path is closer to a straight line. Moreover, a completely random sequence over the alphabet

[1] Note that a similar string geometrization has been considered in [11].

$\{A, T, C, G\}$ has no structure, and therefore its corresponding monotone path can be expected to be much closer to a straight line.

To test this hypothesis, we compare the deviation from linearity of the biosequences of 25 organisms with varying biological complexity, as well as that of random sequences. The number and type of organisms we consider is typical of comparative analysis studies in molecular biology (see, e.g., [7]). We took the biosequences from the genome-scale repository and browser Ensembl Genomes, which is managed by the European Bioinformatics Institute. For each organism, we processed relatively short, randomly selected substrings of DNA in FASTA format.

In our experimental study we first studied the effects of string size on the proposed linearity measures and then selected a suitable string size for our more extensive experiments. This helped save computation time and in turn allowed us to repeat each procedure 1,000 times with different samples of the studied genomes, thus increasing the confidence in the obtained results. Note that some organisms' genomes have billions of letters (whose processing would require a lot of time), while others have genomes that are still uncharted or studied only partially.

Thus, we operated under the assumption that a comparison of the linearity of "relatively small" but "sufficiently large" genome samples will classify organisms in the same *relative* order as a comparison of their whole genomes. As we will see in the following section, our experiments support this claim.

3.2 Effects of Substring Length

We investigated how the length of a biosequence affects its deviation from linearity, and ascertained that deviation increases with the size of the string. However, the rate of increase seems to be independent from the *type* of organism, and therefore an organism's deviation from linearity *relative* to the deviation of other organisms is independent of the length of the biosequences, as long as the length is constant across organisms. These claims are supported by Figure 1, which shows the absolute and normalized maximum and average deviations from linearity of different organisms measured for substrings of increasing length.

The left two graphs display the absolute *adv* and *mdv* and the right two graphs display the normalized *adv* and *mdv* for the graphs on their left. Note that the graphs in the two columns are essentially the same, where the normalized graphs in the right column are the result of "pulling up" the left sides of the graphs in the left column. For all graphs, the substrings are taken randomly from the 26 sources listed in Table 1. The corresponding linearity measures were computed for substrings of length $10,000 \times k$ for $1 \leq k \leq 20$; for each of these lengths, the linearity measures were computed for 200 different substrings taken randomly from each of the 26 sources, and the average values of the 200 trials were plotted. From Figure 1, it can be seen that *adv* and *mdv* are principally independent of length, since for any of the lengths examined the organisms are more or less in the same relative position compared to the other organisms. Clearly, as the length of substrings approaches 0, the measures of deviation from linearity will approach

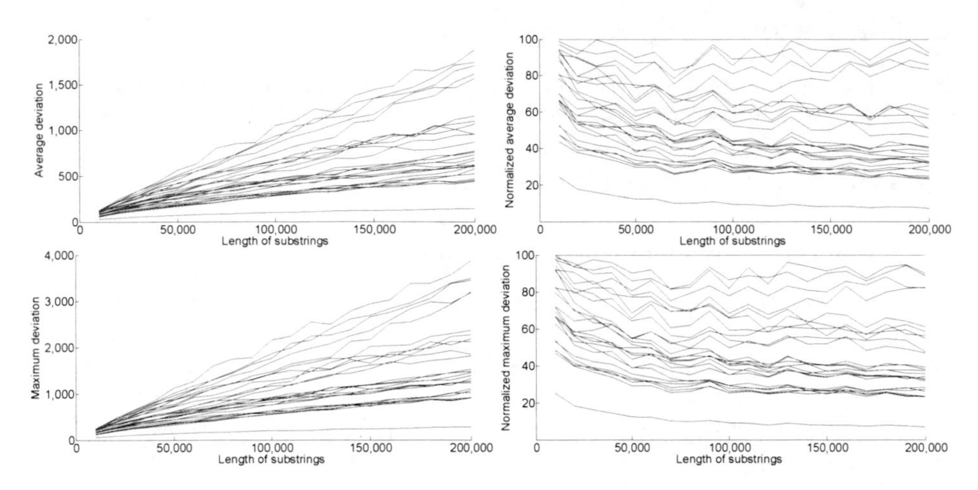

Fig. 1. Absolute and normalized *avd* and *mdv* measured for substrings of increasing length. In all diagrams the lowest line corresponds to randomly generated strings. The relative positions of the other lines match the numerical data of Table 1.

0, and will be more unstable and unreliable. From the normalized graphs, we notice that the initial fluctuations first disappear around substrings of length 50,000. For this reason, we carried out our further experiments (which involve more trials and hence a greater confidence) with substrings of this length.

3.3 Computational Procedure

The performed computational process can be broken up into the following components: selection of samples, computation of linearity measures, and compilation and normalization of data. We used version R2011a of Matlab and several built-in functions to carry out our computations.

Due to the motivation given earlier that fluctuations in the linearity measures first disappear in samples of length 50,000, we randomly selected substrings of length 50,000 from the larger excerpts of genomes. Random sequences have been generated with independent symbols with uniform distribution (25% each) (see [3] for more details). Given such a string, using elementary techniques we compute and store an array of distances from points of a monotone path representing a biosequence to the corresponding straight line. After obtaining an array of 26 samples (25 biosequences and one random sequence), we calculate the linearity measures *adv* and *mdv* for each sample in the array and end up with a 26×2 array. We repeat this procedure $1,000$ times with different random samples, and attain a 3-dimensional array ($26 \times 2 \times 1,000$). We then take the average over the $1,000$ trials, and again obtain a 2-dimensional array, which we normalize for each linearity measure in order to better see the relationships between organisms. Thus, our final product is a 26×2 array with values ranging between 0 and 100, which allows us to easily compare organisms based on the two linearity criteria.

4 Discussion

In this section we provide further details about our experimental work and discuss the obtained results. We comment on results which are obvious to our unarmed eye, hoping that other interesting conclusions could also be drawn by experts with higher expertise in biological sciences.

4.1 General Observations and Comments

The first column of Table 1 gives common names for the organisms which we have examined. The specific strains of Chlamydia, Tuberculosis, Gingivalis, and Streptococcus are Nigg, CCDC5180, W83, and ND03, respectively. All the DNA we processed was from the first chromosomes of the organisms, except for the fruit fly and the yeast, where the DNA was taken from chromosomes 2L and 4, respectively.

Table 1. Summary of experimental results

Common Name	Substrings		Common Name	Substrings		Common Name	Substrings	
	adv	*mdv*		*adv*	*mdv*		*adv*	*mdv*
Random	12.8	12.8	Mouse	62.5	64.9	Chlamydia	29.7	29.3
Human	88.9	89.9	Platypus	41.2	41.7	Tuberculosis	33.8	34.0
Neanderthal	85.5	85.1	Lizard	31.6	32.3	Gingivalis	50.0	48.1
Gorilla	100.0	100.0	Zebrafish	59.6	58.5	Streptococcus	37.2	36.5
Chimp	81.2	81.0	Medaka fish	46.6	47.5	Rice	73.8	76.9
Dog	71.8	74.3	Fruit fly	49.1	49.3	Corn	76.1	80.3
Chicken	59.1	57.2	Sea squirt	32.8	32.2	Cress	44.8	44.4
Marmoset	55.0	55.8	Nematode	51.6	52.4	Soybean	52.2	53.0
Rat	68.2	68.0	Yeast	39.4	39.8			
Maximum pre-normalized value:							554.5	1100.8

Columns 2 and 3 of Table 1 show the results of our experiments when the *adv* and *mdv* were measured for substrings taken from the 25 biosequences and one random sequence. All measures were computed on samples of length 50,000 and are the average of 1,000 trials. Note that for most species, *adv* and *mdv* measured from substrings differ by less than 1%; only for four organisms this difference is more than 2%, but no more than 4.2%. This observation is also supported by Figure 1, which shows that the *adv* graphs are nearly identical in appearance to the *mdv* graphs.

4.2 Distinction between Biosequences and Random Sequences

Our first important conclusion is the distinction between the linearity of biosequences and random sequences. All of our experiments show that biosequences have a higher average and maximum deviation from linearity than random sequences. This difference is very significant. In particular, in Table 1, the normalized *adv* and *mdv* for random sequences are both 12.8, whereas the normalized *adv* and *mdv* for the organism with the smallest deviation are 29.3 and 29.7, respectively. This conclusion is also supported by Figure 1, where the line positioned visibly below the others is the one representing the random sequence.

4.3 Gradient between Primitive and Biologically Complex Organisms

Our experiments also support the hypothesis that the sequences of primitive organisms are closer in linearity to random sequences than the sequences of biologically complex organisms. As most primitive organisms, we consider bacteria and microscopic organisms; we consider plants the next most evolved organisms, followed by fish, reptiles, and other egg-laying vertebrates. Finally, we consider mammals and primates as organisms at the top of the evolutionary ladder. We expected that the graded change in the magnitude of deviation from linearity of different organisms would be in accordance with the aforementioned classification of their biological complexity. Indeed, our experiments support this expectation.

In particular, the Human, Neanderthal, Gorilla, and Chimpanzee have the highest *adv* and *mdv*; the bacterium Chlamydia has the smallest *adv* and *mdv* after the random sequence. The other organisms with the lowest deviations from linearity are two other bacteria, the yeast, sea squirt, and lizard. In the mid-low range are organisms like the fruit fly, medaka fish, and soybean plant, and in the mid-high range are organisms like the zebrafish, chicken, and mouse.

In our considerations, some anomalies and incongruences with expectation are manifested. For example, the *adv* and *mdv* of rice and corn are relatively high – higher, for example, than the *adv* and *mdv* of the mouse and rat.

5 Concluding Remarks

In this paper we introduced a geometric approach for string analysis based on the notions of string linearity and deviation from linearity. Our experiments showed that, unlike some other criteria, ours strongly separate random sequences from biosequences, as well as primitive from biologically complex organisms. These results are in accordance with certain earlier interpretations that biosequences have been evolving towards energy minimization in physical terms, as well as of lowering their information complexity [2,10].

As the proposed quantitative measures seem to be quite robust and reliable in practice, important future tasks are seen in performing systematic extensive experiments on a larger set of biosequences (including in addition results on the values' dispersion, e.g., in terms of standard deviation and confidence intervals), their interpretation and deeper analysis from a biological point of view, and comparison with results obtained by other approaches.

In addition to a more extensive study of the general trends exhibited in the present work, possible future tasks can pursue understanding the meaning and functions (from a biological point of view) of biosequence locations where deviation from linearity achieves local maxima or minima. Certain anomalies from the general trends featured by the experiments could also be addressed.

Acknowledgements. The authors thank the four anonymous referees for their useful remarks and suggestions. In particular, we are indebted to a referee for bringing to our attention ref. [11], of which we were not aware.

References

1. Apostolico, A., Giancarlo, R.: Sequence alignment in molecular biology. Journal of Computational Biology 5(2), 173–196 (1998)
2. Apostolico, A., Cunial, F.: The subsequence composition of polypeptides. Journal of Computational Biology 17(8), 1–39 (2010)
3. Brimkov, B., Brimkov, V.E.: Geometric approach to string analysis: deviation from linearity and its use for biosequence classification (2013), `http://arxiv.org/abs/1308.2885v1`
4. Broox Jr., F.P.: Three great challenges for half-century-old computer science. J. ACM 50, 25–26 (2003)
5. Monod, J.: Chance and Necessity. Collins, London (1972)
6. Nevil-Manning, C., Witten, I.: Protein is incompressible. In: Proc. Conf. Data Compression, p. 257 (1999)
7. Salzburger, W., Steinke, D., Braasch, I., Meyer, A.: Genome desertification in eutherians: can gene deserts explain the uneven distribution of genes in placental mammalian genomes? J. Mol. Evol. 69(3), 207–216 (2009)
8. Sankoff, D., Kruskal, J.B. (eds.): Time Warps, String Edits, and Macromolecules: The Theory and Practice of Sequence Computation. Addison-Wesley, Reading (1983)
9. Schwartz, R., King, J.: Sequences of hydrophobic and hydrophilic runs and alternations in proteins of known structure. Protein Sci. 15, 102–112 (2006)
10. Pande, V., Grosberg, A., Tanaka, T.: Nonrandomness in protein sequences: evidence for a physically driven stage of evolution. Proc. Natl. Acad. Sci. USA 91, 12972–12975 (1994)
11. Pandić, M., Balaban, A.T.: On a four-dimensional representation of DNA primary sequences. J. Chem. Inf. Comput. Sci. 43, 532–539 (2003)
12. Waterman, M.S.: Introduction to Computational Biology. Maps, Sequences and Genomes. Chapman Hall (1995)
13. Weiss, O., Jiménez-Montañgo, M., Herzel, H.: Information content of protein sequences. J. Theoret. Biology 206, 379–386 (2000)
14. White, S., Jacobs, R.: Statistical distribution of hydrophobic residues along the length of protein chains. Biophys. J. 57, 911–921 (1990)

Timed and Probabilistic Model Checking over Phylogenetic Trees

José Ignacio Requeno and José Manuel Colom

Department of Computer Science and Systems Engineering (DIIS),
Universidad de Zaragoza, C/ María de Luna 1, 50018 Zaragoza, Spain
{nrequeno,jm}@unizar.es

Abstract. Model checking is a generic verification technique that allows the phylogeneticist to focus on models and specifications instead of on implementation issues. Phylogenetic trees are considered as transition systems over which we interrogate phylogenetic questions written as formulas of temporal logic. Nonetheless, standard logics become insufficient for the usual practices of phylogenetic analysis since they don't allow the inclusion of explicit time and probabilities. The aim of this paper is to extend the application of model checking techniques beyond qualitative phylogenetic properties and adapt the existing logical extensions and tools to the field of phylogeny. The introduction of time and probabilities in phylogenetic specifications is motivated by a real example.

Keywords: phylogenetic analysis, timed & probabilistic model checking.

1 Introduction

A phylogenetic tree is a description of the evolution process which is discovered via molecular sequencing data and morphological data matrices [1]. Computer science tools have upgraded the capabilities of biologists for their construction as well as for extracting and analyzing the implicit biological messages embedded on them [2,3]. Today, one of the most relevant challenges is the introduction of a generic framework for heterogeneous hypothesis verification over trees.

Model checking is a generic unifying framework that allows the phylogeneticist to focus on tree structures, biological properties and symbolic manipulation of phylogenies described using temporal logic, instead of on implementation issues concerned with verification algorithms [4]. Model checking allows us to uncouple software tools from the definition of properties and it hides the underlying implementation technology [5]. Besides, phylogenetic properties can be exported and evaluated in other structures (i.e., trees or networks) with minimum effort.

Nonetheless, standard logics such as *Computational Tree Logic* (CTL [5]) become insufficient for the usual practices of phylogenetic analysis since they don't allow the inclusion of explicit time and probabilities in the specifications and models. The labeling of the phylogenetic tree sometimes includes extra quantitative information beyond the original propositional information of the states [6].

J. Sáez-Rodríguez et al. (eds.), *8th International Conference on Practical Appl. of Comput. Biol. & Bioinform. (PACBB 2014)*, Advances in Intelligent Systems and Computing 294,
DOI: 10.1007/978-3-319-07581-5_13,

These numerical annotations of the tree changes with every particular study, but they are commonly divided in a) *timed* or distance information, b) *probabilistic* information and c) *raw quantitative* information. The first two ones are mainly related to the labeling of the tree branches [7]. The last one is more general and involves numerical values and comparisons in the atomic propositions.

Hence, the aim of this paper is to analyze the requirements of non-qualitative phylogenetic properties, adapt the existing probabilistic and timed extensions to temporal logics and use the associated model checking tools in the field of phylogeny. The introduction of time and probabilities in phylogenetic specifications is motivated by means of a real example. The paper is divided in six sections. After this introduction, Section 2 explains a real example that motivates the definition of the discrete-time probabilistic logic and structures of Section 3. Next, Section 4 details an algorithm of model checking that computes the probabilities and verifies the logical specifications over the phylogenetic tree. Later, Section 5 shows the experimentation with a temporal and probabilistic model checking tool. Finally, Section 6 briefs the conclusions and draws the future work.

2 Motivation

A phylogenetic tree is a directed graph that offers a realistic model of aggregated evolution in which each vertex represents an inferred state of the evolution characterized by biological sequences (e.g., DNA) [4]. The phylogenies are occasionally enriched with time labels or weights in the edges. This knowledge is useful for learning complex properties about the evolution, for instance, the estimation of the temporal point of divergence between species [8] or the diaspora of human populations [9]. The extension of the phylogenetic properties in [4, Table 1] with time and probabilities increases the expressivity of biological hypothesis.

Take the following disease as a clarifying example. The lactose intolerance in adults is a chronic disease caused by the inhibition of the lactase gene after the breastfeeding and childhood. The inability for processing the milk and its derivations is not homogeneously distributed in the human population. While in some African pastoralist groups of North/East Africa and the northern cultures of Europe their stock breeding tradition and diet motivated an evolutionary adaptation to digest the milk ($> 70\%$ of tolerance), the percentage of acceptation decreases in the rest of areas and ethnic groups [10]. In addition, the phenotype in Europe and Africa appeared at a different epoch and the point mutations that regulate the activation of the lactase persistence are disparate [11]. Some illustrative questions that we desire to ask to the phylogeny, and that are expressed below, require the addition of time to the branches of a population tree. The time allows the estimation of the divergence points between individuals or mutations, while the probability of the lactose persistence in different zones is calculated through the study of the distribution of the point mutations that regulate the phenotype. The questions are:

I What is the rate of lactase persistence in a population? i.e., do their members define a characteristic haplogroup? and in that case,

II Which polymorphism, among the multiple activators and inhibitors of the lactase gene, is the most frequent over there? and finally,
III When did this phenotype approximately start to predominate? i.e., does this date mark a major event in the diet, culture or migration of that population?

These questions ask about the time (dates) and probabilities (frequencies/rates) stored in the branches of the tree. Besides, the deductive process that answers the queries also needs the manipulation of quantitative information. Thus, we must introduce a logic, a transition system and a model checking algorithm capable of expressing and managing these kind of questions. The notion of time introduced here matches with the concept of evolutionary or chronological clock.

3 Discrete-Time Probabilistic Logic and Structure

In this section we are considering a phylogenetic tree enriched with numerical information that tells the probability of selecting a branch descending from an internal node. Therefore, we can analyze properties like: what is the probability of reaching a set of states of the tree from the root? The logic and data structure defined here settle the basis for future updates and extensions for continuous-time systems [12]. Stochastic systems generally use Markov chains as the underlying data structure that provides semantics to the verification process. Discrete-time Markov chains capture the essentials of probabilities between states of the tree and implicitly associates an unit time step to every transition of the system.

Definition 1 (Discrete-time Markov Chain). *A discrete-time Markov chain is a finite transition system represented by a tuple* $M = (S, S_0, \mathbf{P}, L)$, *where 1)* S *is a finite set of states, 2)* $S_0 \subseteq S$ *is the set of initial states, 3)* $\mathbf{P} : S \times S \rightarrow [0,1]$ *is the transition probability matrix that indicates the probability of moving from a state* s_i *to a state* s_j *satisfying* $\Sigma_{s_j \in S} P(s_i, s_j) = 1$, *and 4)* $L : S \rightarrow 2^{AP}$ *is the labeling function that associates each state with the subset of atomic propositions (AP) that are true of it.*

A phylogenetic tree is assimilated to a discrete-time Markov chain making the corresponding association of states to the definition of phylogeny ([4, Def. 3]). The leaves are labeled with the genome information of the population or specie they represent, plus additional data when necessary. Each branch of the phylogeny is labelled with an element $\mathbf{P}(s_i, s_{i+1}) > 0$ of the transition probability matrix. This value gives the probability of moving from state s_i to state s_{i+1} in one time step. The self-loops of the terminal leaves in the Kripke structure of a branching-time phylogeny are represented by a single transition going back to the same state with probability 1 in the transition probability matrix.

For any set of infinite paths Π starting in the initial state s_0, the subset $\Pi(\pi_n)$ selects the paths $\pi \in \Pi$ whose prefix equals to the finite sequence $\pi_n = s_0 s_1 s_2 \ldots s_n$ of length $n+1$ states. The set of infinite sequences sharing the prefix π_n has probability $Pr(\Pi(\pi_n)) = \mathbf{P}_\Pi(\pi_n)$. The probability $\mathbf{P}_\Pi(\pi_n)$ is calculated as the product of probabilities for each intermediate transition, except for paths with unitary length in which case $n = 0$, $\pi_0 = s_0$ and $\mathbf{P}_\Pi(\pi_0) = \mathbf{P}_\Pi(s_0) = 1$. That is, $\mathbf{P}_\Pi(\pi_n) = \mathbf{P}(s_0, s_1) \cdot \mathbf{P}(s_1, s_2) \cdot \ldots \cdot \mathbf{P}(s_{n-1}, s_n)$.

Probabilistic CTL (PCTL) [13,5] helps to formulate conditions on a discrete-time Markov chain. The properties are referred to state formulas (ϕ) or path formulas (Φ). Besides, PCTL allows enriched queries such as $\mathbb{P}_{\sim\lambda}(\Phi)$. Given an initial state s and a comparison $\sim \in \{<, \leq, =, \geq, >\}$, the operator $\mathbb{P}_{\sim\lambda}(\Phi)$ returns true if the probability for a set of paths satisfying Φ is $\sim \lambda$, with $\lambda \in [0, 1]$.

Definition 2 (Probabilistic Computation Tree Logic). *A temporal logic formula ϕ is defined by the following grammar, where $p \in AP$, $k \in \mathbb{N} \cup \{\infty\}$:*

$$\begin{aligned} \phi &::= true \mid p \mid \neg\phi \mid \phi \vee \phi \mid \mathbb{P}_{\sim\lambda}[\Phi] \\ \Phi &::= \mathbf{X}\phi \mid [\phi \mathbf{U}_{\leq k} \phi] \end{aligned} \tag{1}$$

The formulas are checked against a structure M considering all infinite paths $\pi \in \Pi$ from a certain state s_0. Notice that $M, s_0 \models \phi$ means that s_0 satisfies ϕ. The semantics of well-formed formulas is as follows (let $\pi = s_0 s_1 s_2 \ldots$):

- $M, s_0 \models p \Leftrightarrow p \in L(s_0)$,
- $M, s_0 \models \neg\phi \Leftrightarrow M, s_0 \not\models \phi$,
- $M, s_0 \models \phi \vee \psi \Leftrightarrow M, s_0 \models \phi$ *or* $M, s_0 \models \psi$,
- $M, s_0 \models \mathbb{P}_{\sim\lambda}[\Phi] \Leftrightarrow Prob(M, s_0, \Phi) \sim \lambda$,

The calculation of the probability $Prob(M, s_0, \Phi)$ requires the identification of the infinite paths π satisfying the path formula $M, \pi \models \Phi$:

- $M, \pi \models \mathbf{X}\phi \Leftrightarrow M, s_1 \models \phi$
- $M, \pi \models [\phi \mathbf{U}_{\leq k} \psi] \Leftrightarrow \exists 0 \leq i \leq k, \forall 0 \leq j \leq i : (M, s_i \models \psi) \wedge (M, s_j \models \phi)$

This set, $\{\pi \in \Pi \mid M, \pi \models \Phi\}$, can be obtained by the union of finitely many pairwise disjoint subsets $\Pi(\pi_n)$ by [12, Def. 3], each one characterized by the finite prefix π_n of all infinite sequences of the set. Therefore, $Prob(M, s_0, \Phi) = Pr\{\pi \in \Pi \mid M, \pi \models \Phi\} = \Sigma_{\pi_n} Pr(\Pi(\pi_n))$ computes the probability as the summation of probabilities in all possible prefixes π_n by [12, Theor. 1].

The logic supports timed transitions in the $\mathbf{U}$ operator. The notion of time in a Markov chain falls within the concept of state distances. Each state transition of the discrete-time Markov chain involves an unit time step. A mapping between the chronological time and state distances allows the inference of the evolutionary speed in the branches of the phylogenetic tree. The computation of time and probabilities are embedded in the model checking algorithm. Timed variants of the modal operators $\mathbf{F}$ and $\mathbf{G}$ are obtained via $\mathbf{U}$ as $\mathbf{F}_{\sim c}\phi = true\ \mathbf{U}_{\sim c}\phi$ and $\mathbf{G}_{\sim c}\phi = \neg\mathbf{F}_{\sim c}\neg\phi$. Instead of writing the intervals explicitly, sometimes they are abbreviated with comparisons. For example, $\mathbb{P}_{\leq 0.5}[\Phi]$ denotes $\mathbb{P}_{[0,0.5]}[\Phi]$.

By now, we can translate the questions presented in the motivation example of lactose into the PCTL syntax. In a phylogenetic tree, the tips correspond to individuals of disjoint populations whose states are tagged with their DNA and a boolean indicating if they are lactose (in)tolerant. The internal nodes of the inferred ancestors are labeled with their estimated DNA sequence and lactose phenotype as well. The following equation asks if there exists an ancestor ($\mathbb{P}_{>0}$) at distance 3 or above from the initial state ($\mathbf{F}_{\geq 3}$) that is the root of a population with lactase persistence over 70% ($\mathbb{P}_{\geq 0.7}\,[\mathbf{F}_{\geq 0}\,\mathit{lactose\_tolerant}]$). The members of a population, including the leaves and internal nodes, are reached by $\mathbf{F}_{\geq 0}$.

$$\mathbb{P}_{>0}\,[\mathbf{F}_{\geq 3}\,(\mathbb{P}_{\geq 0.7}\,[\mathbf{F}_{\geq 0}\,\mathit{lactose\_tolerant}]\,)] \tag{2}$$

The outer restriction $\mathbb{P}_{>0}\,[\mathbf{F}_{\geq 3}]$ corresponds to the question III of the motivation. It searches for an internal node from which the phenotype starts to be predominant after a certain date since the phylogenetic root. The inner formula $\mathbb{P}_{\geq 0.7}\,[\mathbf{F}_{\geq 0}\,\mathit{lactose\_tolerant}]$ answers the question I about the rate of lactase persistence in a population. Finally, the addition of a genetic marker in this place inside the $\mathbb{P}_{\geq 0.7}$ equation helps to investigate the relation between a polymorphism and phenotype (question II). The evaluation of the formulas needs the algorithm introduced in the next section.

4 Algorithm for PCTL Model Checking

The algorithm for managing and solving PCTL formulas in stochastic systems is mainly identical to that of classic model checking except for the resolution of $\mathbb{P}_{\sim\lambda}[\Phi]$, i.e., the $\mathbf{X}$ and $\mathbf{U}$ operators with probability thresholds. In short, the recursive algorithm of model checking incorporates the new sentence [5]:

$$Sat(\mathbb{P}_{\sim\lambda}\,[\Phi]) = \{s \in S \mid Prob(M, s, \Phi) \sim \lambda\}$$

$\mathbb{P}_{\sim\lambda}[\mathbf{X}\phi]$ formula. In PCTL, the probability of satisfying the next operator requires the probabilities of the immediate transitions from s. It is resolved by $Prob(M, s, \mathbf{X}\phi) = \Sigma_{s' \in Sat(\phi)}\mathbf{P}(s, s')$.

$\mathbb{P}_{\sim\lambda}[\psi\mathbf{U}_{\leq k}\phi]$ formula. The computation of the probability for the until operator depends on the value of k. For $k \in \mathbb{N}$, then $Prob(M, s, \psi\mathbf{U}_{\leq k}\phi)$ is:

$$\begin{cases} 1 & \text{if } s \in Sat(\phi) \\ 0 & \text{if } k = 0 \text{ or } s \in Sat(\neg\phi \wedge \neg\psi) \\ \Sigma_{s' \in S}\mathbf{P}(s, s') \cdot Prob(M, s', \psi\mathbf{U}_{\leq k-1}\phi) & \text{otherwise} \end{cases}$$

When $k = \infty$, the until operator is analogous to the original until operator of CTL with semantics of infinite paths. That is, $Prob(M, s, \psi\mathbf{U}_{\leq\infty}\phi)$ can be rewritten as $Prob(M, s, \psi\mathbf{U}\phi)$ and it equals to:

$$\begin{cases} 1 & \text{if } s \in Sat(\phi) \\ 0 & \text{if } k = 0 \text{ or } s \in Sat(\neg\phi \wedge \neg\psi) \\ \Sigma_{s' \in S}\mathbf{P}(s, s') \cdot Prob(M, s', \psi\mathbf{U}\phi) & \text{otherwise} \end{cases}$$

The time complexity of verifying a PCTL formula ϕ against a discrete-time Markov chain is linear in $|\phi|$ and polynomial in the size of S, with $|\phi|$ the number of logical connectives and temporal operators of the formula. In short, $\Theta(poly|S|) * k_{max} * |\phi|)$ where k_{max} is the maximal step bound of a path subformula $\psi_1 \mathbf{U}_{\leq k} \psi_2$ of ϕ, with $k_{max} = 1$ if it doesn't contain any $\mathbf{U}_{\leq k}$ subformula.

In sum, most of the investigations related with quantitative information are a prolongation of the phylogenetic properties analyzed with boolean temporal logics. Consequently, the inconveniences presented in [4] would appear even more dramatically now. The techniques introduced for the optimization and scalability in classic model checking environments also work for this context [14]. The incorporation of external data bases for storing the labeling of the states is compatible with any kind of atomic propositions, quantitative or not. The vertical and horizontal partitioning of the database table adds an extra dimension of parallelism. Furthermore, the summations in the computation of probability paths, specially with the $\mathbf{U}_{\leq k}$ operator, can be executed in parallel.

5 Model Checking Tools and Experimentation

PRISM [15] is a generic model checking tool capable of handling probabilistic and timed specifications over Markov chains. There exist many other model checking tools [16]. Although the real performance depends on the particular structure of the model and specifications, PRISM offers Java portability, a powerful syntax, and a good scientific community support.

The data set used for this experimentation is synthetic. With this data set, we try to cover the spectrum of small phylogenies and analyze the cost of the evaluation of the lactose property over there. We have created random phylogenetic trees of up to 1000 tips with a Yule speciation model. The DNA sequences have 50 bases with an homogeneous distribution of nucleotides. We have evaluated the lactose formula introduced in the motivation but enriched for the detection of polymorphisms. The underlying objective consists of the identification of a correlated evolution between lactose tolerance and genomic patterns. Other studies such as [10] use cultural information for discovering this coevolution and the influence of a milk-based diet. The utilization of phylogenetic comparative methods and regression techniques establishes the essentials of this approach.

The codification of the phylogeny in PRISM follows the same idea presented for NuSMV in [4]. The probability threshold of the internal $\mathbb{P}_{\geq x}[\mathbf{F}_{\geq 0}\, seq[i] = j]$ ranges from $x \in [0.1, 0.9]$, with $i \in [1, 50]$ the position where we search for the polymorphism and j a certain nucleotide. The Figure 1 plots the time required for the computation of 50×9 formulas corresponding to the expansion of i and x for all the columns of the alignment and probability bounds. All tests have been run on a Intel Core 2 Duo E6750 @ 2.66 GHz with 8 GB RAM and Linux.

PRISM performs well for the verification of the lactose formulas in small phylogenies as it follows a polynomial trend in time with respect to the number of tips. However, it requires the integration of new technologies and solutions to scale for larger phylogenetic trees and specifications. In fact, we desire to find the values of x, i and j for which the verification of the equation returns true. The definition of patterns is a common procedure, which intuitively leads

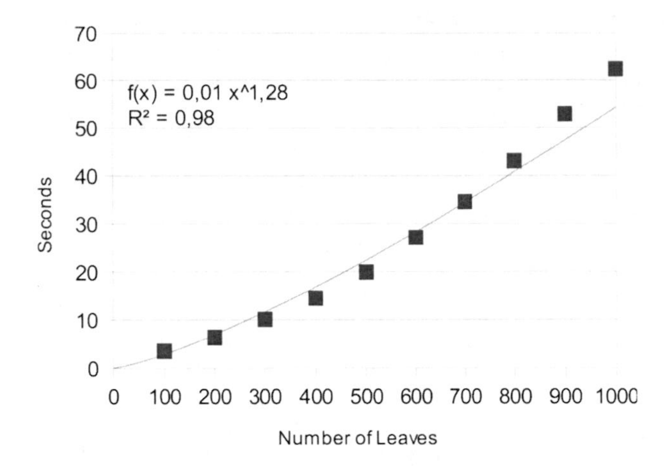

Fig. 1. Time required for the verification of a set of probabilistic formulas

to parametric model checking [17]. Nonetheless, mining for knowledge without prior information requires a more or less thorough exploration of the structure, which can be combinatorial in some or all of its dimensions. Although inherently parallel, the exhaustive inspection of potential solutions involves the test of large sets of formulas and an intensive use of the topology and information of each state. The application of parametric model checking for model exploration is a future extension that will increase the potential of our framework.

6 Conclusions

In this paper we have motivated the extension of phylogenetic analysis via model checking using quantitative information. We have proposed the inclusion of time and probabilities in the branches of the tree because of its natural interpretation in the phylogeny. In particular, we have presented a phylogenetic example based on the lactose (in)tolerance that needs these kind of information. To this end, we have introduced an extended logic and data structure adapted for probabilities and time together with the algorithms and computations for managing them. Our first goal has been the increase of the logical capabilities for querying about the date of appearance and degree of distribution of mutations and phenotypes.

Next, we have experimented with synthetic data in order to prove the feasibility of our approach with existing probabilistic model checking tools. PRISM is a generic model checking tool that performs well for small phylogenies in polynomial time with respect to the number of tips. The tool is independent of the application domain: it automatically verifies any proposition expressed with temporal logic over a model of the system. However, it requires the integration of new technologies and solutions to scale for larger phylogenies and specifications due to the particularities of the phylogenetic analysis. The distribution of the Markov chain structure, the paralellization of the formula verification with the computation of probabilities, and the integration of PRISM with the atomic propositions stored in an external database constitutes our following step. Some of these ideas has been already applied for standard CTL model checking.

This work opens the door for the review of bigger phylogenies with properties similar to the lactose persistence. The modularity of our framework allows the evaluation of hypothesis and the comparison of results for a set of phylogenetic trees by only changing the tree file (the specification of the property remains constant). Finally, the search for the valuations that verify a certain specification leads to an intensive exploration of the formula space or the solution of linear systems. The introduction of parametric model checking for the automatic discovery and mining of phylogenetic information outlines our future work. This work was supported by MICINN [TIN2011-27479-C04-01] and DGA [B117/10].

References

1. Felsenstein, J.: Inferring phylogenies. Sinauer, Sunderland (2003)
2. Yang, Z., Rannala, B.: Molecular phylogenetics: principles and practice. Nat. Rev. Genet. 13(5), 303–314 (2012)
3. Fitch, W.M.: Uses for evolutionary trees. Philos. T. Roy. Soc. B 349(1327), 93–102 (1995)
4. Requeno, J.I., De Miguel Casado, G., Blanco, R., Colom, J.M.: Temporal logics for phylogenetic analysis via model checking. IEEE ACM T. Comput. Bi. 10(4), 1058–1070 (2013)
5. Baier, C., Katoen, J.P.: Principles of model checking. MIT Press, Cambridge (2008)
6. Zmasek, C.M., Eddy, S.R.: ATV: display and manipulation of annotated phylogenetic trees. Bioinformatics 17(4), 383–384 (2001)
7. Rambaut, A.: How to read a phylogenetic tree (August 2013)
8. Barraclough, T.G., Nee, S.: Phylogenetics and speciation. Trends Ecol. Evol. 16(7), 391–399 (2001)
9. Cavalli Sforza, L.L., Feldman, M.W.: The application of molecular genetic approaches to the study of human evolution. Nat. Genet. 33, 266–275 (2003)
10. Holden, C., Mace, R.: Phylogenetic analysis of the evolution of lactose digestion in adults. Hum. Biol. 597–619 (2009)
11. Tishkoff, S.A., et al.: Convergent adaptation of human lactase persistence in Africa and Europe. Nat. Genet. 39(1), 31–40 (2006)
12. Kwiatkowska, M., Norman, G., Parker, D.: Stochastic model checking. In: Bernardo, M., Hillston, J. (eds.) SFM 2007. LNCS, vol. 4486, pp. 220–270. Springer, Heidelberg (2007)
13. Hansson, H., Jonsson, B.: A logic for reasoning about time and reliability. Form. Asp. Comput. 6, 512–535 (1994)
14. Requeno, J.I., Colom, J.M.: Model checking software for phylogenetic trees using distribution and database methods. J. Integr. Bioinformatics 10(3), 229 (2013)
15. Kwiatkowska, M., Norman, G., Parker, D.: PRISM 4.0: Verification of probabilistic real-time systems. In: Gopalakrishnan, G., Qadeer, S. (eds.) CAV 2011. LNCS, vol. 6806, pp. 585–591. Springer, Heidelberg (2011)
16. Jansen, D.N., Katoen, J.-P., Oldenkamp, M., Stoelinga, M., Zapreev, I.: How fast and fat is your probabilistic model checker? an experimental performance comparison. In: Yorav, K. (ed.) HVC 2007. LNCS, vol. 4899, pp. 69–85. Springer, Heidelberg (2008)
17. Bruyère, V., Raskin, J.-F.: Real-time model-checking: Parameters everywhere. In: Pandya, P.K., Radhakrishnan, J. (eds.) FSTTCS 2003. LNCS, vol. 2914, pp. 100–111. Springer, Heidelberg (2003)

mBWA: A Massively Parallel Sequence Reads Aligner*

Yingbo Cui[1,**], Xiangke Liao[1,**], Xiaoqian Zhu[1,**], Bingqiang Wang[2], and Shaoliang Peng[1,**,***]

[1] National University of Defense Technology, Changsha, China
{yingbocui,xkliao,zxq,pengshaoliang}@nudt.edu.cn
[2] BGI, Shenzhen, China
wangbingqiang@genomics.cn

Abstract. Mapping sequenced reads to a reference genome, also known as sequence reads alignment, is central for sequence analysis. Emerging sequencing technologies such as next generation sequencing (NGS) lead to an explosion of sequencing data, which is far beyond the process capabilities of existing alignment tools. Consequently, sequence alignment becomes the bottleneck of sequence analysis. Intensive computing power is required to address this challenge. A key feature of sequence alignment is that different reads are independent. Considering this property, we proposed a multi-level parallelization strategy to speed up BWA, a widely used sequence alignment tool and developed our massively parallel sequence aligner: mBWA. mBWA contains two levels of parallelization: firstly, parallelization of data input/output (IO) and reads alignment by a three-stage parallel pipeline; secondly, parallelization enabled by Intel Many Integrated Core (MIC) coprocessor technology. In this paper, we demonstrate that mBWA outperforms BWA by a combination of those techniques. To the best of our knowledge, mBWA is the first sequence alignment tool to run on Intel MIC and it can achieve more than 5-fold speedup over the original BWA while maintaining the alignment precision.

Keywords: NGS, sequence aligner, BWA, parallelization, MIC coprocessor.

1 Introduction

Mapping sequence reads to a reference genome is central and fundamental for sequence analysis. Many biological applications take sequence alignment as the first step, such as the detection of single-nucleotide polymorphism (SNP) (Li et al., 2009), the study of genome-wide methylation patterns (Cokus et al., 2008). To meet this demand, a number of sequence alignment tools have been developed, including MAQ (Li, et al., 2008), SOAP (Li et al, 2008), BWA (Li et al., 2009), Bowtie (Langmead et al., 2009), etc.

---

* mBWA is under BSD and freely available at http://sourceforge.net/projects/mbwa

** Equal Contributors.

*** Corresponding author.

J. Sáez-Rodríguez et al. (eds.), *8th International Conference on Practical Appl. of Comput. Biol. & Bioinform. (PACBB 2014)*, Advances in Intelligent Systems and Computing 294,
DOI: 10.1007/978-3-319-07581-5_14, © Springer International Publishing Switzerland 2014

Emerging sequencing technologies like the next generation sequencing (NGS) technology, have led to an explosive increase of sequenced data. Taking human genome as an example, for 2nd generation sequencing technology like Illumina Hiseq, each human genome contains billions of bases and each sequenced reads file contains millions of reads. Existing alignment tools cannot cope with the rapid growth of data scale. As a result, intensive computing power is demanded to speed up sequence alignment.

BWA is one of the most widely used alignment tools and it has a fast processing speed with low memory footprint. Before alignment, BWA first constructs an index for reference genome to accelerate alignment, and the index is reusable for the same species. For most sequence alignment tools, including BWA, the most time-consuming part is sequence alignment. One key but neglected feature is that the alignment of each read is independent. Based on this, we propose a multi-level parallelization strategy to accelerate BWA.

Our key contributions are listed as follows:

1) We designed a three-stage pipeline for the sequence alignment process. The three stages include data input, reads aligning and data output. The pipeline increases efficiency by overlapping data IO with the actual alignment process.

2) We ported the alignment kernel of BWA onto the Intel MIC coprocessor by reorganizing data transformation in the kernel. In addition, CPU and MIC perform the alignment cooperately.

The rest of this paper is organized as follows. Section 2 presents background and related work. Section 3 describes the architecture of our massive parallel sequence aligner. Performance evaluation is presented in Section 4. Section 5 concludes the paper.

2 Background and Related Work

In this section, we first review the sequence alignment tools of recent years in section 2.1; then we introduce the Intel MIC coprocessor and the Tianhe-2 supercomputer in section 2.2 and section 2.3.

2.1 Sequence Alignment Tools

Sequence alignment tools take sequenced reads and reference genome as input, map reads to the genome, and output the alignment information of reads against genome. As described in Section 1, the scale of reference genome and sequenced reads data is so massive that most existing alignment tools have to rely on advanced indexing techniques like hash table and suffix index to deal with the large scale data. Most of alignment tools are either based on: hash table and suffix tree.

The first hash table based aligner is BLAST (Altschul et al., 1990), and a variety of optimizations and improvements of BLAST are developed after that, such as SOAP, SeqMap (Jiang et al., 2008), ZOOM (Lin et al., 2008), MAQ, RMAP (Smith et al., 2008), BFAST (Homer et al., 2009)(Li et al., 2010), CloudBurst (Schatz, 2009),

SHRiMP (Rumble et al., 2009), PerM (Chen et al., 2009), Mosaik (https://code.google.com/p/mosaik-aligner/), and GNUMAP (Clement et al., 2010) etc.

The suffix tree based alignment tools can reduce flexible matching to exact matching. A suffix tree is a data structure that stores all suffices of a string, and all identical copies of these strings collapse to a single path in the tree. Because of this, it's easy to migrate from one representation of suffix tree to another, which makes it possible to transform inexact matching to exact matching. The exact matching complexity of a query is linear with the length of the query. However, the memory required for constructing a suffix tree for a genome is so huge that it can hardly be implemented on a commodity server, resulting in the rare use of suffix tree based aligners in the early stage. The appearance of FM-index (Ferragina et al., 2000) solves this problem effectively: the memory consumption is around the size of reference genome which is affordable for most servers. Hence, most today's popular alignment tools are FM-index based, such as BWA, Bowtie, SOAP2 (Li et al., 2009), SEAL (Pireddu, et al., 2011) and each of them has its specific properties (Medina et al., 2012).

2.2 Intel MIC Coprocessor

Intel recently announced the Intel MIC architecture based coprocessor Intel Xeon Phi (Renders, 2012). The coprocessor is equipped with up to 50 cores clocked at 1GHz or higher, and 512-bit SIMD (Single Instruction Multiple Data) capabilities. Due to the large number of cores and wide vector size, each coprocessor can deliver 1063.82 TFlop/s of double-precision (DP). As MIC coprocessor is an X86 SMP-on-a-chip, major parallel programming strategies on CPU can be easily implemented on MIC. There are two major approaches to introduce a MIC in an application:

1) Native Model. The processor and coprocessor both have one copy of the application and run the application natively by communicating with each other in various ways. For example, in native model, all residents of BWA run on CPU and MIC simultaneously, just as on two compute nodes in a network.

2) Offload Model. The application is viewed as running on the processor and offloading selected work to coprocessor. For example, in offload model, only the alignment algorithm search kernel will be executed on MIC, while the rest of BWA remains on CPU.

2.3 The Tianhe-2 Supercomputer

The Tianhe-2 high performance computer system is currently the world's fastest computer according to the TOP500 list[1] in June 2013 and November 2013. The theoretical peak performance of the complete system is 54.9 PFlop/s and the Linpack performance is 33.8 PFlop/s. Tianhe-2 is developed by the National University of Defense Technology (NUDT) as a national project. Most of our evaluations are performed on Tianhe-2.

The system consists of 16,000 compute nodes. Each compute node contains two Intel Xeon E5-2692 CPUs and three Intel Xeon Phi 31S1P coprocessors. Each Xeon Phi 31S1P has 57 cores and 8 GB on-card memory.

---

[1] TOP 500 Lists: `http://www.top500.org/lists/`

3 Design and Implementation

A multi-level parallelization is implemented in mBWA: the first one is a three-stage pipeline with overlapped IO and computation, and the second one is parallelization utilizing Intel MIC Coprocessor.

3.1 A Three-Stage Parallel Pipeline

In order to improve CPU hardware utilization, we design a three-stage parallel pipeline. Both in original BWA and mBWA, reads are mapped from batch to batch, with each round mapping a batch of reads. Each round is processed sequentially in BWA while in mBWA each round is deployed as a three-stage parallel pipeline and overlapped with adjacent round. The three stages are: loading reads, aligning reads and writing results. At the start of each mapping round, the three stages of pipeline are launched at the same time by means of multi-threading, with one thread for loading reads, one thread for writing results and one or more threads for aligning reads. The loading stage loads reads for the next mapping round. The aligning stage aligns the reads of current batch, which are loaded in last mapping round. The writing stage writes alignment results of last mapping round. Thus, the processes of data IO and reads aligning can be overlapped except for the first and last mapping round in each batch, resulting in an improved utilization of CPU.

3.2 Collaborated Parallel Using CPU and MIC

In mBWA, the Intel MIC coprocessor is introduced to accelerate the aligning stage, taking advantage of data independency of each mapping round.

mBWA adopts MIC's offload model. In this model, the data required by the offload region of program must be transferred to MIC in advance. This model supports flat data structure, such as basic variables, arrays and structures that are bitwise copyable (Jeffers et al., 2013). If a data structure allows a simple bit-by-bit copy operation to work properly, it is defined as bitwise copyable (Jeffers et al., 2013). A bitwise copyable data structure should not contain any pointer. Therefore, structures containing pointers or arrays of pointers are not supported by MIC's offload model.

However, most of BWA's key data structures contain pointers or array of pointers, including structs representing FM-index, reads. To solve this problem, we reorganize the data structure of FM-index and reads by storing each bitwise uncopyable structure as an array to make them bitwise copyable. The operations in loading stage and writing stage are modified accordingly for the new data structure.

The 512-bit vector processor in each MIC core is also quite powerful. It can process eight DP or sixteen single-precision (SP) floating point numbers at once. To fully utilize the powerful vector processor, we vectorize the algorithm with the help of compiler by adding several pragma instructs.

Best performance can only be achieved when both MIC coprocessor and CPU are fully exploited. In MIC's "offload" mode, when the offload region of program is running on MIC, CPU is usually stalled to wait for the results transferred from MIC.

As an improvement, we designed an asynchronous data transfer framework between CPU and MIC, which allows CPU to continue working after launching the offload region. In this framework, CPU transfers data to MIC to launch the offload region, and immediately starts its own aligning stage. Thus, CPU and MIC perform aligning stage simultaneously. After CPU and MIC finish their aligning jobs, CPU retrieves the results from MIC and continues other operations. With the asynchronous data transfer framework, the utilization of CPU and MIC is optimized.

4 Evaluation

We evaluate the performance of mBWA in the pipeline strategy and the effectiveness of MIC coprocessor.

4.1 Experiment Setup

In order to evaluate the improvement of pipeline strategy, two versions of BWA are used: the original BWA whose version is 0.5.10 and the pipeline only version of mBWA which contains no optimization strategies expect for the three-stage parallel pipeline. We used an Inspur NF5280M server equipped with one Intel Xeon Phi 3120A coprocessor and the Tianhe-2 supercomputer equipped with three MIC cards on each compute node to evaluate the effectiveness of MIC. The 3120A coprocessor on NF5280M server also has 57 1.1 GHz cores and 6 GB memory on card. Each MIC card in Tianhe-2 has 57 1.1 GHz cores, but 8 GB memory on card.

Our experiments focus on human genome, but the methodology is applicable for all species. The reference genome is hg19 and the reads are Illumina paired-end reads of YH samples whose default length is 100bp.

4.2 Pipeline Strategy

mBWA implements a three-stage parallel pipeline to improve CPU efficiency by hiding the time of data IO. We first evaluate the performance improvement introduced by the pipeline using different number of threads. As depicted in Fig. 1(a), the pipeline only version of mBWA achieved its best performance at 32 threads, and achieved its highest speedup at 8 threads with nearly 2-fold faster. The speedup reached top at 8 threads rather than 32 threads. With the number of threads increasing, the cost of creating and joining threads increases, but the scale of input data remains equally. Consequently, the proportion of computing time in whole time decreases, which leads to the decline of speedup.

Then we evaluated the improvement of pipeline strategy under different batch sizes. The recommended default batch size of BWA is 0x40000 reads, so the size of batch increases by 0x40000 each time. As depicted in Fig. 1(b), pipeline strategy is useful for mBWA and the speedup increases along with the increasing of batch size. As described above, the pipeline strategy only reorganizes the IO of alignment, so the pipeline has no effect on the quality of alignment.

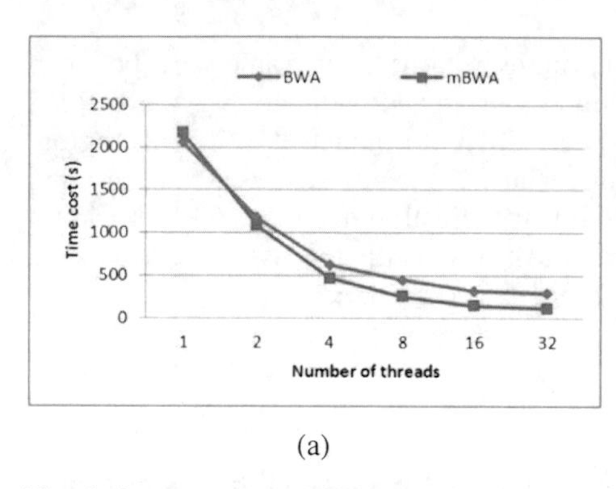

(a)

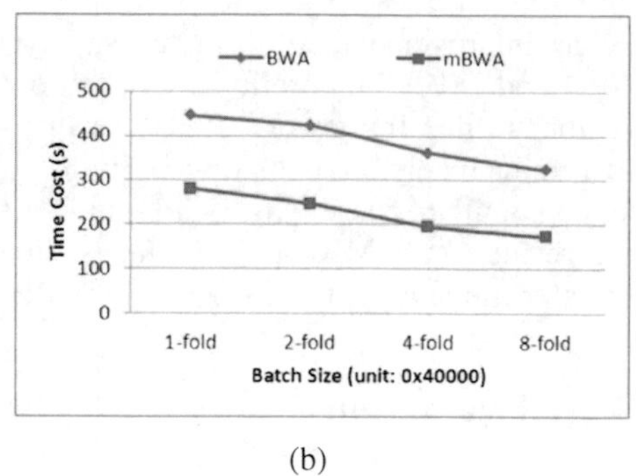

(b)

Fig. 1. Improvements of Pipeline Strategy. (a) Improvement under different number of threads. (b) Improvement under different batch size.

4.3 Effectiveness of MIC

Wc first evaluate the effectiveness of accelerators on Inspur NF5280M. NF5280M blade server is equipped with one MIC card, so we can compare the time cost of BWA and mBWA directly.

BWA runs only on CPU, and mBWA runs both on CPU and MIC. The number of threads on CPU is 32 and the number of threads on MIC is 224. We tested the effectiveness of MIC coprocessor with three-stage pipeline and without three-stage pipeline both. As shown in Table 1, mBWA achieved about 2.6-fold speedup with the acceleration of MIC, and achieved 5.3-fold speedup with additional three-stage pipeline.

Table 1. Time Cost of mBWA & BWA

Application	Pipeline	mBWA	BWA
Time Cost/s	off	104.258	271.923
	on	51.381	

We run mBWA under 56, 112, 168 and 224 threads (empirically selected values) to evaluate the scalability of mBWA on MIC. As depicted in Fig. 2(a), the performance of mBWA on MIC increases with more threads in the beginning and tops at 224 threads. With more and more threads, the cost of threads creating and joining increases, and more data cache and DTLB misses occur, thus resulting in some loss of performance.

We evaluate the performance of mBWA under multiple MIC cards on Tianhe-2. Each compute node of Tianhe-2 is equipped with three coprocessors, so we test the performance on one, two and three cards. The data needed by MIC is all transferred from CPU, no communication between MIC cards. Thus, the performance of mBWA increased linearly with the number of coprocessor increasing, as depicted in Fig. 2(b). Although we reorganize the key data structures and operation of BWA, the accuracy of alignment is not affected, and the experiment results demonstrate this.

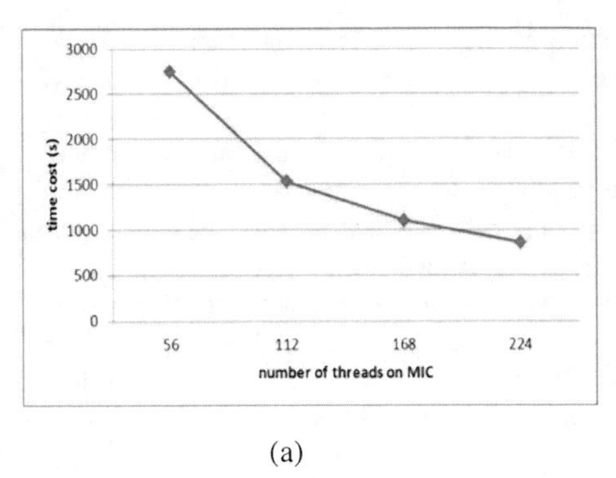

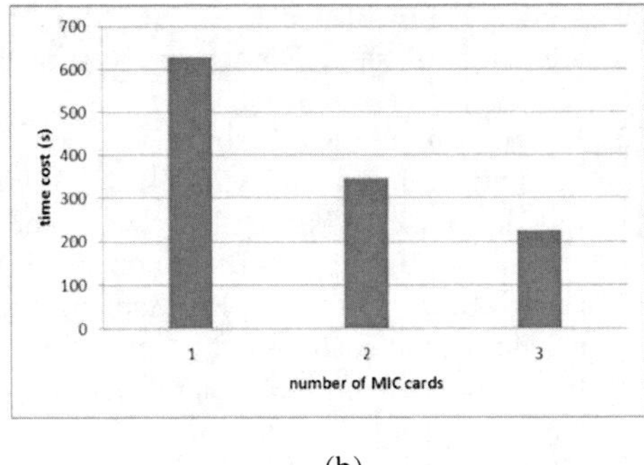

(a) (b)

Fig. 2. Effectiveness of MIC. (a) Scalability of mBWA on MIC. (b) Performance of mBWA under different number of MICs.

5 Conclusions

mBWA is a massively parallel sequence reads aligner, which contains three-stage parallel pipeline strategy and massive parallelization with Intel MIC coprocessor It achieves about 5-fold speedup in one blade server equipped with one MIC card, while maintaining the quality of alignment. As our parallelization approaches are generic, all BWT based application can take use of them.

Acknowledgement. This work is supported by NSF Grant 61272056 and NSF Grant 61133005.

References

1. Altschul, S.F., Gish, W., Miller, W., Myers, E.W., Lipman, D.J.: Basic local alignment search tool. Journal of Molecular Biology 215(3), 403–410 (1990)
2. Chen, Y., Souaiaia, T., Chen, T.: PerM: efficient mapping of short sequencing reads with periodic full sensitive spaced seeds. Bioinformatics 25(19), 2514–2521 (2009)
3. Clement, N.L., Snell, Q., Clement, M.J., Hollenhorst, P.C., Purwar, J., Graves, B.J., Johnson, W.E.: The GNUMAP algorithm: unbiased probabilistic mapping of oligonucleotides from next-generation sequencing. Bioinformatics 26(1), 38–45 (2010)
4. Cokus, S.J., Feng, S., Zhang, X., Chen, Z., Merriman, B., Haudenschild, C.D., Jacobsen, S.E.: Shotgun bisulphite sequencing of the Arabidopsis genome reveals DNA methylation patterning. Nature 452(7184), 215–219 (2008)
5. Ferragina, P., Manzini, G.: Opportunistic data structures with applications. In: Proceedings of the 41st Annual Symposium on Foundations of Computer Science 2000, pp. 390–398. IEEE (2000)
6. Homer, N., Merriman, B., Nelson, S.F.: BFAST: an alignment tool for large scale genome resequencing. PloS One 4(11), e7767 (2009)
7. Jeffers, J., Reinders, J.: Intel Xeon Phi Coprocessor High Performance Programming. Newnes, Boston (2013)
8. Jiang, H., Wong, W.H.: SeqMap: mapping massive amount of oligonucleotides to the genome. Bioinformatics 24(20), 2395–2396 (2008)

9. Langmead, B., Trapnell, C., Pop, M., Salzberg, S.L.: Ultrafast and memory-efficient alignment of short DNA sequences to the human genome. Genome Biol. 10(3), R25 (2009)
10. Li, H., Durbin, R.: Fast and accurate short read alignment with Burrows–Wheeler transform. Bioinformatics 25(14), 1754–1760 (2009)
11. Li, H., Homer, N.: A survey of sequence alignment algorithms for next-generation sequencing. Briefings in Bioinformatics 11(5), 473–483 (2010)
12. Li, H., Ruan, J., Durbin, R.: Mapping short DNA sequencing reads and calling variants using mapping quality scores. Genome Research 18(11), 1851–1858 (2008)
13. Li, R., Li, Y., Fang, X., Yang, H., Wang, J., Kristiansen, K., Wang, J.: SNP detection for massively parallel whole-genome resequencing. Genome Research 19(6), 1124–1132 (2009)
14. Li, R., Li, Y., Kristiansen, K., Wang, J.: SOAP: short oligonucleotide alignment program. Bioinformatics 24(5), 713–714 (2008)
15. Li, R., Yu, C., Li, Y., Lam, T.W., Yiu, S.M., Kristiansen, K., Wang, J.: SOAP2: an improved ultrafast tool for short read alignment. Bioinformatics 25(15), 1966–1967 (2009)
16. Lin, H., Zhang, Z., Zhang, M.Q., Ma, B., Li, M.: ZOOM! Zillions of oligos mapped. Bioinformatics 24(21), 2431–2437 (2008)
17. Ma, B., Tromp, J., Li, M.: PatternHunter: faster and more sensitive homology search. Bioinformatics 18(3), 440–445 (2002)
18. Medina-Medina, N., Broka, A., Lacey, S., Lin, H., Klings, E.S., Baldwin, C.T., Steinberg, M.H., Sebastiani, P.: Comparing Bowtie and BWA to Align Short Reads from a RNA-Seq Experiment. In: Rocha, M.P., Luscombe, N., Fdez-Riverola, F., Rodríguez, J.M.C. (eds.) 6th International Conference on PACBB. AISC, vol. 154, pp. 197–207. Springer, Heidelberg (2012)
19. Pireddu, L., Leo, S., Zanetti, G.: MapReducing a genomic sequencing workflow. In: Proceedings of the Second International Workshop on MapReduce and its Applications, pp. 67–74. ACM (2011)
20. Rumble, S.M., Lacroute, P., Dalca, A.V., Fiume, M., Sidow, A., Brudno, M.: SHRiMP: accurate mapping of short color-space reads. PLoS Computational Biology 5(5), e1000386 (2009)
21. Schatz, M.C.: CloudBurst: highly sensitive read mapping with MapReduce. Bioinformatics 25(11), 1363–1369 (2009)
22. Smith, A.D., Xuan, Z., Zhang, M.Q.: Using quality scores and longer reads improves accuracy of Solexa read mapping. BMC Bioinformatics 9(1), 128 (2008)

Optimizing Multiple Pairwise Alignment of Genomic Sequences in Multicore Clusters*

Alberto Montañola[1], Concepció Roig[1], and Porfidio Hernández[2]

[1] Computer Science Department, Universitat de Lleida, Lleida, Spain
alberto.montanola@udl.cat, roig@diei.udl.cat
[2] Computer Architecture and Operating Systems Department,
Universitat Autònoma de Barcelona, Cerdanyola del Vallès, Spain
Porfidio.Hernandez@uab.es

Abstract. Multiple sequence alignment (MSA), used in biocomputing to study similarities between different genomic sequences, is known to require important memory and computation resources. Determining the efficient amount of resources to allocate is important to avoid waste of them, thus reducing the economical costs required in running for example a specific cloud instance. The pairwise alignment is the initial key step of the MSA problem, which will compute all pair alignments needed. We present a method to determine the optimal amount of memory and computation resources to allocate by the pairwise alignment, and we will validate it through a set of experimental results for different possible inputs. These allow us to determine the best parameters to configure the applications in order to use effectively the available resources of a given system.

Keywords: sequence alignment, shared memory, message-passing, distributed computing, multi-core.

1 Introduction

In biology, a genomic sequence is a character string representing the codification of different basic structural elements known as amino-acids that will compose a specific protein with specific properties. There is the need for several applications to study the differences between different genomic sequences, for example in the aid to study the evolution of species and other usages in genomic research. This comparative, when is performed with two sequences, is known as pair sequence alignment. In some cases comparing two sequences is sufficient, but in other cases there is need of comparing thousands of sequences, then the problem is known as Multiple Sequence Alignment (MSA).

The MSA problem [edgar06] aims to find out the best possible alignment of all sequences. The steps composing the MSA process are illustrated in Figure 1, and can be described as follows:

* This work was supported by the MEyC-Spain under contract TIN 2011-28689-C02-02 and Consolider CSD2007-0050. The CUR of DIUE of GENCAT and the European Social Fund.

J. Sáez-Rodríguez et al. (eds.), *8th International Conference on Practical Appl. of Comput. Biol. & Bioinform. (PACBB 2014)*, Advances in Intelligent Systems and Computing 294,
DOI: 10.1007/978-3-319-07581-5_15, © Springer International Publishing Switzerland 2014

1. Read a list of genomic sequences to be aligned.
2. Pairwise alignment of all possible pairs of input sequences, calculating a similarity score for each pair.
3. Building a phylogenetic guide tree relating the sequences by similarity. The tree indicates the order for the subsequent multiple alignment.
4. Progressive alignment of the final tree to obtain the multiple alignment of all sequences.

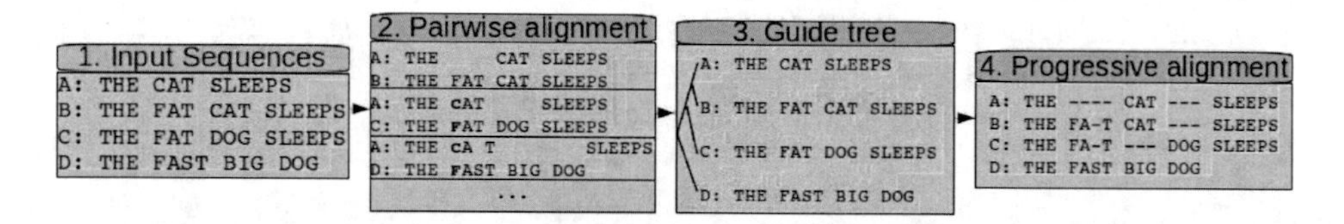

Fig. 1. MSA: Multiple Sequence Alignment steps

In this paper we study the impact of the initial pairwise alignment step when it is applied to a big number of sequences, and we focus on its improvement by studying its performance. The main goal of our work, consists on developing a method to determine the best configuration and tuning parameters that should be used to run the MSA on a specific hardware.

This article extends our previous work [montanola13], where we presented an hybrid programming solution, that combines the shared memory model with a distributed one, by expanding as many parallel threads as possible on a determined number of computing nodes.

In the present work, we carry out an experimental study to evaluate the performance and scalability of our mechanism according to the number and the length of sequences to be aligned. Additionally, we propose a method to estimate the number of computation resources needed to solve the problem, given a set of input sequences, in such a way that an efficient use of resources could be achieved. Finally, we compare the experimental data with the estimated prediction given by our proposed model.

The remainder of the paper is organized as follows: Section 2 reviews related work, Section 3 describes the used multiple pairwise implementation, Section 4 discusses the experimental results and finally, Section 5 presents our conclusions.

2 Related Work

Sequence alignment is a widely used application in the biocomputing community, were several previous efforts have been done in order to solve the problem [daugelaite13]. For the pairwise alignment of two single sequences, authors have proposed methods such as Smith-Waterman [smith81] and Needleman and Wunsch [needleman70].

When more than two sequences are involved, the problem complexity scales into MSA, where algorithms have to compute all possible combinations of pairs of sequences in order to achieve a final multiple alignment. Such implementations are for example T-Coffee [notredame00], Muscle [edgar04] and ClustalW [larkin07] among several others.

In order to improve the performance of these MSA algorithms, some authors have proposed distributed memory solutions based on message passing (MPI), including Parallel-T-Coffee [zola07] and ClustalW-MPI [li03] that have considerably improved their speedup. They coincide in the need to align several pairs of sequences. Thus, they are focused on solving the problem using only a distributed memory paradigm and lack the benefits of a hybrid system (distributed/shared memory).

While previous implementations will run several instances of the Smith-Waterman algorithm in parallel, there are also parallel implementations using shared memory such as [farrar07] and [liu09]. Using both methods may benefit from a better hybrid implementation that could exploit and use the modern parallel systems more efficiently.

3 Multiple Pairwise Implementation

In this section we are going to briefly describe the algorithm used to carry out step 2 of pairwise alignment in the MSA process. The current version of the algorithm is available at *http://lescaffee.sevendeset.org*. We study its impact and efficiency. Furthermore, we are going to present a model used to determine the amount of resources required for its optimal use.

We assume a computing platform based on an hybrid architecture where cores in the same node have a shared memory access while different nodes act in a distributed memory fashion through message-passing. The proposed algorithm distributes the pairwise alignment work by mapping to each available core and node a partition of sequences from all possible combinations, by distributing the total computation time in the best uniform way across the system.

We consider a number of n sequences of similar length to be aligned. Then, the number of pairwise alignments to carry out will be $\frac{n^2-n}{2}$, which corresponds to the total number of pairs. Assuming a balanced distribution of all pairwise alignments into the cores of the system and considering that a single pair alignment is entirely done in a core, the total pairwise computation (t_{pwc}) can be calculated with expression (1), where pwt is the time needed to carry out a single pairwise operation and f is the total number of cores.

$$t_{pwc} = pwt \times \frac{n^2 - n}{2f} \tag{1}$$

The expected communications overhead (t_{comm}) of the message passing stack, for the distribution of their sequences to all nodes and the gathering of all the aligned pairs, can be calculated with expression (2), where ℓ is the average sequence length in bytes, bw the transmission speed in bytes per second and k is the number of nodes of the system.

$$t_{comm} = \frac{\ell \times n \times k}{bw} + \frac{\ell \times (n^2 - n)}{bw} \quad (2)$$

The total computation time t, given in (3), corresponds, at maximum, to the sum of total pairwise computation and the communication time. It has also to be added a value k_o that corresponds to the remaining application overhead used to compute the pairwise tasks and reading sequences from disk. The communication time between the cores in the same node is considered to be negligible.

$$t = t_{pwc} + t_{comm} + k_o \quad (3)$$

Finally, the expected total amount of memory (mem) that will be used by our implementation can be defined by (4), were k_{sm} is the average size of required data structures for one pair of sequences and k_{th} the overhead penalty for each additional thread. In average, we consider the size in memory of two aligned sequences as $3l$, taking into account that the aligned sequences can add gaps inside. Additionally it has to be added the storage of the n original sequences for each of the k nodes.

$$mem = \frac{n^2 - n}{2} \times (3 \times l + k_{sm}) + f \times k_{th} + n \times \ell \times k\ . \quad (4)$$

From equation (3) we can calculate the speedup as the sequential computation time ($pwt \times \frac{n^2-n}{2}$) divided by the parallel computation time t. According to this, the efficiency is calculated with expression (5) as the obtained speedup with respect to the optimum one, that is of number of cores. This permits us to determine the expected number of resources needed for an specific input problem.

$$Efficiency = \frac{pwt \times bw + 2 \times \ell}{pwt \times bw + 2 \times \ell \times f}\ . \quad (5)$$

This means that the efficiency directly depends on the length of the input sequences and the computing time required to align one pair of sequences. From this expression, it can be calculated the recommended number of cores to achieve the desired efficiency.

The accuracy of the performance parameters previously defined, is proved for the implementation of the multiple pairwise alignment. The functionality of these steps is the following:

1. Input sequences are parsed and loaded into the memory by a master task, and the system is queried to determine the total number of nodes, maximum memory available and number of available cores per node. Then we determine some system information, by testing the data transfer speed t_k and calculating a set of benchmarking pairs of sequences in order to calculate t_{pwt}. Then, using equation (5), we calculate the ideal f value to run the application and the required application memory using (4).
2. All possible pairs of sequences are generated and distributed.

3. Each process spawns a worker pool of c threads, where c is the total number of cores available in a given node.
4. Each thread carries out the Smith-Waterman algorithm for each pair of sequences that has been assigned to.
5. All threads and processes are joined and all aligned pairs are merged into a final list.

4 Experimental Results

Different tests were performed in order to analyse the behaviour of our presented implementation, thus proving the correctness of the prediction model used to evaluate and determine the best parameters for a specific input. In these experiments we are measuring the total computation time and memory usage by the application and comparing it to the model that defines the expected behaviour of the experiments. Different sequence sets from 400 to 1400 sequences of lengths varying from 100 to 200 residues were used.

The platform used to perform executions consists of a cluster composed by 24 nodes of Intel Quad Core processors running at 2.4 GHz with 8 Gigabytes of RAM each one. Moreover, the configuration used for each execution consists on running a fixed number of MPI processes, one per node, expanding as many threads as available cores on the node. For example, one execution may consists on expanding 5 MPI processes in 5 nodes with 4 threads each one, thus achieving a total number of 20 threads.

In all samples, the average processing time is calculated by measuring the total time required by the application since it loads the sequences from disk until all the results are written back to disk by the master process. The total memory usage is measured on each node by measuring all memory allocations by providing replacements to the standard allocation library. The displayed results show the total system memory by adding the memory measures of each node.

By applying the previously seen expression (4), considering $k_{sm} = 400bytes$ and depreciating k_{th}, we determine the expected amount of memory consumption for 400 to 1400 sequences of 229 residues. These results are compared with experimental ones on Table 1. As can be seen, the values that we calculated can successfully predict the behaviour of the algorithm in the experimentation.

Figure 2 shows in logarithmic scale the execution time, in seconds, with the following graphs: t_{pwc}: the execution of Smith-Watermanm, t_{comm}: the communications time, and $total$: the total time. The simulated graphic is generated from data gathered from the prediction algorithm using the after-mentioned equation models. As it can be observed, the simulated graphs are a good prediction of the behaviour of the algorithm. Communications overhead stays stable on both graphs, being constant in the simulation because the time to transfer $\frac{n^2-n}{2}$ pairs of sequences is fixed indifferently on the number of nodes.

Figure 3 shows in logarithmic scale the computation time, in seconds of step 4, corresponding to the execution of Smith-Waterman in each core on the left, and the average time inverted in the communications layer of MPI on the right

Table 1. Comparison of experimental memory usage versus the predicted one

Input n	Memory MB	
	Experimental	Predicted
400	94	92
600	196	196
800	343	341
1000	539	527
1200	762	755
1400	1038	1025

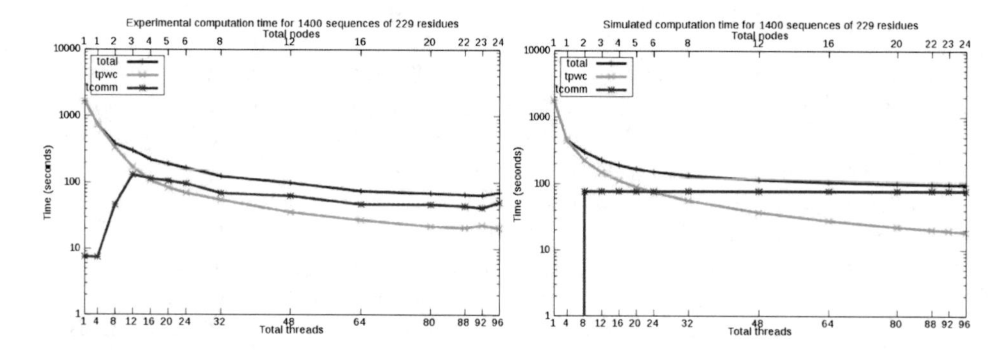

Fig. 2. Experimental and simulated execution time of different algorithm steps

graph. The number of sequences varies from 400 to 1400, with a fixed length of 229 residues. While studying the pairwise step, as it can be observed the computation time increases in proportion of the number of sequences to be aligned. In all the cases, increasing the number of threads entails a reduction of computation time. From this graph we can calculate the speedup, which is linear, thus, the algorithm correctly scales, meaning the computation balance is correctly distributed among the nodes. On the other hand, it can be seen that the time invested in communications is proportional to the number of sequences as it is also established in equation (2). However, it can be observed in the experiment that this time is stabilized from 8 nodes to 24.

Finally, Figure 4, shows the sum of total computation time required by the expression (3). We can see that the time stabilizes to a certain point were adding more nodes will not improve the overall time, thus we can see that the impact of the communications are affecting negatively to the Efficiency endangering the scalability of the algorithm. From the prediction expression (5) we determined that the best number of nodes for 400 sequences is 20, between 600 and 1000 is 24, for 1200 is 28, and lastly for 1400 is 32 cores. Thus, we can see that experimental results are not so far from the expected ones.

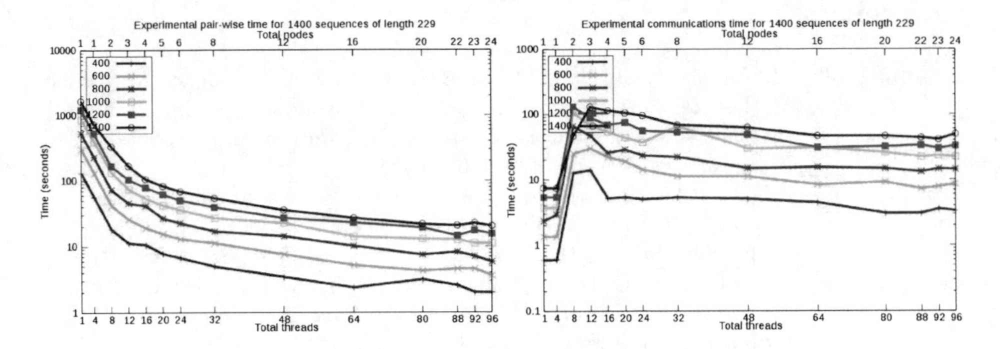

Fig. 3. Pairwise time and communications overhead varying the number of sequences

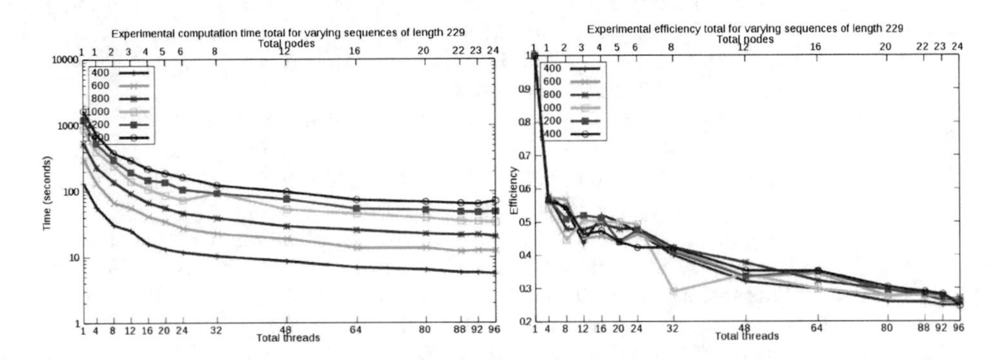

Fig. 4. Total pairwise time and Efficiency varying the number of sequences

5 Conclusions

The pairwise sequence alignment, as a key step in the Multiple Sequence Alignment problem, will influence in the overall performance of the problem, being the step with more computing resources required. As an essential step towards MSA, we should find new methodologies and techniques to improve and to parallellize it. Furthermore, we need a way to predict which configuration is the best one for a specific input, in order to be able to run the given algorithm with the best efficient parameters, to use the resources in the best possible way.

This paper presents a new implementation based on the Smith-Waterman pairwise algorithm, that aims to study its scalability and behavior and finds a way to predict the best parameters that should be used. Thus, we present an hybrid implementation using a threading library combined with a message passing one to exploit the available resources as best as possible. Finally, we obtain a set of results that prove that the algorithm behavior can be predicted by the defined equations.

References

[montanola13] Montañola, A., Roig, C., Hernández, P.: Pairwise Sequence Alignment Method for Distributed Shared Memory Systems. In: 21st Euromicro International Conference on Parallel, Distributed, and Network-Based Processing (PDP), pp. 432–436 (2013)

[edgar06] Edgar, R.C., Batzoglou, S.: Multiple sequence alignment. Current Opinion in Structural Biology 16(3), 368–373 (2006), http://dx.doi.org/10.1016/j.sbi.2006.04.004 ISSN 0959-440X

[daugelaite13] Daugelaite, J., O' Driscoll, A., Sleator, R.D.: An Overview of Multiple Sequence Alignments and Cloud Computing in Bioinformatics. ISRN Biomathematics 2013, Article ID 615630, 14 pages (2013)

[notredame00] Notredame, C., Higgins, D.G., Heringa, J.: T-Coffee: A novel method for fast and accurate multiple sequence alignment. J. Mol. Biol. 302(1), 205–217 (2000), PMID: 10964570 (PubMed - indexed for MEDLINE)

[smith81] Smith, T.F., Waterman, M.S.: Identification of Common Molecular Subsequences. Journal of Molecular Biology 147, 195–197 (1981), doi:10.1016/0022-2836(81)90087-5

[needleman70] Needleman, S.B., Wunsch, C.D.: A general method applicable to the search for similarities in the amino acid sequence of two proteins. Journal of Molecular Biology 48(3), 443–453 (1970), doi:10.1016/0022-2836(70)90057-4; PMID 5420325

[larkin07] Larkin, M.A., Blackshields, G., Brown, N.P., Chenna, R., McGettigan, P.A., McWillian, H., Valentin, F., Wallace, I.M., Wilm, A., Lopez, R., Thompson, J.D., Gibson, T.J., Higgins, D.G.: ClustalW and ClustalX version 2. Bioinformatics 23(21), 2947–2948 (2007)

[edgar04] Edgar, R.C.: MUSCLE: multiple sequence alignment with high accuracy and high throughput. Nucleic Acids Research 32(5), 1792–1797 (2004)

[zola07] Zola, J., Yang, X., Rospondek, S., Aluru, S.: Parallel T-Coffee: A Parallel Multiple Sequence Aligner. In: Proc. of ISCA PDCS-2007, pp. 248–253 (2007)

[li03] Li, K.B.: ClustalW-MPI: ClustalW analysis using distributed and parallel computing. Bioinformatics 19, 1585–1586 (2003)

[farrar07] Farrar, M.: Striped Smith-Waterman speeds database searches six times over other SIMD implementations. Bioinformatics 23(2), 156–161 (2007), doi:10.1093/bioinformatics; first published online (November 16, 2006)

[liu09] Liu, Y., Maskell, D.L., Schmidt, B.: CUDASW++: optimizing Smith-Waterman sequence database searches for CUDA-enabled graphics processing units. BMC Research Notes 2, 73 (2009)

High Performance Genomic Sequencing: A Filtered Approach

German Retamosa[1], Luis de Pedro[1], Ivan Gonzalez[1], and Javier Tamames[2]

[1] High Performance and Computing Networking Department, University Autonoma of Madrid, Spain
{german.retamosa,luis.depedro,ivan.gonzalez}@uam.es
[2] National Biotechnology Research Center, CSIC, Spain
jtamames@cnb.csic.es

Abstract. Protein and DNA homology detection systems are an essential part in computational biology applications. These algorithms have changed over the time from dynamic programming approaches by finding the optimal local alignment between two sequences to statistical approaches with different kinds of heuristics that minimize former executions times. However, the continuously increasing size of input datasets is being projected into the use of High Performance Computing (HPC) hardware and software in order to address this problem. The aim of the research presented in this paper is to propose a new filtering methodology, based on general-purpose graphical processor units (GP-GPUs) and multi-core processors, for removing those sequences considered irrelevant in terms of homology and similarity. The proposed methodology is completely independent from the homology detection algorithm. This approach is very useful for researchers and practitioners because they do not need to understand a new algorithm. This design has been approved by the National Biotechnology Research Center of Spain (CNB).

Keywords: Comparison and alignment methods, BLAST, High Performance Computing, GP-GPU.

1 Introduction

Genomics and genetics are one of the biology specialties more relevant today [1]. Genomics start-ups are the best candidates of such relevance and considered to be one of the most profitable companies [2]. Furthermore, thanks to the research carried out until now, it is possible to obtain a better understanding of the human and other species genomes. However, this technical revolution evolves step-by-step with some short-term dangers to be taken into account. Stepping up the computational services of research centers in order to fit their current scientific needs must be interpreted as an urgent wake-up call and a technical challenge known as the *big data* paradigm [3]. This new and growing paradigm is the answer of a number of challenges such as the capture, curation, storage, search, sharing, analysis and visualization in terms of data sets whose size and complexity is beyond the ability of commonly used software algorithms.

J. Sáez-Rodríguez et al. (eds.), *8th International Conference on Practical Appl. of Comput. Biol. & Bioinform. (PACBB 2014)*, Advances in Intelligent Systems and Computing 294,
DOI: 10.1007/978-3-319-07581-5_16,

Starting with the objective of improving these algorithms with High Performance Computing (HPC) techniques, researchers and practitioners have currently two different approaches. On the one hand, there are some applications able to speed up the performance of older algorithms, ie GPU-BLASTP [10] and GPU-NCBI-BLAST [11]. These applications are based on refactoring and re-implementing the source code of the corresponding algorithms according to the HPC architectures. The usual approach consists of acquiring a deeper knowledge about these algorithms with the corresponding investment of time and effort. On the other hand, there are an existing line of research based on filtering and removing those sequences considered irrelevant according to a given model [9] with Smith-Waterman [7] and Aho-Corasick [8] algorithm as core elements. Such approach is really interesting because it guarantees an autonomous behavior between the filtering model and the sequencing algorithm and provides a useful application to perform daily genomic studies in less time, without additional efforts to integrate the filtering module or to understand how it works and it is even more accurate in some particular cases.

Finally, this paper aims to follow the line of research about filtering algorithms by using of NCBI BLAST [4] as a widely used reference sequencing algorithm for the filtering proposed methodology to be applied. Furthermore, it is structured into three main blocks. Section 2 provides a generic introduction of the most important homology detection algorithm NCBI BLAST [4]. Section 3 provides the proposed design with an overview of the hitting process and describes the theoretical model of the filter implementation. Section 4 evaluates the performance and accuracy of a non-optimized prototype implementation with different simulated examples and make some conclusions and future work about the research carried out in section 5.

2 Related Works

Homology detection algorithms represents the technical approach of most of the genomic research. These applications help to researchers and practitioners saving time and facilitating the obtained results interpretation. In particular, NCBI BLAST [4] is known as the most used algorithm and use statistical methods to make a link between sequences and relevance. BLAST (Basic Local Alignment Search Tool) [4] is based exclusively on statistical methods with hit-and-extend heuristics for proteins and DNA, is divided into the following steps:

1. Remove low-complexity regions or sequence repeats in the query sequence [12].
2. For each k-letter in the query sequence:
   (a) Using substitution matrix such as BLOSUM62 [16] or PAM250 [15], list all matching words with scores greater than a given significance threshold.
   (b) Organize the remaining high-scoring words into an efficient search tree.
3. Scan the database sequences for exact matches with the remaining high-scoring words.

4. Extend the exact matches to high-scoring segment pair (HSP).
5. According to a specific cutoff score, list all of the HSPs in the database whose score is high enough to be considered.
6. Evaluate the significance of the HSP score by exploiting the Gumbel extreme value distribution (EVD) and Karlin-Altschul equation.
7. Make two or more HSP regions into a longer alignment by using of methods like Poisson (original BLAST [4]) or Sum-of-Scores (BLAST2 [13]).
8. Report every match whose expect score is lower than a threshold parameter.

In contrast with other algorithms, the parallellization of BLAST and its variants is nearly impossible to obtain due to its algorithmic complexity. Therefore, HPC research carried out until now [10,11] has focused in two particular stages, hitting and extension.

3 Methodology

Using HPC techniques, the proposed filtering method is based on the assumption that the number of hits between two sequences tends to be, according to their length, directly proportional to the relevance of their alignment. Unlike former filtering approaches [9], this proposed methodology is best suited to any kind of residues, either for proteins, nucleotides or for metagenomes. First of all, it begins with the definition of hit and then explains the multiple scenarios where hits acquire different relevance. Secondly, it extends the previous definition of hit describing the hit proximity concept and how the more hits appear together the more relevant the alignment will be.

3.1 Hit Definition

As we see in Section 2, most of the homology detection algorithms for proteins, nucleotides or metagenomes are based on the search of hits, ie BLAST [4]. However, depending on each particular case, the definition of a hit can be slightly different.

On the one hand, proteins and metagenomes sequencing define a hit as the occurrence of two aligned residues with a positive score given by a specific substitution matrix [14][15]. A substitution matrix is a bidimensional table which describes loosely the rate at which one residue in a sequence changes to other residue state over time and thus, it provides the similarity between them. Some examples of substitution matrix are BLOSUM62 [16], PAM250 [15], etc. As stated in [17], there some research lines that bases the definition of a substitution matrix and its inner data in terms of empirical and experimental analysis. However, there are some other research based on physico-chemical properties of amino acids [19,20] that have stayed outside the scope of this paper.

Equation 1 is the expression of the hit score between two residues. $M_{i,j}$ is the probability that amino acid i transforms into amino acid j and $p_i p_j$ are its individual frequencies. Therefore, we can remark that the likelihood or correlation between two residues ($S(i, j)$), i.e. the score between them, follows a logarithmic distribution where the more similarity between the aligned residues, the higher significance score will be achieved and vice versa.

$$S(i,j) = \log \frac{p_i M_{i,j}}{p_i p_j} = \log \frac{M_{i,j}}{p_j} = \log \frac{observed\ frequency}{expected\ frequency} \quad (1)$$

On the other hand, nucleotide sequencing defines a hit in two different ways. Firstly, there are some reference sequencing algorithms, i.e. NCBI BLAST, that define a nucleotide hit as an exact match between two aligned residues [14,15]. In this particular case, the non-existence of substitution matrix simplifies the hitting process and increases the estimation reliability in terms of sequence similarity. However, there are some other research that can take into account nucleotide transversions and transitions to create customized substitution matrix [21].

3.2 Hits Density Principle

As discussed on previous sections, hitting process is a preliminary way to predict the relevance of an alignment between two sequences with a highly parallellizable nature and algorithmic complexity of ($O(N^4)$) where N is the number of *l-mers* per sequence founded in every database (target and query) and described in algorithmic terms as follows:

```
1. Iterate over a list of query sequence
for query_sequence in query_sequence_list:
    for query_sequence_lmer in query_sequence:
        2. Iterate over a list of database sequence
        for db_sequence in db_sequence_list:
            for db_sequence_lmer in db_sequence:
                3. Check matches
                if db_sequence_lmer == query_sequence_lmer:
                    match++
```

As mentioned before, another key point is that all this filtering process and its corresponding output can be redirected to any statistical analysis tool such as BLAST [4], BWA [5], BLAT [6]. These examples are complex applications that invest large amounts of time evaluating the relevance of each sequence. The understanding of the hit density principle begins with the hit proximity concept.

Our first step to deal with hit proximity is the frequency analysis of a hit through comparing long and short sequences. Intuitively, hit encountering in longer sequences will have a higher probability of success, i.e. more hits, than in shorter sequences. However, according to the Karlin-Altschul formula, we can conclude that the sequence length has a little impact on the significance between two sequences. According to the Section 2, every homology detection algorithm has its own significance function. The purpose of this function is to determine, by using of statistical or experimental methods, the relevance of an alignment between two sequences. In particular, the hit proximity and density principle has been endorsed by Karlin-Altschul formula from BLAST [4] algorithm. This formula, shown in Equation 2, determines the significance of an alignment in terms of searching space, raw score and some other scaling factors obtained from experimentation.

$$E = \frac{kmn}{\exp^{\lambda S}} \tag{2}$$

The analysis of this equation has identified some scaling factors, such as k and λ, as constant values and has determined that the raw score S is calculated from the mixture of statistical and experimental values. Therefore, the search space of the target and query sequence, represented by mn, is the key point to determine the relevance between two sequences. Hence, once reviewed all elements of the equation, we can conclude that hit proximity reduce the search space between two sequence and thus, increases the relevance between them. However, there is a strong contradiction between hitting and phylogenetic research that we must consider. According to a phylogenetic study [18], all homology detection algorithms that rely exclusively on the closest BLAST [4] hit should be interpreted with great caution because it does not always implies a phylogenetic proximity. This phylogenetic distance is understood as the distance obtained from a morphological matrix between two species according to their evolutionary behavior. In this case, the proposed hit proximity concept can also contribute with new levels of phylogenetic similarity because it is able to consider groups of hits instead of individual ones.

4 Evaluation

This assessment procedure, performed joined by the National Biotechnology Research Center of Spain and the University Autonoma of Madrid, begins with a complete definition of tests to perform, identifying for each one its relevance in terms of performance and accuracy. The genomes used during these tests were obtained from the National Centre for Biotechnology Information (NCBI) database and consists of the following examples, *Anaplasma marginale* genome (AMA) [26] is a tick-borne livestock pathogen member of the proteobacteria group, *Escherichia coli* genome (ECS) [25] is an human pathogen member of the proteobacteria group, *Buchnera aphidicola* genome (BAP) [24] is a symbiotic bacteria member of proteobacteria group and *Pseudomonas putida* genome (PPU) [23] is a versatile saprophytic soil bacterium that has been certified as a biosafety host for the cloning of foreign genes and member of proteobacteria group.

4.1 Functionality Tests

Functionality tests are a fine-grained validation procedure which aims to determine that the obtained results are correct and therefore, to certify the compatibility of the proposed methodology with other reference applications.

The inner implementation of these tests is based on the comparison between the obtained results from the proposed filtering method integrated with some reference applications and these applications in a standalone mode. This resulting likelihood percentage, measured from every alignment, is based on two different levels.

This first level of test aims to check the existence of a small protein fragments into the complete reference genome. Genomes like AMA or ECS have been selected to be compared to their own fragments, such as AMA1981 or ECS2480. In all these tests, the accuracy of the prototype has been 100% with a total inclusion of fragments and even more, due to the fine grain of the proposed system and the corresponding reduction of the search space by splitting and parallelizing the reference algorithm, new alignments, undetectable for conventional algorithms, have been detected. However, there is still some pending studies with the objective of analyzing the relevance of this new alignments.

The second level of test is the most realistic approach to scientific environments because it checks the findings of best alignments between two different and complete genomes. To support this stage, genomes like ECS, BAP, PPU and a reference algorithm like NCBI BLAST has been selected, concluding with an accuracy between 70 - 80%.

Either the increasing accuracy of the first level of test or the more differences in the second level has a simple explanation, the accuracy effects of the search space variable in heuristic homology detection algorithms like NCBI BLAST. As reference papers [4] discussed, the scoring system and thus, the relevance of an alignment, depends on the search space; therefore, the bigger search space value, the less accurate the algorithm will be. Some multi-core algorithms, such as MPI BLAST [22] or our proposed methodology, have demonstrated this statistical dependence by splitting the original dataset in blocks for individual processing, reducing the search space and increasing the accuracy of results.

4.2 Performance Test

Performance test are part of the second stage of this evaluation process which aims to mix the obtained accuracy in the previous phase with a good performance in terms of execution time.

On the one hand, the hardware architecture used in these test has consisted of an Intel Xeon E5506 with 8 physical cores and without hyperthreading features, 24 GBytes of RAM memory and 2 Graphical Processor Units (GPU) NVidia cards GTX590 with their own Scalable Link Interface (SLI) connection. SLI aims to increase the computational performance of applications by connecting multiple GPU in parallel. This technology employs several algorithms for distributing and load-balancing dynamically tasks between GPU cores at driver level. More specifically, the proposed implementation has been tested on a two GPU SLI-equipped system. As table 1 shows, the proposed methodology with the non-optimized prototype has reached a notable improvement with speed ups up to 3x, i.e. three times faster than the reference algorithm NCBI BLAST.

Table 1. Protein Performance Simulated Test

Protein	Database	NCBI BLAST Time	Prototype Time	Speed Up
ECS	NCBI Non-Redundant	79 minutes	49 minutes	2x
AMA	NCBI Non-Redundant	47 minutes	25 minutes	2x
BAP	NCBI Non-Redundant	20 minutes	7 minutes	3x

5 Conclusions

Based on the well-known algorithms such as BLAST [4] and the HPC techniques, new design ideas, statistical and numerical methodologies and implementations about filtering models have been analyzed for improving these algorithms in terms of execution time and accuracy. The highly parallellizable architecture of NVidia GPUs could achieve a massive sequence processing with significance improvements in commodity hardware. The proposed methodology provides a novel approach with regard to former solutions [9]. This design is able to reach a filtered model compatible with any kind of sequence, either for proteins, nucleotides or metagenomes and achieving an autonomous behavior from the original sequencing algorithm.

Based on the promising results, future work will be focused on completing and optimizing the implementation of the proposed model on GP-GPUs and carry out with a more extended and more detailed testing plan in terms of accuracy and performance.

Acknowledgements. The authors wish to thank the project CEMU-2013-14 supported by Universidad Autonoma de Madrid for providing the necessary means to carry out the proposed filtering system.

References

1. National Human Genome Research Institute. Why are genetics and genomics important to my health (2014), `http://www.genome.gov/19016904`
2. Weiss, B.: Genomics companies ripe for flurry of mergers. The Wall Street Journal (2013), `http://www.marketwatch.com/story/genomics-companies-ripe-for-flurry-of-mergers-2013-04-16`
3. Humphries, C.: A Hospital Takes Its Own Big-Data Medicine. MIT Technology Review (2013), `http://www.technologyreview.com/news/518916/a-hospital-takes-its-own-big-data-medicine/`
4. Altschul, S.F., Gish, W., Miller, W., Myers, E.W., Lipman, D.J.: Basic Local Alignment Search Tool. J. Mol. Biol. 215, 403–410 (1990)
5. Burrows, M., Wheeler, D.J.: A block sorting lossless data compression algorithm. Technical Report 124, Digital Equipment Corporation, Palo Alto, California (1994)
6. Kent, W.J.: BLAT the BLAT-like alignment tool. Gen. Res. 12, 656–664 (2002)
7. Smith, T.F., Waterman, M.S.: Identification of common molecular subsequences. Journal of Molecular Biology 147, 195–197 (1991)
8. Aho, A.V., Corasick, M.J.: Efficient string matching: An aid to bibliographic search. Comm. ACM 18(6), 333–340 (1975)

9. Nordin, M., Rahman, A., Yazid, M., Saman, M., Ahmad, A., Osman, A., Tap, M.: A Filtering Algorithm for Efficient Retrieving of DNA Sequence. International Journal of Computer Theory and Engineering 1(2), 1793–8201 (2009)
10. Xiao, S., Lin, H., Feng, W.: Accelerating Protein Sequence Search in a Heterogeneous Computing System. In: Parallel & Distributed Processing Symposium, IPDPS (2011)
11. Vouzis, P.D., Sahinidis, N.V.: GPU-BLAST: using graphics processors to accelerate protein sequence alignment. Bioinformatics 27(2), 182–188 (2011)
12. Wootton, J.C., Federhen, S.: Statistics of local complexity in amino acid sequences and sequence databases. Comput. Chem. 17, 149–163 (1993)
13. Tatusova, T.A., Madden, T.L.: BLAST 2 Sequences, a new tool for comparing protein and nucleotide sequences. FEMS Microbiol. Lett. 174, 247–250 (1999)
14. Dayhoff, M.O., Schwartz, R.M., Orcutt, B.C.: A Model of Evolutionary Change in Proteins. In: Dayhoff, M.O. (ed.) Atlas of Protein Sequence and Structure, vol. 5(3), pp. 345–352 (1978)
15. Henikoff, S., Henikoff, J.G.: Amino acid substitution matrices from protein blocks. Proceedings of National Academic Science USA 89, 10915–10919 (1992)
16. Altschul, S.F.: Amino acid substitution matrices from an information theoretic perspective. J. Mol. Biol. 219, 555–565 (1991)
17. Eddy, S.R.: Where did the BLOSUM62 alignment score matrix come from? Nature Biotechnology (2004)
18. Liisa, B., Koski, G., Golding, B.: The Closest BLAST Hit Is Often Not the Nearest Neighbor. Journal Molecular Evolution 52, 540–542 (2001)
19. Risler, J.L., Delorme, M.O., Delacroix, H., Henaut, A.: Amino acid substitutions in structurally related proteins. A pattern recognition approach. J. Mol. Biol. 204, 1019–1029 (1988)
20. Johnson, M.S., Overington, J.P.: A structural basis for sequence comparisons. An evaluation of scoring methodologies. J. Mol. Biol. 233, 716–738 (1993)
21. Ziheng, Y., Yoder, A.D.: Estimation of the Transition/Transversion Rate Bias and Species Sampling. J. Mol. Evol. 48, 274–283 (1999)
22. Darling, A., Carey, L., Feng, W.: The Design, Implementation, and Evaluation of mpiBLAST. In: Proc. of the 4th Intl. Conf. on Linux Clusters, p. 14 (2003)
23. Cornelis, P.: Pseudomonas: Genomics and Molecular Biology. Caister Academic Press (2008) ISBN 1-904455-19-0
24. Douglas, A.E.: Nutritional interactions in insect-microbial symbioses: Aphids and their symbiotic bacteria Buchnera. Annual Review of Entomology 43, 17–38 (1998)
25. Karch, H., Tarr, P., Bielaszewska, M.: Enterohaemorrhagic Escherichia coli in human medicine. Int. J. Med. Microbiol. 295(67), 405–418 (2005)
26. Brayton, K.A., Kappmeyer, L.S., Herndon, D.R., Dark, M.J., Tibbals, D.L., et al.: Complete genome sequencing of Anaplasma marginale reveals that the surface is skewed to two superfamilies of outer membrane proteins. Proc. Natl. Acad. Sci. USA 102, 844–849 (2005)

Exceptional Single Strand DNA Word Symmetry: Universal Law?

Vera Afreixo[1], João M.O.S. Rodrigues[2], and Carlos A.C. Bastos[2]

[1] CIDMA - Center for Research and Development in Mathematics and Applications, Department of Mathematics, University of Aveiro, 3810-193 Aveiro, Portugal
[2] Signal Processing Lab., IEETA and Department of Electronics Telecommunications and Informatics, University of Aveiro, 3810-193 Aveiro, Portugal

Abstract. Some previous studies point to the extension of Chargaff's second rule (the phenomenon of symmetry) to words of large length. However, in random sequences generated by an independent symbol model where the probability of occurrence of complementary nucleotides is the same, we expect that the phenomenon of symmetry holds for all word lengths. In this work, we measure the symmetry above that expected in independence contexts (exceptional symmetry), for several organisms: viruses; archaea; bacteria; eukaryotes. The results for each organism were compared to those obtained in control scenarios. We created a new organism genomic signature consisting of a vector of the measures of exceptional symmetry for words of lengths 1 through 12. We show that the proposed signature is able to capture essential relationships between organisms.

Keywords: DNA, Single strand symmetry, Exceptional symmetry, Effect size measure.

1 Introduction

The detailed analysis of some bacterial genomes led to the formulation of Chargaff's second parity rule, which asserts that complementary nucleotides occur with similar frequencies in each of the two DNA strands [12,7,13,5]. Extensions of this rule state that the frequencies of inverted complementary words (such as AAC and GTT) should also be similar. Chargaff's second parity rule and its extensions have been extensively confirmed in bacterial and eukaryotic genomes, including recent results (e.g [11,4,3,10,8,14,5,9,1,2]).

The views about the origins and biological significance of Chargaff's second parity rule and its extensions are conflicting [15]. For example, Forsdyke and Bell (2004) argue that this symmetry results from DNA stem-loop secondary structures [6]. However, Albrecht-Buehler (2007) argues that the presence of Chargaff's second parity rule and its extensions are due to the existence of certain mechanisms of inversions, transpositions, and inverted transpositions [3].

If a sequence is randomly generated using an independent symbol model that assigns equal probability to complementary nucleotides, then it is expected that the extensions of Chargaff's second parity rule will hold. In this case, however, words other than inverted complements (e.g. AAG, TTC, CAA) will also be equally likely. In real

J. Sáez-Rodríguez et al. (eds.), *8th International Conference on Practical Appl. of Comput. Biol. & Bioinform. (PACBB 2014)*, Advances in Intelligent Systems and Computing 294,
DOI: 10.1007/978-3-319-07581-5_17,

sequences, we found that the similarity between the frequency of each word and that of the corresponding inverted complement is stronger than the similarity to the frequency of any other word.

We will analyze not only the symmetry phenomenon (similarity between the frequencies of symmetric pairs) but also the exceptional symmetry phenomenon. This exceptional symmetry will be evaluated by a relative measure that corresponds to the ratio of the goodness of fit of the symmetry hypothesis and the goodness of fit of the uniformity in equivalent composition group hypothesis. An equivalent composition group is composed by words with equal expected frequencies under independence and Chargaff's second parity rule hypothesis. We focus our study on the analysis of several species and compare the results of the exceptional symmetry with those obtained in the control scenarios.

Based on the exceptional symmetry measure, we created a kind of organism genomic signature by using a vector with the first 12 measures of exceptional symmetry (word lengths from 1 to 12). Our results show that the genomic signature has the potential to discriminate between species groups.

2 Methods

Let $\mathscr{A}$ be the set $\{A, C, G, T\}$ and let π_S denote the occurrence probability of symbol $S \in \mathscr{A}$. Chargaff's second parity rule states the equality between the occurrence of complementary nucleotides: $\pi_A = \pi_T$ and $\pi_C = \pi_G$.

We define a symmetric word pair as the set composed by one word w and the corresponding inverted complement word w', with $w'' = w$. The extensions of Chargaff's second parity rule state that all symmetric word pairs have similar occurrence frequencies.

We call equivalent composition groups (ECG) to the sets of words with length k which contain the same number of nucleotides A or T. Every symmetric word pair is a subset of an ECG, which contains several distinct symmetric word pairs. Note that, for k-mers (word of length k) we have $k+1$ ECGs and we denote the i-th ECG by G_i, $0 \leq i \leq k$.

When all words in each ECG have similar frequency, we have a particular single strand symmetry phenomenon that we call uniform symmetry. A random sequence generated under independence with $\pi_A = \pi_T$ and $\pi_C = \pi_G$ is obviously expected to exhibit uniform symmetry. We expect that, in a natural DNA sequence, the frequency of a word is generally more similar to the frequency of its inverted complement than to the frequencies of other words in the same ECG. We call this exceptional symmetry.

We use uniform symmetry as the reference to evaluate the possible exceptional symmetry (non uniform symmetry) of a DNA sequence.

2.1 Exceptional Symmetry Measure

To measure exceptional symmetry in a global way we propose the following ratio

$$R_s = \frac{X_u^2 + \tau}{X_s^2 + \tau}. \tag{1}$$

where $X_u^2 = \sum_{i=0}^{k} X_u^2(G_i)$ with $X_u^2(G_i)$ being the chi-square statistic used to evaluate the uniformity in ECG G_i, and $X_s^2 = \sum_{i=0}^{k} X_s^2(G_i)$ with $X_s^2(G_i)$ being the chi-square statistic to evaluate the symmetry in ECG G_i. $\tau > 0$ is a residual value to avoid an indeterminate ratio in the presence of exact uniform symmetry.

We observe that R_s statistic does not depend on the sample size dimension (n), but depends on the degree of freedom of X_u^2 and X_s^2. This measure depends, in an indirect way, on the word length. X_u^2 has $df_u = 4^k - (k+1) - 1$ degrees of freedom and X_s^2 has $df_s = 4^k/2 - 1$ degrees of freedom.

In order to obtain an effect size measure able to compare the symmetry effect of all k-mers we create the following measure

$$VR = \sqrt{\frac{df_s}{df_u} R_s}. \tag{2}$$

The equivalence between symmetric word pairs may be evaluated using Cramér's coefficient. If a DNA sequence reveals symmetry, it is worthwhile to evaluate the existence of exceptional symmetry. If VR takes on values $\gg 1$ and there is equivalence between symmetric word pairs, we conclude that there is exceptional word symmetry in the sequence being analyzed.

Note that, VR measure can be described as the ratio of two Cramér's coefficients which may be considered an effect size measure.

2.2 Simulated Data

Based on human genome characteristics, we simulated random sequences with 2.8×10^9 base pairs. In the first scenario, called "random (mean by nucleotide type)", we simulated random nucleotide sequences considering Chargaff's second parity rule using the nucleotide composition $\pi_A = \pi_T = 0.2955$ and $\pi_C = \pi_G = 0.2045$. In the second scenario, called "random (nucleotide composition)", we simulated random nucleotide sequences only considering the nucleotide composition $\pi_A = 0.2953$, $\pi_T = 0.2957$, $\pi_C = 0.2044$ and $\pi_G = 0.2045$ (nucleotide frequencies in the human genome).

For each scenario we generated a total of 100 sequences of nucleotides under the independent symbol assumption. The first scenario results from the uniform symmetry hypothesis.

2.3 DNA Sequences

In this study, we used the complete DNA sequences of 27 organisms obtained from the National Center for Biotechnology Information (NCBI; ftp://ftp.ncbi.nih.gov/genomes/); The species used in this work are listed in Tab. 1 and were downloaded in January 2014.

All genome sequences were processed to obtain the word counts. The words were counted considering overlapping. We use the word lengths from 1 to 12.

3 Results

Figure 1 presents the mean exceptional symmetry profiles for the four types of organisms and the profiles for the random control scenarios. We observe that for all the

Table 1. List of organisms whose DNA was used in this study

organism	group
Homo sapiens	eukarya (animalia)
Macaca mulatta	eukarya (animalia)
Pan troglodytes	eukarya (animalia)
Mus musculus	eukarya (animalia)
Rattus norvegicus	eukarya (animalia)
Danio rerio	eukarya (animalia)
Apis mellifera	eukarya (animalia)
Arabidopsis thaliana	eukarya (plantae)
Vitis vinifera	eukarya (plantae)
Saccharomyces cerevisiae	eukarya (fungi)
Candida albicans	eukarya (fungi)
Plasmodium falciparum	eukarya (protozoon))
Helicobacter pylori	bacteria
Streptococcus mutans GS	bacteria
Streptococcus mutans LJ23	bacteria
Streptococcus pneumoniae	bacteria
Escherichia coli	bacteria
Aeropyrum camini	archea
Aeropyrum pernix	archea
Caldisphaera lagunensis	archea
Candidatus Korarchaeum	archea
Nanoarchaeum equitans	archea
NC001341	virus
NC001447	virus
NC004290	virus
NC008724	virus
NC011646	virus

organism profiles and for all word lengths there is a global exceptional symmetry tendency: all groups of organisms under study present higher VR values than in the random sequences profiles. Bacteria and archea groups have the most similar profiles.

The simulated data from the first scenario represents the control groups for the symmetry phenomenon without exceptional symmetry. The random sequence generated under the independence hypothesis and with a small bias from Chargaff's second parity rule (second scenario) shows no exceptional symmetry, moreover it shows a weak opposite effect for short word lengths, $VR<1$ (see Fig. 1).

By comparing the VR measures for each cellular organism to the measures of the random data, we observe that all the studied cellular organisms present exceptional single strand DNA word symmetry, for all word lengths.

Some of the viruses studied here do not present significant exceptional symmetry for short word lengths, since VR values are lower than the upper bound (mean +2SE) of the confidence interval obtained in the uniform symmetry context (see Fig. 2). The shapes of the exceptional symmetry profiles for NC001341, NC001447 and NC004290 viruses are similar. The NC008724 and NC011646 exceptional symmetry profiles exhibit a valley in intermediate word lengths.

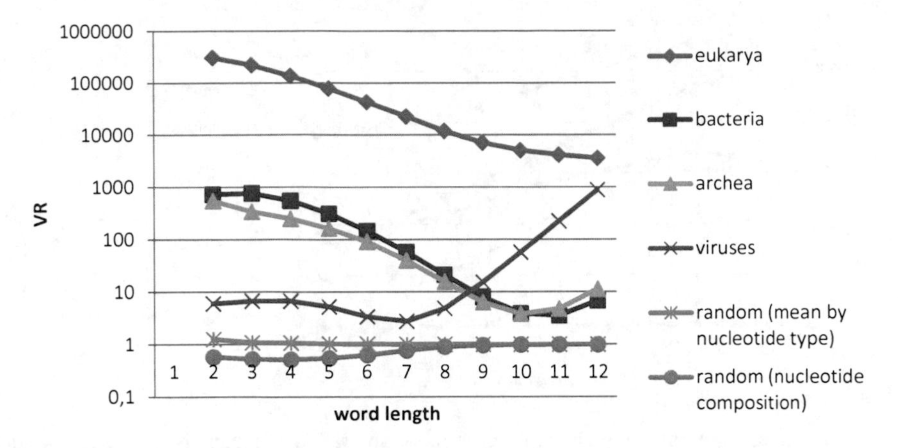

Fig. 1. VR mean values for eukarya, bacteria, archea, viruses and random (mean values from first and second scenarios)

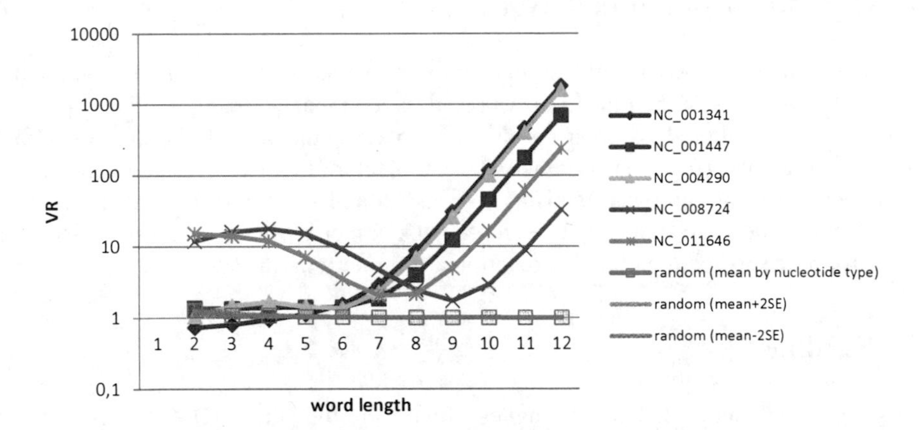

Fig. 2. VR values for viruses; random (mean and mean $\pm$ 2 $\times$ standard error from first scenario)

The highest variation coefficient inside each group (eukarya, bacteria, archea and viruses) for all studied word lengths is obtained in eukarya. Figure 3 shows the subgroup profiles for: animalia, plantae, fungi and protists. The highest exceptional symmetry values were obtained for animalia.

Note that *Rattus norvegicus* shows the highest exceptional symmetry values when compared with the other species in this study. We also observed that the maximum exceptional value for cellular organisms was obtained between word lengths 2 and 5, and for viruses it was obtained for the longer word lengths under study.

We observed also that exceptional symmetry is not strongly related to the length of the organism's genome. For each organism we obtained the average of the VR values (over the 12 word lengths) and the genome size. The Pearson correlation coefficient between mean VR values and genome size is 0.41.

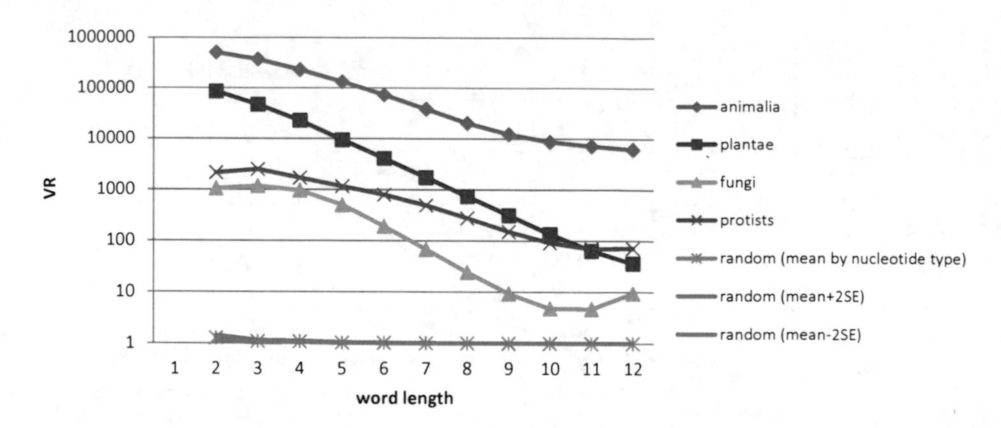

Fig. 3. VR values for animalia, plantae, fungi, protists and random (mean and mean $\pm$ 2 $\times$ standard error from first scenario)

4 Conclusion and Future Work

We created a kind of organism genomic signature which is an exceptional symmetry profile (the vector with the first 12 measures of exceptional symmetry).

Based on exceptional symmetry profile, we observed that all cellular organisms under study present exceptional symmetry: we conjecture that exceptional symmetry is an universal law of cellular organisms. We also found some viruses with a VR profile showing a behavior opposite to exceptional symmetry, $VR<1$, in short words. The animals group showed the highest exceptional symmetry in this study.

5 Funding

This work was supported by Portuguese funds through the CIDMA - Center for Research and Development in Mathematics and Applications, IEETA - *Institute of Electronics and Telematics Engineering of Aveiro* and the Portuguese Foundation for Science and Technology ("FCT–Fundação para a Ciência e a Tecnologia"), within projects PEst-OE/MAT/UI4106/2014 and PEst-OE/EEI/UI0127/2014.

References

1. Afreixo, V., Bastos, C.A.C., Garcia, S.P., Rodrigues, J.M.O.S., Pinho, A.J., Ferreira, P.J.S.G.: The breakdown of the word symmetry in the human genome. Journal of Theoretical Biology 335, 153–159 (2013)
2. Afreixo, V., Garcia, S.P., Rodrigues, J.M.O.S.: The breakdown of symmetry in word pairs in 1,092 human genomes. Jurnal Teknologi 66(3), 1–8 (2013)
3. Albrecht-Buehler, G.: Inversions and inverted transpositions as the basis for an almost universal "format" of genome sequences. Genomics 90, 297–305 (2007)

4. Baisnée, P.-F., Hampson, S., Baldi, P.: Why are complementary DNA strands symmetric? Bioinformatics 18(8), 1021–1033 (2002)
5. Forsdyke, D.R.: Evolutionary Bioinformatics. Springer, Berlin (2010)
6. Forsdyke, D.R., Bell, S.J.: Purine loading, stem-loops and Chargaff's second parity rule: a discussion of the application of elementary principles to early chemical observations. Applied Bioinformatics 3(1), 3–8 (2004)
7. Karkas, J.D., Rudner, R., Chargaff, E.: Separation of B. subtilis DNA into complementary strands. II. template functions and composition as determined by transcription with RNA polymerase. Proceedings of the National Academy of Sciences of the United States of America 60(3), 915–920 (1968)
8. Kong, S.-G., Fan, W.-L., Chen, H.-D., Hsu, Z.-T., Zhou, N., Zheng, B., Lee, H.-C.: Inverse symmetry in complete genomes and whole-genome inverse duplication. PLoS One 4(11), e7553 (2009)
9. Mascher, M., Schubert, I., Scholz, U., Friedel, S.: Patterns of nucleotide asymmetries in plant and animal genomes. Biosystems 111(3), 181–189 (2013)
10. Okamura, K., Wei, J., Scherer, S.W.: Evolutionary implications of inversions that have caused intra-strand parity in DNA. BMC Genomics 8, 160 (2007)
11. Qi, D., Cuticchia, A.J.: Compositional symmetries in complete genomes. Bioinformatics 17(6), 557–559 (2001)
12. Rudner, R., Karkas, J.D., Chargaff, E.: Separation of B. subtilis DNA into complementary strands, I. biological properties. Proceedings of the National Academy of Sciences of the United States of America 60(2), 630–635 (1968)
13. Rudner, R., Karkas, J.D., Chargaff, E.: Separation of B. subtilis DNA into complementary strands. III. direct analysis. Proceedings of the National Academy of Sciences of the United States of America 60(3), 921–922 (1968)
14. Zhang, S.-H., Huang, Y.-Z.: Limited contribution of stem-loop potential to symmetry of single-stranded genomic DNA. Bioinformatics 26(4), 478–485 (2010)
15. Zhang, S.-H., Huang, Y.-Z.: Strand symmetry: Characteristics and origins. In: 2010 4th International Conference on Bioinformatics and Biomedical Engineering (iCBBE), pp. 1–4 (June 2010)

Mutation Analysis in *PARK2* Gene Uncovers Patterns of Associated Genetic Variants

Luísa Castro[1], José Luís Oliveira[1,2], and Raquel M. Silva[1]

[1] IEETA, University of Aveiro, Campus Universitário de Santiago, 3810-193 Aveiro, Portugal
[2] DETI, University of Aveiro, Campus Universitário de Santiago, 3810-193 Aveiro, Portugal
{luisa.castro,jlo,raquelsilva}@ua.pt

Abstract. We present a comparative analysis of *PARK2* genetic variants based on genotype data from HapMap. We focused our study on the association between missense mutations and all other variations within the same gene to uncover patterns of hidden genetic variation. Alzheimer's disease (AD) and Parkinson's disease (PD) are the main neurodegenerative diseases and represent a growing health concern worldwide, with the increase in the elderly population. Mutations in several genes have been associated with either AD or PD, and the number of novel genetic variants characterized is expanding rapidly with the introduction of next generation sequencing technologies. Most of these variants, however, are of unknown consequences as their effect might be mediated through association with additional mutations. Our results show that significant correlation between genetic variants exists and their co-occurrence might contribute to previously unidentified risk increase.

Keywords: neurodegenerative diseases, mutation analysis, SNP association.

1 Introduction

Ageing is one of the main medical challenges of modern societies, as the average lifespan increases and consequently the incidence of age-related disorders. Among these, neurodegenerative diseases represent a great economic and social burden worldwide, considering that thousands of new patients are reported every year [1]. Alzheimer's disease (AD) and Parkinson's disease (PD) are common and complex neuropathologies with both familial and sporadic cases [2, 3].

Here we describe a comparative study of mutations in genes associated with neurodegenerative diseases. We have narrowed our analysis to the associations between missense mutations (substitutions that change the encoded amino acid) and other single nucleotide polymorphisms (SNPs) within the *PARK2* gene. Mutations are often associated with disease and some are important for diagnosis or prognosis, which motivates the search for genetic variants with clinical relevance. In particular, there is evidence that mutations in several genes are associated with PD [2] and AD [3].

PARK2 is one of the largest human genes, spanning approximately 1.38 Mb of the genome, and mutations in this gene are implicated in autosomal recessive early-onset PD [2]. Parkin is an E3 ubiquitin ligase that attaches ubiquitin to specific protein

J. Sáez-Rodríguez et al. (eds.), *8th International Conference on Practical Appl. of Comput. Biol. & Bioinform. (PACBB 2014)*, Advances in Intelligent Systems and Computing 294,
DOI: 10.1007/978-3-319-07581-5_18, © Springer International Publishing Switzerland 2014

substrates, modulating their stability and activity. Parkin functions in mitophagy and has a neuroprotective role, through the interaction with the mitochondrial serine/threonine kinase PINK1 [4].

Multiple risk loci have been identified through genome-wide association studies (GWAS), and the emergence of the next generation sequencing technologies will contribute to further developments, namely, enabling the detection of rare genetic and structural variants, as well as gene–gene and gene–environment interactions [5]. How most genetic variants influence the risk or contribute to the pathophysiology of neurodegenerative disorders is not yet known, and it is difficult to identify underlying genetic variation by considering each mutation alone, as inter- and intra-genic interactions might occur.

Motivated by these questions, we have analyzed mutations in a key gene of Parkinson's disease, *PARK2*. Through comparative analyses between single nucleotide polymorphisms within the same gene and using Fisher's exact test for independence, we found that several SNPs show patterns of significant correlation. Surprisingly, association between missense and adjacent intronic variants was common, which suggests that previously unconsidered variants can be pathogenic. In particular, variants that can alter transcript expression and alternative splicing might contribute to disease pathophysiology and are underestimated by common approaches.

2 Methods

Genes with a role in Parkinson disease were selected according to the Parkinson Disease Mutation Database [2]. In this study, our analysis was narrowed to the *PARK2* gene as it has known three-dimensional structure [4] allowing the correlation of SNPs with their 3D location in the protein. Phased haplotype data for *PARK2* was imported from HapMap PhaseIII/Rel#3, May 2010 [6]. Data includes a total of 954 variations for 2022 haplotypes, from 11 different populations (table 1).

Using the BioMart tool from Ensembl release 74 - December 2013 [7] we retrieved the missense mutations from the SNPs present in the HapMap dataset. Four mutations were identified as missense in the dataset of *PARK2* gene: rs1801334, rs1801582, rs9456735 and rs1801474.

Table 1. Populations of origin of the 2022 genotypes used for the mutations analysis

Description	Identifier	Size
African ancestry in Southwest USA	ASW	126
Utah residents with Northern and Western European ancestry (CEPH collection)	CEU	234
Han Chinese in Beijing, China	CHB	168
Chinese in Metropolitan Denver, Colorado	CHD	170
Gujarati Indians in Houston, Texas	GIH	176
Japanese in Tokyo, Japan	JPT	172
Luhya in Webuye, Kenya	LWK	180
Mexican ancestry in Los Angeles, California	MEX	104
Maasai in Kinyawa, Kenya	MKK	286
Toscans in Italy	TSI	176
Yoruba in Ibadan, Nigeria	YRI	230

The genotypes dataset was divided into 11 subgroups according to the individuals' population of origin. Each population dataset was studied separately in order to control for population stratification possibly present in the data. Within all given subgroups (gene-population), comparative analysis was performed for each of the four missense mutations and all other variants of the gene. Each subgroup of genotypes was divided in two sets according to the two nucleotides present in the missense position. Populations with a set containing less than 30 haplotypes for a particular allele were not taken into account for subsequent analysis.

Missense mutations rs1801334 and rs9456735 from *PARK2* were not considered as they did not accomplish this requisite for any population set. For missense mutation rs1801582 only CEU, GIH, LWK, MKK and YRI sets satisfied this condition and for missense mutation rs1801474 only CHB, CHD and JPT fulfilled the requirement. For any two sets of genotypes within the same population subgroup we computed the allele frequencies that were then compared between the two sets. For example, to evaluate the association between SNP rs6942109 and the missense mutation rs1801474 in individuals from the CEU population, we compared the proportion of allele T in the group of haplotypes with C in the missense position with the proportion of allele T in the group of haplotypes carrying allele G in the missense position. This procedure was adopted to evaluate the association of each of the 953 SNPs with the missense mutation rs1801474 and, separately, with the missense mutation rs1801582. Fisher's exact test [8] was used for inferring which SNPs had allele frequencies dependent on the set they were, i.e., dependent on the nucleotide present in the missense position. In the dataset used there are some SNPs with a low or null minor allele frequency which implies a small expected number of haplotypes in the group with this allele. This and unequally distribution of alleles in the groups to be compared turns Fisher's exact test more suitable than chi-square test or other common independence tests.

For each missense position, 953 pair-wise comparison tests were performed to evaluate the association between the missense and each one of the other SNPs in the dataset. In order to control for multiple testing we employed a conservative procedure aiming at reducing the number of false positives (with respect to the non-null hypothesis). The critical level for significance was set at 0.01 and Bonferroni correction was applied, with a resultant significance level for each individual test of 10^{-5}. All analyses were done using MATLAB (R2011a, MathWorks, Natick, MA, USA).

In the following results and discussion the SNPs with significant association with the missense positions rs1801582 and rs1801474 of the *PARK2* gene are presented.

3 Results

We found evidence of associated SNPs for two missense variants in the PARK2 gene, namely, rs1801582 (V380L) and rs1801474 (S167N). Five distinct populations (CEU, GIH, LWK, MKK and YRI) show statistically significant associations for rs1801582 (table 2).

Table 2. *PARK2* SNPs associated with missense mutations rs1801582 and rs1801474. For each population, the p-values below the threshold obtained from the multiple comparisons correction ($p < 10^{-5}$) are displayed, as well as the distances in base pairs to the missense position. All SNPs are intronic.

rs1801582						
		CEU	GIH	LWK	MKK	YRI
SNP	Dist	p-val	p-val	p-val	p-val	p-val
rs9458229	-35116		6.9E-09			
rs3798964	-34906		1.5E-18			
rs7752498	-28149		8.1E-10			
rs6942109	-21161	8.3E-10	1.9E-20		2.5E-07	
rs7766508	-14536	1.5E-08	5.1E-25		4.0E-13	2.7E-15
rs9347502	-6701	3.7E-16	2.5E-28	5.2E-06	1.7E-15	9.7E-19
rs12215325	-5407				4.4E-13	3.9E-19
rs3890730	-3903	2.0E-20	2.5E-38	2.1E-12	6.6E-22	5.4E-28
rs4587148	-2148	6.4E-17	4.8E-42			
rs9355894	-1313	7.7E-36	7.7E-44	9.7E-34	2.2E-52	8.2E-40
rs4315992	813	9.3E-16	4.8E-42			
rs4574609	1049	1.1E-13	1.8E-29	3.0E-21	3.0E-25	1.2E-25
rs4131770	3053		1.4E-09		1.1E-06	
rs9346851	4564	2.2E-22	2.5E-38	2.2E-19	2.5E-26	1.4E-16
rs11964284	4771	1.6E-38	6.1E-46	4.5E-28	1.8E-39	5.7E-23
rs9365285	9313		7.6E-07		2.1E-06	
rs9365286	11783	1.3E-14	3.7E-36			
rs9355897	15494	9.5E-35	7.0E-31	3.1E-16	1.9E-22	1.6E-09
rs9347505	16660	5.1E-24	6.1E-24	2.7E-17	1.0E-25	4.2E-20
		13	**18**	**8**	**13**	**10**

rs1801474				
		CHB	CHD	JPT
SNP	Dist	p-val	p-val	p-val
rs12205861	-23384			3.5E-07
rs2000752	-48776		3.7E-06	
rs2023081	-30886		8.6E-06	
rs7765100	-16867		4.6E-06	4.7E-07
rs9295184	-12478	2.0E-11	4.6E-11	5.0E-14
rs11961127	-11867	1.2E-36	3.5E-41	1.6E-36
rs9347599	-10305	1.4E-41	7.2E-43	2.0E-37
rs9355988	-10194	3.3E-09	2.2E-10	2.3E-08
rs12529283	-6591	4.5E-13	6.6E-12	1.9E-16
rs1954943	-5185	3.3E-09	5.3E-10	2.2E-08
rs10455789	-2211	3.3E-09	5.3E-10	2.2E-08
rs9355387	-935	6.9E-06		
rs10945803	631	3.3E-09	5.3E-10	9.3E-09
rs9458481	2449	2.0E-11	2.5E-09	5.0E-14
rs2023073	9974	7.7E-06		
rs2023071	10921	1.5E-08	2.8E-09	5.5E-08
rs4546464	11421	7.8E-08	6.2E-09	2.0E-06
rs6921002	12848			7.7E-06
rs1954939	44814			3.9E-06
rs6922813	46692			4.4E-07
rs9365393	51364			3.9E-06
rs9355994	59547			2.0E-06
rs9355391	66808			4.2E-06
rs6942198	67825			4.2E-06
rs9346911	134917			1.9E-06
		13	**14**	**21**

In the genotyped positions from CEU individuals, results for the 13 associated variants indicate that, when the base in rs1801582 is a G, 4 SNPs possess only one possible allele (30 vs 204 haplotypes with allele C) (Fig.1 (a)). The same pattern occurs for allele C, although for a distinct set of SNPs. These results indicate differentially distributed genotypes, depending on the allele found in a given position. A similar behavior is verified for the GIH population, but for 18 SNPs, where 4 variations have one possible allele in the set with base G in the missense position, and 5 in the other set (Fig.1 (b)). One of these variants is linked to the missense position in both CEU and GIH datasets. When nucleotide C (G) is in rs1801582, the SNP rs11964284 always carries base T (C). In the LWK population, 8 positions show evidence of association with the missense variant (table 2) but only one allele is fixed, occurring for the set where base G is in the missense position (not shown). Individual genotypes from population MKK show a group of 13 SNPs associated with the missense variant rs1801582. In the set where the missense SNP carries nucleotide C, one mutation shows only one possible allele, while for the other genotype set the fixed pattern includes three positions (not shown). Finally, results for individuals from YRI suggest 10 variations associated with the missense position (table 2).

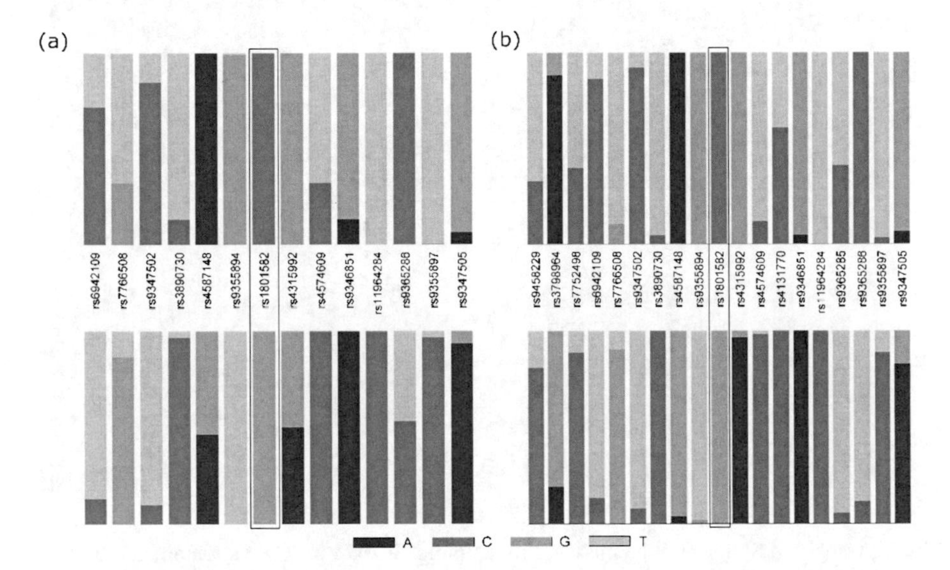

Fig. 1. Associated SNPs with the missense mutation rs1801582 in *PARK2*. Independence tests were performed to infer on the association of each of the 953 available SNPs with the missense but only the SNPs presented were statistically significant ($p < 10^{-5}$ for individual tests). (a) Thirteen variations revealed association to rs1801582 in CEU individuals. Allele frequencies for these SNPs are presented as the missense position carries allele C (204 haplotypes, up), or allele G (30, down). Ancestral alleles, including the missense position, correspond to TGCCAGGGAATCTA. (b) In individuals from population GIH, the results suggest 18 variations in association with the missense position. Allele frequencies of the associated mutations are displayed for the set with allele C in rs1801582 (124 haplotypes, up) and for the genotypes with base G in position rs1801582 (52, down). Ancestral alleles are CGCTGCCAGGGACATTCTA. In both panels missense mutations are highlighted.

For the missense position rs1801474, 3 populations show relevant results. In the CHB population, haplotypes that carry allele T also carry only one possible nucleotide in 12 of the 13 associated SNPs. As a consequence, a fixed sequence (GGTGTTTCTGTT) is observed for the 68 haplotypes with base T in rs1801474, in comparison with no fixed sequence for the 100 carrying allele C (Fig.2 (a)). Similarly, for the population CHD, in 14 positions significantly associated with rs1801474, 9 SNPs show only one possible allele in the set when the variant base is a T (63 vs 107 haplotypes) (data not shown). From the genotypes of JPT individuals, the results provide evidence that 21 positions are associated with the missense variant. From those, and in the set where the missense variant is T, 11 positions present only one possible allele (Fig.2 (b)). Throughout the 3 populations mentioned, several SNPs with significant association with the missense variation rs1801474 are shared. In particular, a set of 6 SNPs has a fixed sequence when the missense carries allele T: rs9295184, rs9355988, rs12529283, rs1954943, rs10455789 and rs10945803, which are all intronic variants.

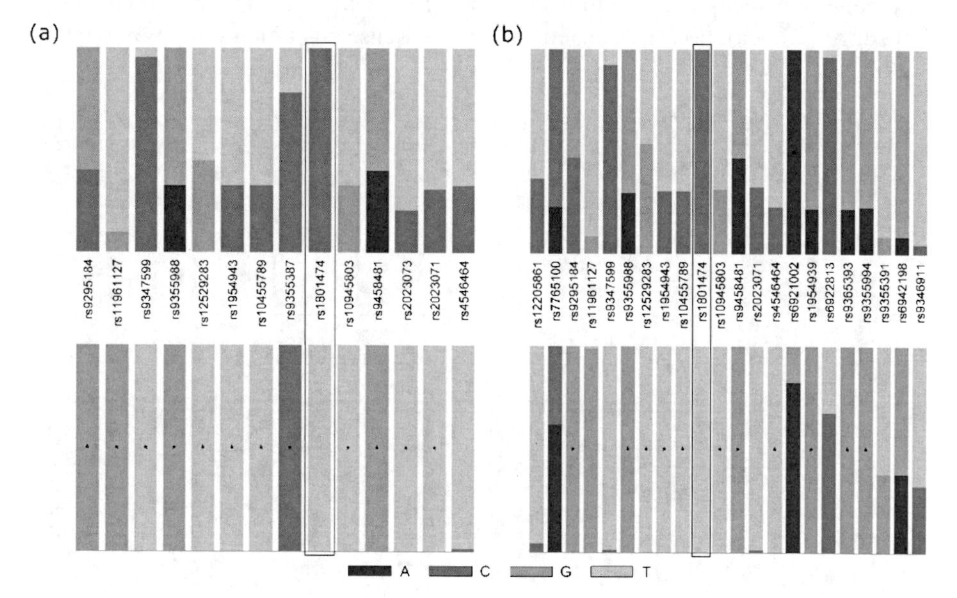

Fig. 2. Associated SNPs with the missense mutation rs1801474 in *PARK2*, obtained as in Fig.1. (a) The analysis made suggest evidence of association of the missense position with 13 variations in population CHB. In the upper histogram, the allele frequencies of the associated SNPs are shown for the set carrying allele C in the missense position (100 haplotypes). The lower histogram shows allele frequencies for the genotypes carrying allele T in rs1801474 (68) and a set of 12 positions, signed by stars on the bars, carry only one possible allele. Ancestral alleles are GTCGTTTCCTGTTT. (b) Results provide evidence of 21 variations associated with rs1801474 in JPT genotypes. Allele frequencies for the 21 SNPs are presented as the missense position carries allele C (104, up) or allele T (68, down). For the set carrying allele T in rs1801474, 11 SNPs, signed by stars, carry only one possible allele. Ancestral alleles are TAGTCGTTTCTGTTAGTAGGAT. In both panels missense mutations are highlighted.

Linkage disequilibrium scores for the SNPs were extracted from Haploview [9]. We observed that the LOD (logarithm of odds ratio) score follows a similar trend to the natural logarithm of the p-values obtained in our analysis. However, the SNP associations found cannot be explained by proximity (table 2), as some variants are found several kilobases apart, while recombination rates are not expected to vary in such short distances. Point mutations affecting individual positions are the most plausible cause for such heterogeneity of associations with neighboring SNPs [10].

Our results extend beyond the concept of linkage, as our analysis show positions with strong association scores intermingled with mutations in linkage equilibrium with the missense mutation. Some of these positions that are in strong association with the missense mutation (low p-values or high LOD values) are not allowing variability and these findings are discussed below.

4 Discussion

In this study we performed mutation analysis in the PARK2 gene. We have focused on the effect that missense mutations have on other single nucleotide polymorphisms within the gene. In both reported cases there was evidence of association for a number of SNPs and the missense alleles. Moreover, in some examples, the presence of a specific nucleotide in one allele has a fixation effect on the associated SNPs, that is, these positions no longer accommodate all possible genetic variants observed for the other allele. In general, positions that are in strong association may result from constraints in the 3D protein structure or from the loss of deleterious variants by natural selection (positive selection).

Several pathogenic variants have been described in *PARK2* to date, including rearrangements of single or multiple exons in Parkinson's disease patients that are associated with an earlier age at onset (EOPD) [1]. These structural variants account for the majority of *PARK2* mutations and, although the gene is composed of 12 exons, most rearrangements locate in the region between exon 2 and exon 4 [11]. Studies carried out in a Korean population indicate that, although exon deletions are frequent in *PARK2* in EOPD patients, no exon rearrangements were detected in *SNCA*, *PINK1*, and *PARK7* genes [11]. Also, structural variants might be underestimated as these mutations were not specifically screened for in initial studies and, in addition, conventional sequencing is unable to detect exon rearrangements in heterozygosity [11].

It is interesting that the significant associations found for the *PARK2* gene are within the region surrounding exon 4 and 10. Six protein-coding transcripts are described in the Ensembl genome browser for the *PARK2* gene, and two lack exon 4 (PARK2-001 and PARK2-005). The association between the missense variant rs1801474 in exon 4 and several SNPs in the adjacent introns 3 and 4 suggests that these positions might mediate an alternative-splicing event resulting in exon 4 skipping, probably to prevent the deleterious impact of the missense mutation S167N. Although S167N is considered benign by several pathogenicity prediction tools (PolyPhen, SIFT), recent structural studies indicate that this position is comprised in a previously undescribed domain that coordinates metal ions and forms an interface with the catalytic domain, thus, important to the function of the Parkin protein [4, 12].

Additional studies will be performed in protein models containing relevant mutations, generated by homology modeling, to predict the structural impact of the mutations.

Genome-wide association studies have identified many risk loci, yet, a greater than expected number of disease-associated SNPs are located in non-coding regions (intronic or intergenic) [13]. This points to an underestimated role of regulatory regions in conferring risk to particular genetic variants, either by changing splicing, expression levels, or sub cellular localization. Splicing variants can be difficult to identify exclusively by sequence analysis, and include mutations that can result in exon skipping, intron retention, gain or loss of a splice site, or enhancement of cryptic splice sites [14]. The increasing recognition of the role of alternative splicing in pathology is leading the development of a new field of spliceopathies, and our study should further explore this possibility. Future work will also extend the analyses to other genes associated with neurodegenerative disorders, such as *LRRK2*, *PINK1* or *MAPT*.

Acknowledgements. This work was supported by the project Neuropath (CENTRO-07-ST24-FEDER-002034), co-funded by QREN "Mais Centro" program and the EU.

References

1. Corti, O., Lesage, S., Brice, A.: What genetics tells us about the causes and mechanisms of Parkinson's disease. Physiol. Rev. 91(4), 1161–1218 (2011)
2. Nuytemans, K., et al.: Genetic etiology of Parkinson disease associated with mutations in the SNCA, PARK2, PINK1, PARK7, and LRRK2 genes: a mutation update. Hum. Mutat. 31(7), 763–780 (2010)
3. Schellenberg, G.D., Montine, T.J.: The genetics and neuropathology of Alzheimer's disease. Acta Neuropathol. 124(3), 305–323 (2012)
4. Trempe, J.F., et al.: Structure of parkin reveals mechanisms for ubiquitin ligase activation. Science 340(6139), 1451–1455 (2013)
5. Scholz, S.W., et al.: Genomics and bioinformatics of Parkinson's disease. Cold Spring Harb. Perspect. Med. 2(7), a009449 (2012)
6. Altshuler, D.M., et al.: Integrating common and rare genetic variation in diverse human populations. Nature 467(7311), 52–58 (2010)
7. Flicek, P., et al.: Ensembl 2013. Nucleic Acids Res. 41(Database issue), D48–D55 (2013)
8. Boedigheimer, M.: Fisher's Exact Test. MATLAB Central File Exchange (2008)
9. Barrett, J.C., et al.: Haploview: analysis and visualization of LD and haplotype maps. Bioinformatics 21(2), 263–265 (2005)
10. Hartl, D.L., Clark, A.G.: Principles of Population Genetics, 4th edn. S. Associates, Sunderland (2007)
11. Kim, S.Y., et al.: Phase analysis identifies compound heterozygous deletions of the PARK2 gene in patients with early-onset Parkinson disease. Clin. Genet. 82(1), 77–82 (2012)
12. Riley, B.E., et al.: Structure and function of Parkin E3 ubiquitin ligase reveals aspects of RING and HECT ligases. Nat. Commun. 4, 1982 (2013)
13. Hindorff, L.A., et al.: Potential etiologic and functional implications of genome-wide association loci for human diseases and traits. Proc. Natl. Acad. Sci. USA 106(23), 9362–9367 (2009)
14. Baralle, D., Baralle, M.: Splicing in action: assessing disease causing sequence changes. J. Med. Genet. 42(10), 737–748 (2005)

Heterogeneous Parallelization of Aho-Corasick Algorithm

Shima Soroushnia, Masoud Daneshtalab, Juha Plosila, and Pasi Liljeberg

Department of Information Technology, University of Turku, Finland
{shisor,masdan,juplos,pakrli}@utu.fi

Abstract. Pattern discovery is one of the fundamental tasks in bioinformatics and pattern recognition is a powerful technique for searching sequence patterns in the biological sequence databases. The significant increase in the number of DNA and protein sequences expands the need for raising the performance of pattern matching algorithms. For this purpose, heterogeneous architectures can be a good choice due to their potential for high performance and energy efficiency. In this paper we present an efficient implementation of Aho-Corasick (AC) and PFAC (Parallel Failureless Aho-Corasick) algorithm on a heterogeneous CPU/GPU architecture. We progressively redesigned the algorithms and data structures to fit on the GPU architecture. Our results on different protein sequence data sets show 15% speedup comparing to the original implementation of the PFAC algorithm.

Keywords: Heterogeneous, GPU, Pattern Matching, Aho-Corasick, CUDA, OpenMP.

1 Introduction

In the last decade, the exponential growth of biological information, the need to understand the complex interactions that determine the biological processes and the diversity and interconnectedness of organisms at the molecular level have driven the need for high performance computational biology [1]. Heterogeneous architectures that provide different processing elements with different characteristics on the same machine are becoming mainstream in high performance computing platforms and seem able to cope with these requirements [2]. Such architectures integrate different multi-core CPUs with many-core accelerators (GPGPUs) and while running an application, some parts can execute on CPU while the other parts of the same application are running on the GPU side to benefit from the both processing elements features.

Aho-Corasick [3] is one of the widely used pattern matching algorithms in computational biology applications due to its linear complexity and ability to match multiple patterns at the same time. The algorithm works by building a state machine out of patterns and search through the data using the state machine. It can find all the

J. Sáez-Rodríguez et al. (eds.), *8th International Conference on Practical Appl. of Comput. Biol. & Bioinform. (PACBB 2014)*, Advances in Intelligent Systems and Computing 294,
DOI: 10.1007/978-3-319-07581-5_19, © Springer International Publishing Switzerland 2014

occurrence of the patterns inside the data in a single pass of data and that is the reason for the linear complexity. We developed a high performance parallel implementation of the AC algorithm for a heterogeneous computing environment. For this purpose, we use the message passing interface (OpenMP) for parallelizing the building of state machine on shared memory multi-core CPU architectures and CUDA for fine-grain parallelization of searching on the GPU.

To increase the performance of Aho-Corasick algorithm, PFAC [4] which is the highly parallelized version of Aho-Corasick on GPGPUs has been introduced. In PFAC algorithm, the state machine is built without any failure transition and it is placed on the global memory. For searching through the data, a thread will be generated for each single character in the dataset. The total number of threads is equal to the length of text in dataset. Each thread needs to check the automaton in global memory and processes its character to see if it can find a valid transition, otherwise the thread will be terminated.

In this paper, we attempt to speed up the PFAC algorithm by rearranging the data placements in the GPU memories. We also used the heterogeneous programming concept for parallelizing the building of the state machine using OpenMP.

2 Related Works

Many research papers on accelerating pattern matching using GPUs have been published recently as they are becoming increasingly popular for various applications. Different implementations to the Aho-Corasick algorithm have been proposed to increase the efficiency and throughput of multiple pattern matching problems.

Tumeo et al. [5] has introduced a parallel version of AC algorithm for network based applications that process TCP/IP packets. They assigned a single TCP/IP packet to each CUDA thread, loading it in the shared memory, and then divide the packet in further chunks assigned to each thread of the thread block. Another GPU parallelization of AC algorithm has introduced in [6]. They tested their implementation assuming the case when all data reside initially in the GPU memory and the results are to be left in this memory. In this implementation, AC DFA (Deterministic Finite Automaton) is stored in the texture memory. Threads cooperatively read all the data needed to process a block, store this data in shared memory, and finally read and process the data from the shared memory. They extended their idea in [7] and tested the AC and Boyer-Moore [8] in two cases. In the GPU-to-GPU case, they considered several refinements to their base GPU implementation and measured the performance gain from each refinement. For the host-to-host case, they analyzed different strategies to reduce the communication cost between the host and the GPU. Tran et al. [9] has developed a memory efficient parallelization of the AC algorithm by placing the state transition table on texture memory and loading the data chunks from global memory to the shared memory. They gained 15 times speedup comparing to the serial execution of the algorithm. Gnort [10] is another GPU implementation of the AC algorithm for Network Intrusion Detection Systems (NIDS). Each packet is copied to the shared memory of the

Multiprocessor and stream processors search different parts of the packet concurrently. Another implementation based on Memory Page Size is introduced by Peng et al. [11]. They split the large scale text into $\frac{TextSize}{PageSize}$ size-fixed pages. All the data are transferred to the GPU memory and different threads on GPU process different pages using the automaton which is bound to texture memory.

As an experiment of parallelizing the AC algorithm on heterogeneous CPU-GPU platform, Tumeo et al. [1] has implemented the algorithm on GPU clusters for the use of DNA analysis applications. They developed a master/slave implementation, where the master MPI process distributes the work to the various nodes and GPUs, and the slaves perform the effective computation. The GPU parallelization is based on partitioning the input text in multiple chunks and assigning each chunk to a single CUDA thread. To catch the patterns that go over the border of a single chunk, the chunks can be overlapped for the size of the longest pattern in the dictionary. The state transition table is bound to the texture memory. Also Tumeo et al. [12] have extended their previous work [5] on heterogeneous architecture by developing a master/slave scheduler, where the master MPI process distributes the work, and the slaves perform the effective computation.

PFAC (parallel failure less Aho-Corasick) [13] is a highly parallelized implementation of the AC algorithm without any failure transition. In the PFAC algorithm, each character in the text is assigned to a thread and each thread is only responsible for finding the patterns starting at its own starting position. The state transition table is placed in the texture memory and data is preloaded to the shared memory. The PFAC algorithm is then used for implementing serial and parallel Bayesian spam filtering [14]. In this work, we have modified the AC algorithm by rearranging the placement of data in different GPU memories comparing to PFAC. We also include heterogeneous programming concept by using OpenMP interface for building the state machine on a multi-core CPU.

3 Aho-Corasick Algorithm

A multi-string matching problem is defined as "having a set of patterns P and we would like to detect all the occurrences of any of the patterns in P in a text stream T". The Aho-Corasick solution for this problem is to build a Finite State Automaton named pattern matching machine and run this machine on the text to find any existence occurrence of the patterns inside the text. It is also easy to implement a state machine as a lookup table where the rows represent states and the columns represent input character, so that each element of matrix shows the next state for the corresponding alphabet in the current state. Having the information about the current state and next input character, the machine can determine whether the input character causes a failure transition. If not, then it makes a transition to the state corresponding to the character.

The complexities of constructing a pattern matching machine and searching the text are linear to the total length of given patterns and the length of a text, respectively. Figure 1 shows an example of how the algorithm works.

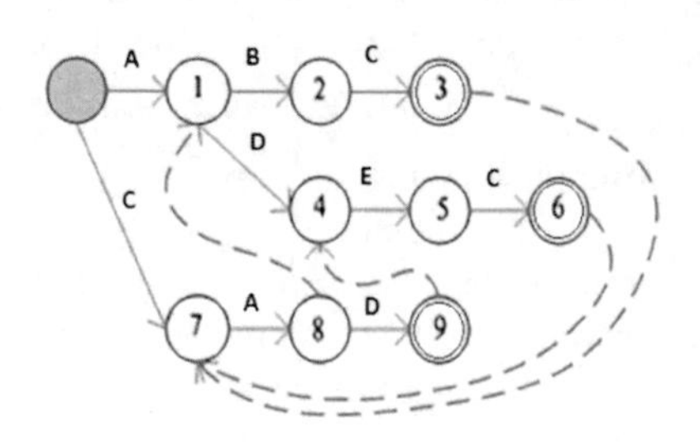

Fig. 1. Aho-Corasick State machine and transition table for finding keywords: "ABC", "ADEC","CAD"

	Final	A	B	C	D	E
0	0	**1**	**0**	**7**	**0**	**0**
1	0	**0**	**2**	**0**	**4**	**0**
2	0	**0**	**0**	**3**	**0**	**0**
3	1	*7*	*7*	*7*	*7*	*7*
4	0	**0**	**0**	**0**	**0**	**5**
5	0	**0**	**0**	**6**	**0**	**0**
6	1	*7*	*7*	*7*	*7*	*7*
7	0	**8**	**0**	**0**	**0**	**0**
8	0	*1*	*1*	*1*	**9**	*1*
9	1	*4*	*4*	*4*	*4*	*4*

4 Implementation

Most of the previous parallelizations of Aho-Corasick algorithm are based on the method that input stream is divided to the chunks and each chunk is assigned to a separate thread to search through. An example can be seen in Fig.2. It illustrates how the data parallelization method can speed-up the application.

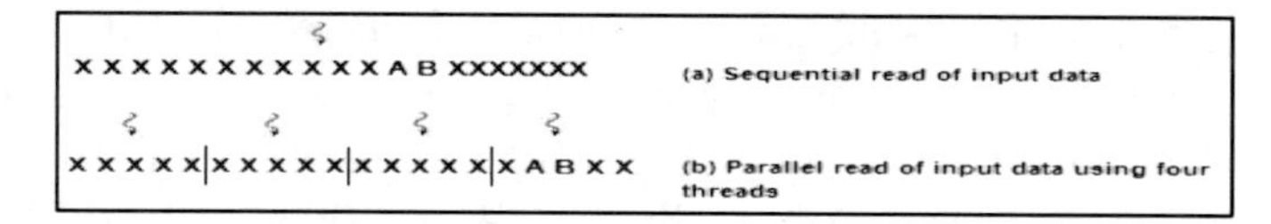

Fig. 2. Sequential vs. parallel search

In theory, finding the pattern in parallel with four threads will be four times faster than doing it sequentially. However, this implementation will face problem when patterns are occurred in the boundary of data as shown in Fig.3.

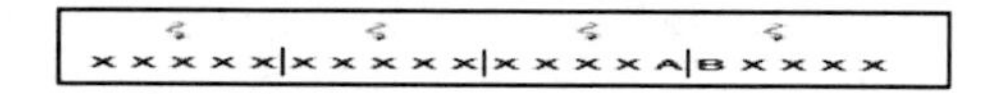

Fig. 3. Boundary Problem

In this case, the only way to solve this problem is to assign overlapped data chunks to the threads and the size of this overlap should be equal to the longest keyword. Fig.4 shows how data overlapping can solve the boundary problem.

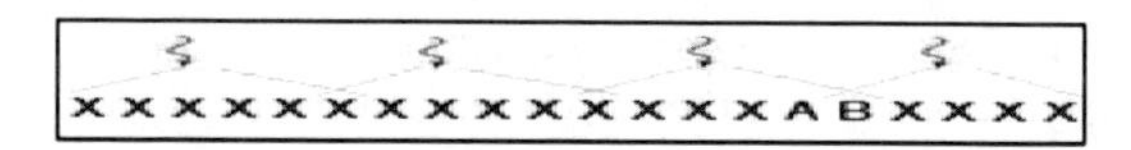

Fig. 4. Threads with overlapped data to search through

But this approach will lead to significant performance reduction, specifically in GPU implementation that memory access is the main issue and has considerable effect on the overall throughput of the system.

To overcome this problem, the PFAC algorithm is introduced [13]. In this algorithm, every character in the input stream is assigned to a thread which is only responsible to find the keywords beginning at its own starting position. If a thread does not have any valid transition for the character, the thread will be terminated immediately. But in case of having a valid transition, thread will continue reading the next character.

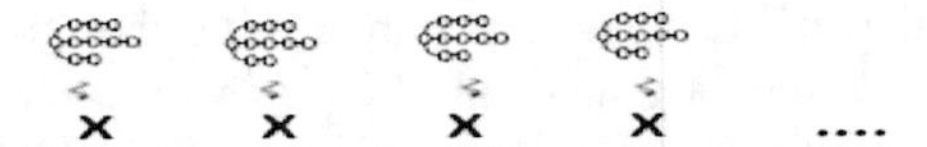

Fig. 5. PFAC algorithm. Generating a thread for each character in the input stream.

One of the most important features of PFAC algorithm is that it removes the failure transition from the Aho-Corasick state machine by assigning a thread for each character. Each thread will only find its own keyword that starts in its own starting position, so there is no need to have failure transition anymore. Also the boundary problem that is introduced in the chunk based approach is somehow solved here as we are not assigning the data chunks to the threads.

The implementation of the PFAC algorithm is in a way that the state transition table is placed in the texture memory, which is a part of global memory with an internal 6KB cache in each streaming processor. Then the input data will be first transferred to the GPU's shared memory and then searching through the data will be started.

Although PFAC eliminates the boundary problem between threads, there are still boundary problems while prefetching data to the GPU's shared memories and patterns that occur in the boundary of data chunks that are going to be loaded to the shared memory will not be detected without use of overlapping method again [15].

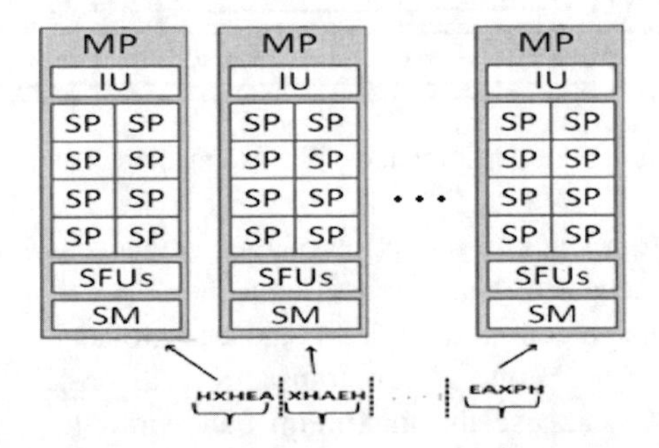

Fig. 6. Boundary Problem in PFAC algorithm

An example is shown in Fig.6. Pattern "AX" in the boundary of first two segments will not be found without overlapping the data chunks.

Another issue about the PFAC algorithm is the memory accesses. Considering the fact that each thread needs to check the next transition for its input character from transition table which is placed in the texture memory, and also very limited size of texture cache, for most of the input characters we need to have two memory accesses. One will happen when reading the character and the other one when checking the transition table for that character. This problem will become more serious when we have large set of patterns to search for, which leads to have a larger transition table and more global memory accesses.

Our solution for the aforementioned problem is to store the transition table in the shared memory of each streaming processor. Threads then start reading the input file as it is explained in the PFAC algorithm. For limiting the global memory access to once per thread, we use the fact that threads inside each streaming processor can communicate with each other via the shared memory. We allocate a small part of the shared memory for communication between the threads in each streaming processor, so that whenever a thread finds its next transition by checking the transition table from the shared memory, it will be assigned the value of next thread for continuing its search. This trend will be continued till each thread finds the keyword starting at the thread's starting position or the termination of the thread due to not having the next transition. An example of this method is illustrated in the Fig.7. Assume that we are searching for pattern "AB", all the threads will read their value from the global memory and start searching for the next transition by checking the transition table in the shared memory. As the thread with the value "A" will have a next transition and needs "B" to be its new value, it can have the value of its next thread instead of reading the value from the global memory.

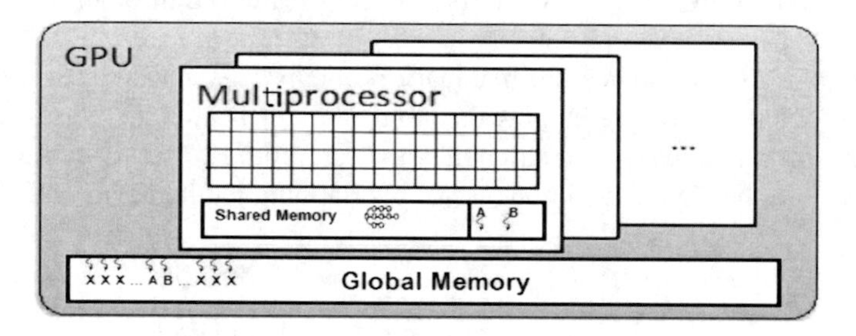

Fig. 7. Storing the transition table of State machine in shared memory

This implementation leads to reduction of memory accesses to once per character. Also, as the threads access to contiguous global memory locations, memory accesses will be coalesced, that means the device coalesces global memory loads and stores issued by threads of a warp into as few transactions as possible to minimize DRAM bandwidth. Hence we can use the maximum bandwidth for fetching data from the global memory and reducing the global memory latency as much as possible.

5 Experimental Results

In this section we will discuss the performance of our heterogeneous implementation with respect to the original implementation of PFAC algorithm. The result for the modified PFAC algorithm without including the OpenMP part is also presented here to have a better overview of our modification results.

We selected a few sets of protein databases for different animals to assess the presented method. Execution time and system throughput for different implementations on NVIDIA Tesla C2075 and Intel® Xeon® Processor E5-2620 are shown in Table 1.

Table 1. Execution time for each implementation

	Modified PFAC + OpenMP	Modified PFAC	Original PFAC	Sequential
Horse Protein Sequence (50MB)	6.20	6.63	6.97	45.87
Mouse Protein Sequence (27.4MB)	6.64	7.36	7.81	44.36
ZebraFish Protein Sequence (22.1MB)	6.33	7.66	7.90	55.00

Results show the average of 15% speedup for our implementation comparing to the original PFAC implementation. Figure 8 shows the system throughput calculated using the formula $\frac{8n}{T(Total)}$ in Gigabits per second (Gbps).

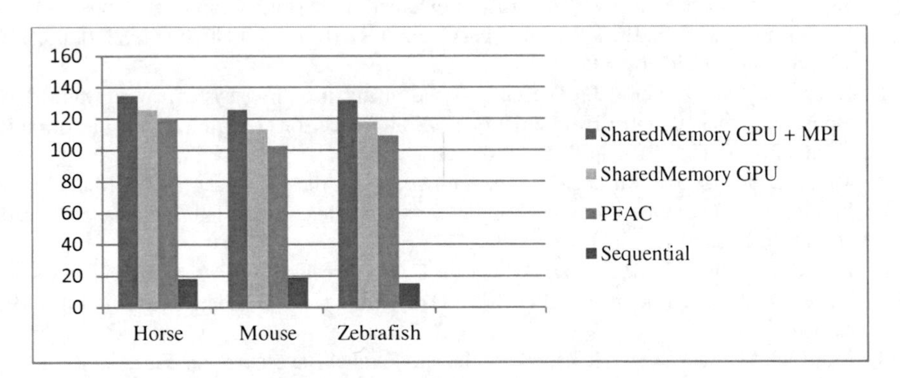

Fig. 8. System throughput (Gbps)

6 Conclusion

We have presented different software implementations of Aho-Corasick algorithm on GPU, CPU, and heterogeneous CPU-GPU platforms, and analyze their performance. We discussed the problems of previous implementations and proposed a new method by redesigning the algorithm and data structures to fit on the GPU architecture. We compared our implementation with sequential version of Aho-Corasick and also PFAC algorithm for finding different animals protein sequences. The experimental results show the average of 15% speedup comparing the original PFAC algorithm.

References

[1] Villa, O., Tumeo, A.: Accelerating DNA analysis applications on GPU clusters. In: 2010 IEEE 8th Symposium on Application Specific Processors (SASP), pp. 71–76 (2010)
[2] Dios, F., Daneshtalab, M., Ebrahimi, M., Carabaño, J.: An Exploration of Heterogeneous Systems. In: 8th IEEE 8th International Symposium on Reconfigurable Communication-centric Systems-on-Chip (ReCoSoC), pp. 1–7 (2013)
[3] Aho, A.V., Corasick, M.J.: Efficient String Matching: An Aid to Bibliographic Search. ACM 18, 333–340 (1975)
[4] Tsai, S.-Y., Liu, C.-H., Chang, S.-C., Shyu, J.-M., Lin, C.-H.: Accelerating String Matching Using Multi-threaded Algorithm on GPU. IEEE (2010)
[5] Villa, O., Sciuto, D., Tumeo, A.: Efficient Pattern Matching on GPUs for Intrusion Detection System, pp. 87–88. ACM (May 2010)
[6] Zha, X., Sahni, S.: Multipattern String Matching on A GPU. In: IEEE, pp. 277–282 (2011)
[7] Zha, X., Sahni, S.: GPU-to-GPU and Host-to-Host Multipattern String Matching on A GPU. Computer and Information Science and Engineering. University of Florida, Florida (2011)
[8] Moore, J.S., Boyer, R.S.: A Fast String Searching Algorithm. Communications of the ACM 20, 762–772 (1997)
[9] Lee, M., Hong, S., Shin, M., Tran, N.-P.: Memory Efficient Parallelization for Aho-Corasick Algorithm on a GPU. In: IEEE 14th International Conference on High Performance Computing and Communications, pp. 432–438 (2012)
[10] Vasiliadis, G., Antonatos, S., Polychronakis, M., Markatos, E.P., Ioannidis, S.: Gnort: High Performance Network Intrusion Detection Using Graphics Processors. In: Lippmann, R., Kirda, E., Trachtenberg, A. (eds.) RAID 2008. LNCS, vol. 5230, pp. 116–134. Springer, Heidelberg (2008)
[11] Chen, H., Shi, S., Peng, J.: The GPU-based string matching system in adavanced AC algorithm. In: 10th IEEE International Conference on Computer and Information Technology (CIT 2010), pp. 1158–1163 (2010)
[12] Villa, O., Chavarría-Miranda, D.G., Tumeo, A.: Aho-Corasick String Matching on Shared and Distributed-Memory Parallel Architectures. IEEE Transactions on Parallel and Distributed Systems 23, 436–443 (2012)
[13] Liu, C.-H., Chien, L.-S., Chang, S.-C., Lin, C.-H.: Accelerating Pattern Matching Using a Novel Parallel Algorithm on GPU. IEEE Transactions on Computers 62(10), 1906–1916 (2013)
[14] Motwani, M., Saxena, A., Haseeb, S.: Serial and Parallel Bayesian Spam Filtering using Aho-Corasick and PFAC. International Journal of Computer Applications 74, 9–14 (2013)
[15] Rasool, A., Khare, N., Agarwal, C.: PFAC Implementation Issues and their Solutions on GPGPU's using OpenCL. International Journal of Computer Applications 72, 52–58 (2013)
[16] Villa, O., Tumeo, A.: Accelerating DNA analysis applications on GPU Clusters. In: IEEE 8th Symposium on Application Specific Processors (SASP), pp. 71–76 (2010)
[17] Tsai, S.-Y., Liu, C.-H., Chang, S.-C., Shyu, J.-M., Li, C.-H.: "Accelerating String Matching Using Multi-threaded Algorithm on GPU. In: IEEE Globecom (2010)
[18] Moore, J.S., Boyer, R.S.: A fast string searching algorithm. Communications of the ACM 20, 762–772 (1997)

High-Confidence Predictions in Systems Biology Dynamic Models

Alejandro F. Villaverde[1], Sophia Bongard[2], Klaus Mauch[2], Dirk Müller[2], Eva Balsa-Canto[1], Joachim Schmid[2], and Julio R. Banga[1]

[1] Bioprocess Engineering Group, IIM-CSIC. Eduardo Cabello 6, 36208 Vigo, Spain
{afvillaverde,ebalsa,julio}@iim.csic.es
[2] Insilico Biotechnology AG, Meitnerstraße 8, 70563 Stuttgart, Germany
{sophia.bongard,klaus.mauch,dirk.mueller,
joachim.schmid}@insilico-biotechnology.com

Abstract. Obtaining reliable predictions from large-scale dynamic models is a challenging task due to frequent lack of identifiability. This work presents a methodology for obtaining high-confidence predictions in biotechnological applications using metabolite time-series data. To preserve the complex behaviour of the network while reducing the number of estimated parameters, model parameters are combined in sets of meta-parameters, obtained from correlations between metabolite concentrations and between biochemical reaction rates. Next, an ensemble of models with different parameterizations is constructed and calibrated. Convergence of model outputs (consensus) is used as an indicator of confidence. Computational tests were carried out on a metabolic model of Chinese Hamster Ovary (CHO) cells. Using noisy simulated data, averaged ensemble predictions with high consensus were found to be more accurate than either predictions of individual ensemble models or averaged ensemble predictions with large variance. The procedure provides quantitative estimates of the confidence in model predictions and enables the analysis of sufficiently complex networks as required for practical applications in biotechnology.

Keywords: Systems biology, Metabolic engineering, Cell line engineering, Dynamic modeling, Ensemble modeling, Consensus.

1 Introduction

Mathematical modeling is a fundamental task in systems and computational biology, with important applications in biotechnology [9]. Among other features, models allow monitoring the state of unmeasured variables and making predictions about system behaviour for a larger number and broader variety of conditions than can be efficiently tested in experiments. The construction and calibration of models of large, complex systems is a particularly challenging task. Uncertainties appear at different stages of the process, limiting the confidence in the resulting predictions [14]. Shortage of experimental data can easily lead to poor identifiability. When the

J. Sáez-Rodríguez et al. (eds.), *8th International Conference on Practical Appl. of Comput. Biol. & Bioinform. (PACBB 2014)*, Advances in Intelligent Systems and Computing 294,
DOI: 10.1007/978-3-319-07581-5_20, © Springer International Publishing Switzerland 2014

number of parameters is larger than what can be actually determined from data, the calibration procedure can sometimes—when allowed by the model structure—yield a perfect fit between model predictions and measurements. However, there is a danger of overfitting in this situation, i.e., the model is being trained to fit in detail the noise contained in the data instead of actually learning the system dynamics. This problem entails the risk that model predictions will be wrong for altered experimental conditions. The problem of dealing with uncertainty in cellular network modeling was reviewed in [7]. In that review, the use of ensembles—sets of models with different structures and/or parameter values—was considered as a powerful and generally applicable approach for reducing prediction errors. However, it was also acknowledged that the concept was not sufficiently matured yet. An early example of the use of an ensemble approach in biological models was presented in [8], which was limited to ensembles of topologies of Boolean networks. Tran et al [12] extended the approach to the dynamic case, building an ensemble of metabolic models that reached the same steady state and applying it to the central carbon metabolism of *E. coli*. A related application was presented in [10]. For a review of metabolic ensemble modeling see [11]. The use of the consensus as an indication of the reliability of the predictions was explored by Bever [3], who computed time-dependent probability distributions of protein concentrations in artificial gene regulatory networks and introduced the concept of consensus sensitivity, finding that consensus among ensemble models was a good indicator of high-confidence predictions. Recently, further steps were taken with the introduction of the concept of "core prediction": a property that must be fulfilled if the model structure is to explain the data, even if the individual parameters are not accurately identified [4].

The present paper deals with the problem of evaluating and, if possible, increasing the confidence in the predictions made by kinetic metabolic models. It is assumed that the model structure—the topology of the metabolic network—is known. Actually, this assumption is not a requirement of the proposed methodology, which may be applied to ensembles of models with different topologies. However, in the present work the uncertainty in the predictions is due only to uncertainty in the parameter values. To overcome uncertainty, an ensemble of models with different parameterizations is built. As a preceding step to improve identifiability and to reduce overfitting, the initial model parameters are grouped in modules of meta-parameters, which are used during calibration. Then a measure of consensus among model outcomes is introduced and used to quantify the confidence in the predicted metabolite concentrations.

2 Methods

The methodology aims at adapting the kinetics of interrelated reaction pathways. Highly correlated trajectories of simulated concentrations and reaction rates point at functional dynamic relations, which can be adjusted by the parameters that correspond to the correlated time courses of concentrations and fluxes. We will refer to these sets of parameters as meta-parameters and use them for improving identifiability and reducing the risk of overfitting. Starting from an ODE model with generic initial parameter settings P_0, sets of model parameters are obtained by calculating the Pearson Correlation Coefficients (PCC) between simulated concentration time courses for all balanced metabolites, here called c, as well as between all simulated rates, here called r, described by the kinetic model. This yields two correlation arrays, C_r and C_c:

$$C_r = \begin{pmatrix} PCC_{r1,r1} & \cdots & PCC_{r1,rm} \\ \vdots & \ddots & \vdots \\ PCC_{rm,r1} & \cdots & PCC_{rm,rm} \end{pmatrix}, C_c = \begin{pmatrix} PCC_{c1,c1} & \cdots & PCC_{c1,cn} \\ \vdots & \ddots & \vdots \\ PCC_{cn,c1} & \cdots & PCC_{cn,cn} \end{pmatrix} \quad (1)$$

To define meta-parameters, a set of metabolites is selected that is of particular interest, e.g. because their time course over a fermentation process has been measured. Depending on the metabolite of interest, rates that represent an input of the utilization or formation pathway are selected. Rates which are highly correlated (PCC > 0.8) with the selected rate r_i are then included in a vector $r_{i\ correlated}$. Similarly, the substrates c_j present in the selected rate reaction r_i are chosen, and the elements of C_c (i.e. concentrations) which are highly correlated with c_j are included in a vector $c_{j\ correlated}$. Thus for each concentration-rate pair (c_j, r_i) two vectors $c_{j\ correlated}$ and $r_{i\ correlated}$ are obtained. Parameters appearing in the equations of the rates included in $r_{i\ correlated}$ are classified into two meta-parameter subsets: one for substrate parameters and one for product parameters. From identical meta-parameters, just one parameter set is kept to avoid co-linearity. For each meta-parameter MP_{ij} an optimization parameter k_{ij} is defined, which multiplies every parameter in the set. This reduces the size of the optimization problem: instead of estimating all the parameters included in a meta-parameter, they are kept to their nominal values, and only k_{ij} needs to be estimated, $MP_{ij\ optim} = MP_{ij} \times k_{ij}$. Meta-parameters have been grouped into modules, which are sets of meta-parameters designed to manipulate the dynamics of one phenotypic property. The property of interest determines the choice of meta-parameters. Determination of the parameter sets has been conducted using the modelling and simulation software Insilico Discovery™ (Insilico Biotechnology AG, Stuttgart, Germany). The routine has a low computational cost: for the model under examination, calculations were carried out in less than three minutes.

The parameter values are estimated by calibrating the models in the ensemble with the enhanced scatter search (eSS) parameter estimation method [6] included in the AMIGO toolbox [2]. It is a state of the art metaheuristic that has shown competitive performance with large-scale multimodal problems [13,15]. This metaheuristic can be regarded as a hybrid technique which combines global and local search, the latter in order to accelerate convergence to nearby optima; in this application the local method chosen is the Matlab routine *fmincon*. When there is lack of identifiability, different sets of parameter values can fit the data equally well. Because eSS is a stochastic technique, several optimization runs with different random initializations result in different solutions, each of which is a member of the ensemble. Models in the ensemble share the same structure but have different parameterizations. Each meta-parameter module leads to a model with a different parameter set for optimization purposes. In turn, each of these models can give rise to several models by changing the parameter values. The resulting ensemble includes (i) models calibrated with the meta-parameter approach, and (ii) models that were calibrated with the original parameter set. After creating an ensemble of models with different meta-parameter modules and/or different parameter values, it is used for estimating the reliability of the predictions. The behaviour of each model in a different experimental condition is simulated. Convergence of the model outputs (consensus) is then taken as an indicator of the confidence in the prediction (see section 3 for justification). Let n_p be the number of data points, n_m the number of models in the ensemble and n_s the number of states (metabolite concentrations) for which a prediction is made. The prediction of the concentration of metabolite j at time i made by model k is A_{ijk}. The average prediction of all the models for every state and time instant is the $n_p \times n_s$ array:

$$\overline{A_{ij}} = \frac{1}{n_m} \sum_{k=1}^{n_m} A_{ijk} \tag{3}$$

The dissensus d_{ijk} of every model k with respect to the average prediction for a metabolite j at every time instant i can be encapsulated as a $n_p \times n_s \times n_m$ array,

$$d_{ijk} = \frac{\left| A_{ijk} - \overline{A_{ij}} \right|}{\max_k (A_{ijk})} \tag{4}$$

The dissensus of every model for every state along the time horizon is

$$d_{jk} = \frac{\sum_{i=1}^{n_p} d_{ijk}}{n_p} \tag{5}$$

And the ensemble dissensus is the dissensus among all the models for a particular state along the time horizon,

$$d_j = \frac{\sum_{k=1}^{n_m} d_{jk}}{n_m} \tag{6}$$

From which we define the ensemble consensus for each state as $c_j = 1 - d_j$, and the global consensus for all the states as

$$c = \frac{\sum_{j=1}^{n_s} c_j}{n_s} \tag{7}$$

The dissensus and consensus metrics defined above have values between 0 and 1, which makes their interpretation easy. For example, if $c_1 = 0.85$ and $c_2 = 0.98$, the ensemble consensus for metabolite 2 is higher than for metabolite 1, and therefore the methodology suggests that the predictions about the concentrations of metabolite 2 are more reliable than for metabolite 1.

3 Results and Discussion

The methodology was applied to a metabolic model of Chinese Hamster Ovary cells (CHO), which are used for recombinant protein production in fermentation processes. We simulated a batch process with resting cells and a time horizon of 300 hours. The fermenter medium contains glucose as main C-source, and leucine and methionine as representative amino acids that are taken up. Lactate is modeled as key by-product of the fermentation process. A synthesized protein serves as main product of the fermentation process. Glucose, lactate, product protein, leucine, and methionine are assumed to be measured in the fermenter. Aspartate, malate, pyruvate, oxaloacetate, ATP, and ADP are assumed to be available as intracellular metabolite measurements. The model comprises 34 metabolites (whose concentrations are the state variables, numbered in Table 1) and 32 reactions, and includes protein product formation, Embden-Meyerhof-Parnas pathway (EMP), TCA cycle, a reduced amino acid metabolism, lactate production, and the electron transport chain. The kinetic ODE model comprises 117 parameters in total. Rate kinetics follow the description in [5], and the flux distribution is obtained by applying Flux Balance Analysis (FBA). The objective function for FBA is defined to achieve a maximal product formation. Uptake and production rate constraints are chosen such to be close to measured stationary fluxes in [1]. The remaining parameters that need to be identified are the K_m values in the Michaelis-Menten kinetics and the elasticities in the linlog kinetics. Five different meta-parameter modules were generated and used for parameter estimation:

- Central metabolism module: MPs from concentration and rate pairs related to the EMP and TCA cycle, describing the dynamics of glucose, leucine, methionine, and 2-oxoglutarate.
- Fermenter module: describing the dynamics of glucose, lactate, methionine, leucine, and product protein.
- Energetics module: TCA cycle and energy-related metabolites, describing the dynamics of glucose, pyruvate, methionine, leucine, and ATP.
- Uptake and energetics module: substrate-associated meta-parameters describing the dynamics of glucose, methionine, leucine, ATP, product protein, and lactate.
- Uptakes module: describing the dynamics of glucose, methionine, and leucine.

Synthetic measurement data was generated using a reference model and was used in lieu of experimental data. Care was taken to ensure resemblance to realistic experimental conditions. For each of the 13 metabolites that are typically measured, 12 sampling times were assumed and Gaussian noise was added with a standard deviation of 20%. Using the eSS algorithm for parameter estimation included in AMIGO, we carried out 10 optimizations of the full model without meta-parameters

Table 1. List of state variables (metabolite concentrations) in the CHO model

Nr	Variable name	Nr	Variable name
1	Glucose (fermenter)	18	H_out (mitochondria)
2	Lactate (fermenter)	19	CoQ (mitochondria)
3	Leucine (fermenter)	20	H_in (mitochondria)
4	Methionine (fermenter)	21	Pyruvate (cytosol)
5	Product protein (fermenter)	22	Phosphoenolpyruvate (cytosol)
6	Glutamate (mitochondria)	23	NADH (cytosol)
7	NAD (mitochondria)	24	Glycerate3-phosphate (cytosol)
8	2-Oxoglutarate (mitochondria)	25	NAD (cytosol)
9	NADH (mitochondria)	26	Glucose (cytosol)
10	Glutamine (cytosol)	27	Malate (cytosol)
11	ADP (cytosol)	28	2-Oxoglutarate (cytosol)
12	Glutamate (cytosol)	29	Aspartate (cytosol)
13	ATP (cytosol)	30	ATP (mitochondria)
14	Aspartate (mitochondria)	31	Orthophosphate (mitochondria)
15	Oxaloacetate (mitochondria)	32	ADP (mitochondria)
16	Malate (mitochondria)	33	Leucine (cytosol)
17	CoQH_radical (mitochondria)	34	Methionine (cytosol)

(117 parameters) starting from different initial random guesses. The lower and upper bounds for the parameters were 1/5 and 5 times the nominal values. All of the optimizations succeeded in obtaining a good fit to the data, yielding 10 solutions that differed significantly; this non-uniqueness suggests lack of practical identifiability. Following this initial calibration with the original parameter set, we re-parameterized the model using a meta-parameter approach. As before, the optimizations started from 10 different initial guesses for each module, whose values were chosen randomly within the parameter bounds. They were carried out twice, resulting in a total of $2 \times 10 \times 5 = 100$ optimizations. Reducing the number of optimized parameters (from the original 117 parameters to between 5 and 12 meta-parameters, depending on the module) improved the identifiability of the model, and the 100 optimizations yielded only 22 different solutions (that is, the fraction of different solutions obtained in the optimizations was reduced from 100% to 22%). The ensemble of models was created with these 22 parameterizations plus the 10 resulting from the optimization of the original model, that is, a total of 32 models. All of these models provided a near-perfect fit to the pseudo-experimental data used for calibration, as shown in Fig. 1. The fact that the models calibrated with the meta-parameter modules (which had

between 5 and 12 free parameters to be optimized) managed to fit the data equally well as the models calibrated with the full original set (117 parameters) supports the idea that the models calibrated with the original parameter set had a clear risk of overfitting. The decrease in the number of optimal solutions found with the meta-parameter approach suggests that its use reduced this risk. The reduction in the number of estimated parameters led also to a decrease of the computation times of the optimization procedure, which were reduced by one or even two orders of magnitude.

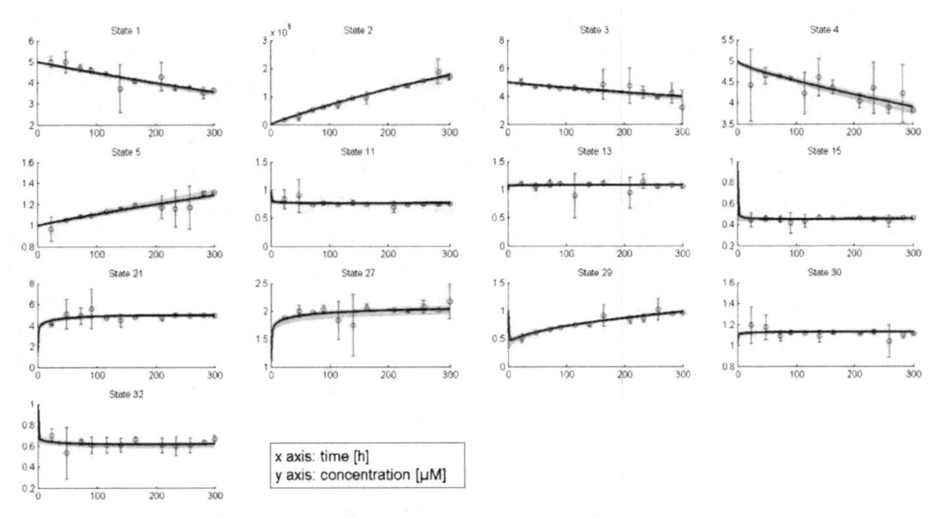

Fig. 1. Pseudo-experimental data (red circles with error bars), individual predictions of the 32 fitted models (grey lines) and their average—i.e., ensemble—predictions (blue lines) for the 13 measured states (metabolite concentrations). The pseudo-experimental data was used in the parameter estimation procedure.

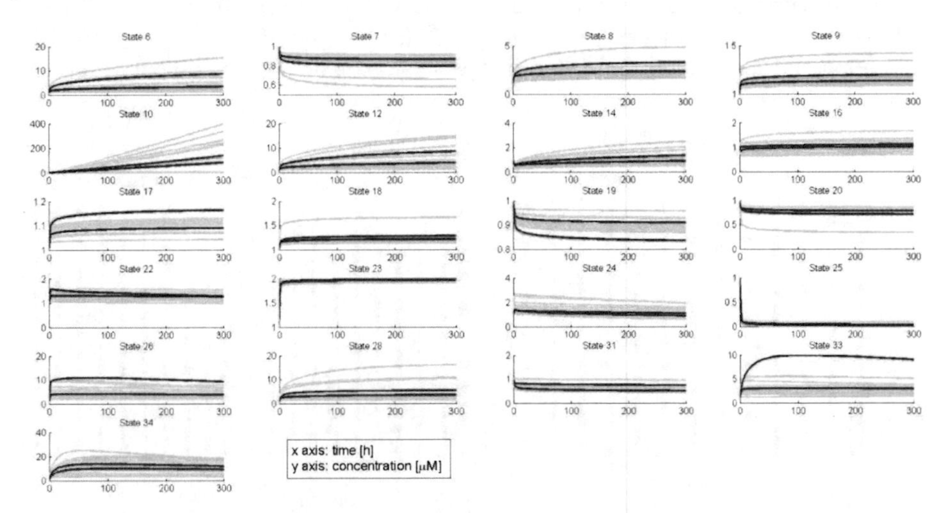

Fig. 2. True model output with nominal parameters (red lines), individual predictions of the 32 fitted models (grey lines), and their average—i.e., ensemble—predictions (blue lines), for the 21 unmeasured states (metabolite concentrations).

Next we evaluated the ability of the calibrated models and the ensemble to reproduce the concentrations of the remaining 21 metabolites, which are typically not measured in practice. Consensus in the predictions among models in the ensemble was calculated for each of the measured and unmeasured metabolites as in equation (5). As expected, in general the measured metabolites elicited more consensus than the unmeasured ones, with consensus ranges of $0.9816 < c_j < 0.9981$ for the former and $0.7988 < c_j < 0.9935$ for the latter. Fig. 2 shows the predictions of every model in the ensemble (in grey) together with the real output (in red) and the average or ensemble prediction (in blue), for the 21 unmeasured states. Results in this figure confirm that consensus is a good indicator of the confidence in the predictions. For example, predictions of the state variable number 23 (NADH) elicit a very high consensus (0.9935), which leads to the assumption that its consensus prediction is very close to the real output, as is indeed the case. Conversely, variable number 10 (glutamine) has the lowest consensus (0.7988), and the resulting prediction is quite far from the real output. Therefore, if we were thinking of measuring additional metabolites, we should concentrate our efforts on those metabolites with the larger dissensus, such as glutamine, and we could safely avoid additional measurements and rely on the ensemble prediction for metabolites such as NADH. Indeed, consensus and prediction error were found to be significantly anti-correlated, with a correlation coefficient of -0.58 ($p < 3 \cdot 10^{-4}$). Fig. 3a plots the consensus c_j and prediction errors e_j for the 13 measured variables. Since the models were fitted to these data, the prediction errors are very low and the consensus is very high for all the variables. Fig. 3b shows the corresponding values for the 21 unmeasured variables; a vertical green line was drawn to separate the 10 variables that elicit the higher consensus from the rest. For these 10 variables the prediction errors are very low, which is not always the case for the other variables. Thus, ranking the predictions according to their consensus is a useful way of deciding which metabolites should be measured and which can be left unmeasured because we have enough confidence in the predictions.

Another application of the consensus approach is for assessing the confidence in predictions of the system's behaviour under different conditions than those used for calibration. For this application it can be the case that no experimental data are

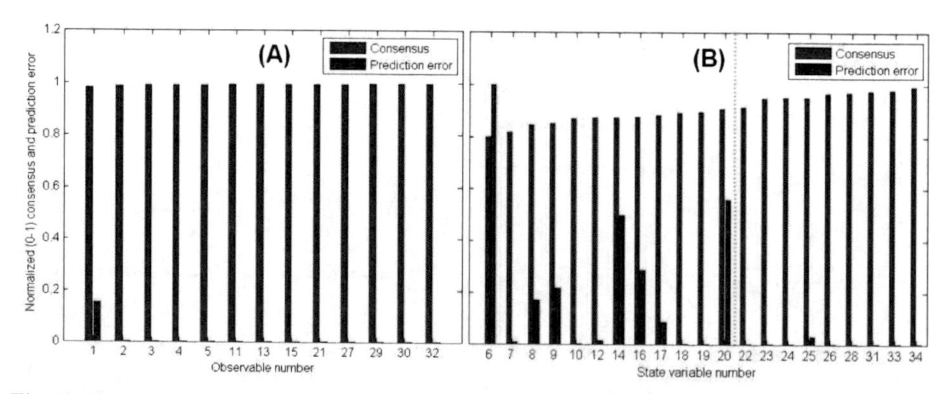

Fig. 3. Bar plots of the consensus c_j and prediction errors e_j for the 13 measured variables ("observables", panel A), and for the 21 unmeasured variables (panel B). The vertical line in panel B demarcates the 10 states with the highest consensus.

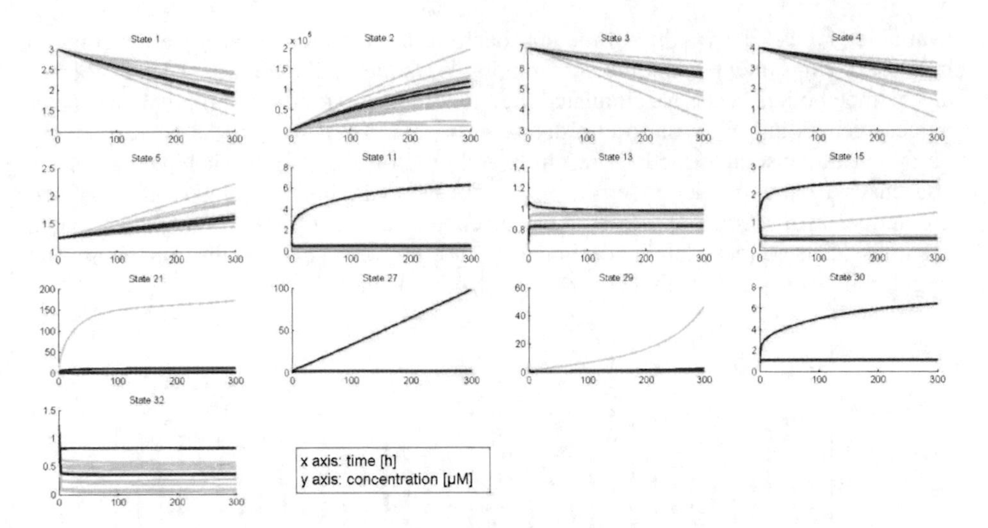

Fig. 4. True model output with nominal parameters (red lines), individual predictions of the 32 fitted models (grey lines) and their average—i.e., ensemble—predictions (blue lines) for 13 states (metabolite concentrations) in the CHO model, under different conditions than those used in the model calibration. The 13 states shown are those for which pseudo-experimental measurements were available.

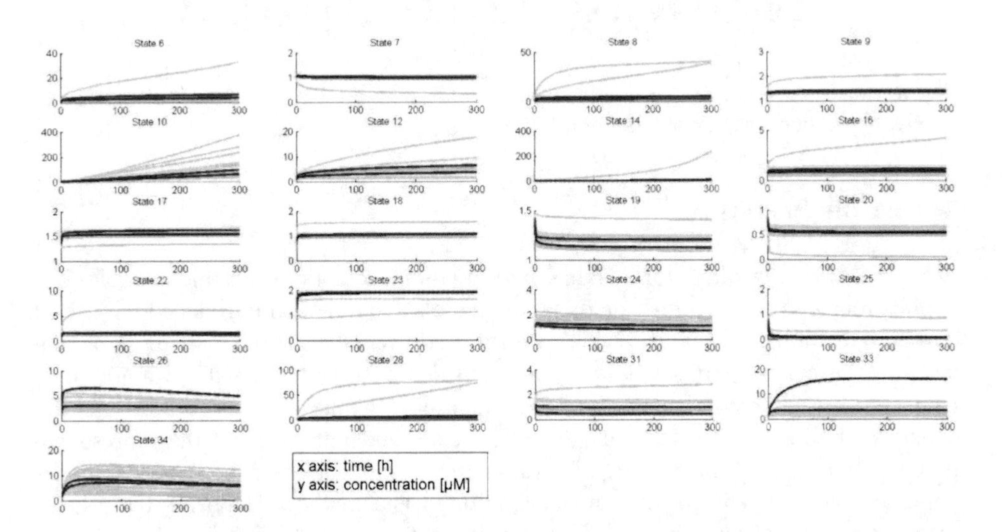

Fig. 5. True model output with nominal parameters (red lines), individual predictions of the 32 fitted models (grey lines) and their average—i.e., ensemble—predictions (blue lines) for 21 states (metabolite concentrations) in the CHO model, under different conditions than those used in the model calibration. The 21 states shown are those for which pseudo-experimental measurements were not available.

available. To test the validity of the approach in this type of scenario we generated a new set of experimental conditions by randomly changing the initial concentrations of all the metabolites. Then we simulated the model's behaviour with nominal parameter values, generating a new artificial dataset, and compared it with the corresponding output of the ensemble models; results are shown on Figs. 4 and 5. Fig. 6 plots the consensus c_j and prediction errors e_j for the 34 variables. Again, consensus and prediction error were found to be anti-correlated, although less strongly than in the previous scenario (correlation coefficient: -0.30, $p < 0.084$). Thus, the conjecture that larger consensus parallels higher accuracy also holds in this case.

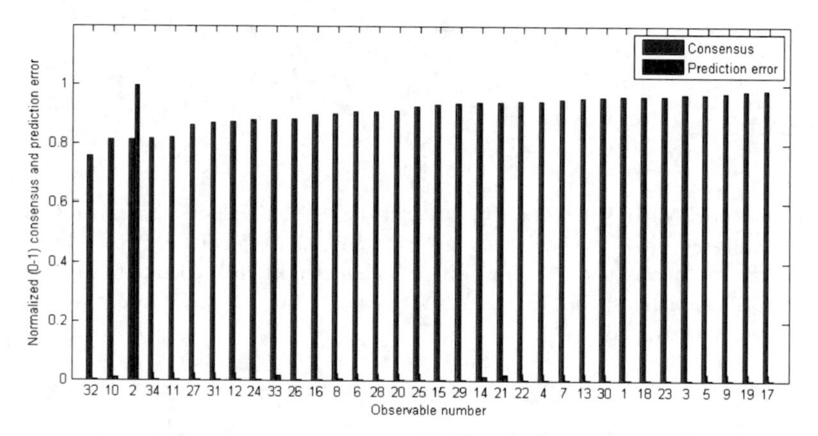

Fig. 6. Bar plots of the consensus c_j and prediction errors e_j for the 34 state variables under different conditions than those used for calibration

4 Conclusions

We presented a method for making high-quality predictions for biotechnological applications by use of large-scale dynamic models. The method introduced the use of two computational techniques, which we referred to as the meta-parameter approach and the consensus approach. Its use has been demonstrated using the example of a batch process for recombinant protein production in Chinese Hamster ovary (CHO) cells. We built a metabolic model for CHO cells and implemented the represented process. We reduced the number of parameters by application of the meta-parameter approach without reducing the model structure. The method decreased the risk of over-fitting and accelerated the parameter estimation procedure. Several variants of meta-parameters served as basis for the application of the consensus approach. The consensus approach is based on an ensemble modeling framework, where the coincidence among predictions of the models in the ensemble is an indication of the reliability of these predictions. Using a measure of the relative distance between the average prediction of all models and the true value, the procedure quantified the degree of confidence in the predictions of the time evolution of each of the state variables (metabolite concentrations in the example application). This quantitative indication can be used to decide where it is most efficient to make an additional measurement effort, that is, which metabolites should be actually measured and for which we can rely on the model simulations. Future applications can be seen for models applied to problems in microbial metabolic engineering as well as in cell line engineering of mammalian cell lines.

Acknowledgements. This work was supported by the EU project "BioPreDyn" (EC FP7-KBBE-2011-5, grant 289434) and by the Spanish MINECO and the European Regional Development Fund (ERDF; project "MultiScales", DPI2011-28112-C04-03). We further would like to thank Michael Rempel for valuable technical support.

References

1. Ahn, W.S., Antoniewicz, M.R.: Towards dynamic metabolic flux analysis in CHO cell cultures. Biotechnol. J. 7(1), 61–74 (2012)
2. Balsa-Canto, E., Banga, J.R.: AMIGO, a toolbox for Advanced Model Identification in systems biology using Global Optimization. Bioinformatics 27(16), 2311–2313 (2011)
3. Bever, C.: Selecting high-confidence predictions from ordinary differential equation models of biological networks. PhD Thesis, MIT (2008)
4. Cedersund, G.: Conclusions via unique predictions obtained despite unidentifiability – new definitions and a general method. FEBS J. 279(18), 3513–3527 (2012)
5. Chassagnole, C., Noisommit, N., Schmid, J.W., Mauch, K., Reuss, M.: Dynamic modeling of the central carbon metabolism of E. coli. Biotechnol. Bioeng. 79(1), 53–73 (2002)
6. Egea, J.A., Martí, R., Banga, J.R.: An evolutionary method for complex process optimization. Comp. Oper. Res. 37(2), 315–324 (2010)
7. Kaltenbach, H.M., Dimopoulos, S., Stelling, J.: Systems analysis of cellular networks under uncertainty. FEBS Letters 583(24), 3923–3930 (2009)
8. Kauffman, S.: A proposal for using the ensemble approach to understand genetic regulatory networks. J. Theor. Biol. 230(4), 581–590 (2004)
9. Kremling, A., Saez-Rodriguez, J.: Systems biology—anengineering perspective. J. Biotechnol. 129, 329–351 (2007)
10. Kuepfer, L., Peter, M., Sauer, U., Stelling, J.: Ensemble modeling for analysis of cell signaling dynamics. Nature Biotechnol. 25, 1001–1006 (2007)
11. Tan, Y., Liao, J.C.: Metabolic ensemble modeling for strain engineers. Biotechnol. J. 7, 343–353 (2012)
12. Tran, L.M., Rizk, L.M., Liao, J.C.: Ensemble Modeling of Metabolic Networks. Biophys. J. 95(12), 5606–5617 (2008)
13. Villaverde, A.F., Ross, J., Morán, F., Balsa-Canto, E., Banga, J.R.: Use of a generalized Fisher equation for global optimization in chemical kinetics. J. Phys. Chem. A 115(30), 8426–8436 (2011)
14. Villaverde, A.F., Banga, J.R.: Reverse engineering and identification in systems biology: strategies, perspectives and challenges. J. R. Soc. Interface 11, 20130505
15. Villaverde, A.F., Egea, J.A., Banga, J.R.: A cooperative strategy for parameter estimation in large scale systems biology models. BMC Syst. Biol. 6, 75 (2012)

A Parallel Differential Evolution Algorithm for Parameter Estimation in Dynamic Models of Biological Systems

D.R. Penas[1], Julio R. Banga[1], P. González[2], and R. Doallo[2]

[1] BioProcess Engineering Group, IIM-CSIC, Spain
{davidrodpenas,julio}@iim.csic.es
[2] Computer Architecture Group, University of A Coruña, Spain
{pglez,doallo}@udc.es

Abstract. Metaheuristics are gaining increased attention as efficient solvers for hard global optimization problems arising in bioinformatics and computational systems biology. Differential Evolution (DE) is one of the most popular algorithms in that class. However, the original algorithm requires many evaluations of the objective function, so its application to realistic computational systems biology problems, like those considering parameter estimation in dynamic models, results in excessive computation times. In this work we present a modified DE method which has been extended to exploit the structure of parameter estimation problems and which is able to run efficiently in parallel machines. In particular, we describe an asynchronous parallel implementation of DE which also incorporates three new search heuristics which exploit the structure of parameter estimation problems. The efficiency and robustness of the resulting method is illustrated with two types of benchmarks problems (i) black-box global optimization problems and (ii) calibration of systems biology dynamic models. The results show that the proposed algorithm achieves excellent results, not only in terms of quality of the solution, but also regarding speedup and scalability.

Keywords: Computational Systems Biology, Parallel Metaheuristics, Differential Evolution.

1 Introduction

Global optimization methods are playing an increasingly important role in computational biology [1], bioinformatics [2] and systems biology [3]. The field of systems biology uses mathematical models and computational resources to understand function in complex biological systems. The problem of parameter estimation in dynamic models is especially challenging to solve due to its NP-hard nature [4]. Many research efforts have focused on developing metaheuristic methods which are able to locate the vicinity of the global solution in reasonable computation.

In order to reduce the computational cost of these methods, a number of researchers have studied parallel strategies for metaheuristics [5, 6]. In the area of

J. Sáez-Rodríguez et al. (eds.), *8th International Conference on Practical Appl. of Comput. Biol. & Bioinform. (PACBB 2014)*, Advances in Intelligent Systems and Computing 294,
DOI: 10.1007/978-3-319-07581-5_21,

computational systems biology, parallel methods have already shown promising results [7, 8].

Differential Evolution (DE) [9] is probably one of the most popular heuristics for global optimization, and it has been successfully used in many different areas [10–12]. However, in most realistic applications, DE needs a significant number of evaluations to achieve an adequate quality solution. This may imply too much time to obtain an acceptable result. Thus, in order to improve the runtime of the classical DE algorithm, two main strategies have been explored in this work. On the one hand, including a selected local search may enhance the classical DE through intensification, drastically reducing the number of evaluations required. On the other hand, exploiting parallelism can help improving the computational time needed for the evaluation of the objective functions.

The organization of this paper is as follows. Section 2 presents a brief overview of the Differential Evolution algorithm and explains the additional heuristics for parameter estimation that have been added. Section 3 describes the asynchronous strategy proposed to parallelize an Island Differential Evolution algorithm. Section 4 evaluates the performance of the new method considering two sets of benchmark problems, demonstrating its efficiency and scalability. Finally, Section 5 summarizes our findings.

2 Improving Search in Differential Evolution Algorithms

Differential Evolution (DE) is an iterative mutation algorithm where vector differences are used to create new candidate solutions. Starting from a initial population matrix composed of NP D-dimensional solution vectors (individuals), DE attempts to achieve the optimal solution iteratively through changes in its vectors. For each iteration, new individuals are generated in the population matrix through operations performed between individuals of the matrix (mutation - F), with old solutions replaced (crossover - CR) only when the fitness value of the objective function is better than the current one.

A population matrix with optimized individuals is obtained as output of the algorithm. The best of these individuals are selected as solution close to optimal for the objective function of the model. However, in some real applications, such as parameter estimation in dynamic models, the performance of the classical sequential DE (`seqDE`) is not acceptable due to the large number of objective function evaluations needed. As a result, typical runtimes for realistic problems are in the range from hours to days. In order to improve the computational effort required by the `seqDE`, three enhancements that exploit the special structure of these parameter estimation problems have been included in this work:

- The search is performed in a logarithmic space, which results in a more suitable exploration of the space of parameters when they are positive and potentially span through several orders of magnitude [13]
- A local method is introduced to achieve a fast local convergence, therefore, reducing the number of objective function evaluations required. The local

Algorithm 1. Differential Evolution algorithm with Local Search and Tabu List - `seqDE_LS`

```
input  : A population matrix P with size D x NP
output: A matrix P whose individuals were optimized

repeat
  for each element x of the P matrix do
    a⃗, b⃗, c⃗ ← different random individuals from P matrix
    for k ← 0 to D do
      if random point is less than CR then
        Ind⃗(k) ← a⃗(k) + F(b⃗(k) - c⃗(k));
      end
    end
    if Evaluation(Ind⃗) is better than Evaluation(P(x)⃗) then
      Replace_Individual(P,Ind⃗);
    end
  end
  if local solver condition then
    Best = Extract_Best_Point(P, TabuList);
    newBest = Run_Local_Solver(Best);
    if success condition then
      Insert_New_Best_Solution(P, newBest);
    end
  end
until Stop conditions;
```

search moves from solution to solution in the space of candidate solutions by applying local changes, until a solution considered optimal is found or a time bound is elapsed. In this work, a suitable local solver (N2SOL [14]) is called every L iterations of the `seqDE`. This local solver is particularly effective for parameter estimation problems [13]

- One drawback of the previous local search is that it tends to become stuck in suboptimal regions or on plateaus where many solutions are equally fit. As a means to avoid this problem, a search space *tabu list* is also implemented in this work. Tabu search enhances the performance of local methods by using memory structures that describe the visited solutions or user-provided sets of rules. If the vicinity of a potential solution has been previously visited within a certain short-term period or if it has violated a rule, it is marked as *tabu* so that the algorithm does not consider that possibility repeatedly.

Algorithm 1 shows the pseudocode for the new DE scheme (`seqDE_LS`) proposed in this work, with local search and the use of a tabu list. This scheme represents a suitable compromise between diversification (global search) and intensification (local search), which is at the core of most modern metaheuristics.

Algorithm 2. Asynchronous island-based parallel DE - `asynPDE_LS`

```
input  : A population matrix P
output: A matrix P whose individuals were optimized
pendingMigration=0, counterMigration=0;
Scatter population P into NP processors (Plocal submatrix);
repeat
    for subpopulation do
        Crossover, mutation and evaluation operations
    end
    if migration condition then
        Selection_Individuals(Plocal, &migrationSet);
        MPI_ISend(&migrationSet, remoteDestination, &requestSend);
        Create_Buffer_Memory(receptionSet, pendingMigration);
        MPI_IRecv(&receptionSet(pendingMigration-counterMigration), remoteOrigin,
        &requestRecv(pendingMigration));
        pendingMigration ++;
    end
    while pendingMigration-counterMigration > 0 do
        MPI_Test(&requestRecv(counterMigration), complete);
        if complete then
            Replacement_Individuals(&Plocal, &receptionSet(pendingMigration -
            counterMigration ));
            Remove_Buffer_Memory(&receptionSet);
            counterMigration ++ ;
        else
            break;
        end
    end
    if local solver condition then
        Local solver and tabu list operations
    end
until Stopping condition;
Gather all subpopulation into matrix P ;
```

3 Asynchronous Island-Based MPI Parallel Implementation

Three different parallel models for metaheuristics can be found in the literature [15]: *Master-slave* model, *Island* model, and *Cellular* model. The solution explored in this work follows the island model approach that divides the population matrix in subpopulations (*islands*) where the algorithm is executed isolately. Sparse individual exchanges are performed among islands to introduce diversity into the subpopulations preventing search from getting stuck in local optima.

The parallel island DE algorithm operates on N independent subpopulations of individuals, each one located in a separate processor. Phases such as selection, recombination and mutation are performed only within each subpopulation, which implies absence of collaboration between processes. After m generations, a migration phase links the subpopulations: selected individuals from each island are communicated to another subpopulation. The migration operation and the checking of the termination criteria imply exchange of communications between processes.

The simplest implementation of the parallel island DE is a synchronous algorithm. The drawback of the synchronous algorithm is that processors are idle during a significant amount of time, while they are waiting for each other during the migration steps. The penalty may be significant when the local search and tabu list is introduced in the DE algorithm since the local solver leads processes into a more computationally unbalanced scenario.

The `asynPDE_LS` algorithm proposed in this work (Algorithm 2), developed using the MPI library, avoids these stalls by implementing a variation of the classical parallel island DE. Whenever a process reaches the migration phase, it sends a set of individuals to the selected remote process using an asynchronous communication (`MPI_ISend` and `MPI_IRecv`). Then, the process in the migration phase checks (`MPI_Test`) if the message with the new individuals of a remote process has already arrived to its memory buffer (`receptionSet`). However, if the new solutions have not arrived yet, the process proceed with the next evaluation. After each evaluation the process searches for the reception of data missed in previous migrations, however avoiding stalls if the messages are not ready. Thus, the new parallel algorithm does not implement straightforwardly the serial one, but it avoids idle processes.

4 Experimental Results

In order to evaluate the efficiency of the proposed cooperative asynchronous algorithm (`asynPDE_LS`), different experiments have been carried out. Its behavior, in terms of convergence and total execution time, was compared with the sequential classic version of DE (`seqDE`), the sequential DE with local search, tabu list and logarithmic search enabled (`seqDE_LS`), and its synchronous parallel version (`synPDE_LS`).

There are many configurable parameters in `seqDE`, such as the mutation scaling factor (F) or the crossover constant (CR), whose selection may have a great impact in the algorithm performance. In parallel island DE algorithms, new parameters have to be also considered, such as migration frequency(μ), island size(λ), communication topology (Tp) between processors, or selection policy (SP) and replacement policy (RR) in the migration step. In addition, the proposed parallel algorithm was tested using two different local configurations (LC): "homogeneous" when the same value of CR and F is used in all processes, and "heterogeneous" when each process has different combination of CR and F parameters. The values used during these experiments were F=$\{.9, .7\}$ and CR=$\{.9, .7, .2\}$.

Table 1. Quality value test in BBOB benchmarks. Configuration: DE/best/1 in f8 and DE/rand/2 in f17, μ=5%, F=0,9, λ=NP/PE.

					Speedup PE (n. of proc.)				
Function	**NP/D**	**CR**	**Algorithm**	**Tp/LC**	**5**	**10**	**20**	**40**	**60**
f8-Rosenbrock	100D/24	0,8	asynPDE_LS	Ring/Homo	4,9	10,8	24,8	38,7	45,2
			synPDE_LS	Ring/Hete	8,7	17,3	27,4	31,4	30,2
f17-Schaffers F7	150D/6	0,8	asynPDE_LS	Ring/Hete	4,9	12,7	24,6	59,7	87,9
			synPDE_LS	Star/Hete	7,6	16,0	29,0	56,4	68,3

The performed experiments use two sets of benchmark problems: the Black-Box Optimization Benchmarking (BBOB) data set [16], and a set of parameter estimation problems in systems biology.

4.1 BBOB Data Set Results

The experiments over the BBOB data set were carried out to evaluate the efficiency of the proposed parallelization. Thus, they were performed in the Intel Sandy Bridge nodes of CESGA SVG Linux cluster [17] and with local solver option disabled. The original population matrix was divided in subpopulations among the processors (λ=NP/PE) in order to share the sequential computational load, and to evaluate its scalability when the number of resources increases.

Both algorithms (synPDE_LS and asynPDE_LS) were executed on this testbed composed of 24 noiseless benchmark functions to be minimized. Up to 41 different experiments, for each of these functions, have been thoroughly performed using different combinations of parameters (PE, Tp, LC, etc). Due to space limitations, only two benchmark functions and particular configurations have been selected to illustrate the experimental results.

Table 1 shows the speedup with quality solution for these two functions and the configurations that eventually obtained the best speedups. With a large number of processors, asynPDE_LS demonstrates its better scalability over synPDE_LS. Note that both the synchronous and asynchronous implementations achieve superlineal speedups. This is due to the good behavior of the parallel algorithms in terms of quality of the solution, where less evaluations are needed to come to the required tolerance when more processors take part in the search. It is also noticeable that, in most cases, the heterogeneous strategies are superior to homogeneous ones, since they benefit from search diversification.

4.2 Systems Biology Parameter Estimation Results

Three challenging parameter estimation problems from the domain of computational system biology were considered. These problems are known to be particularly hard due to their ill-conditioning and non-convexity [18]:

- *Circadian* problem: parameter estimation in a dynamic model of the circadian clock in the plant *Arabidopsis thaliana*, as presented in [19]. The

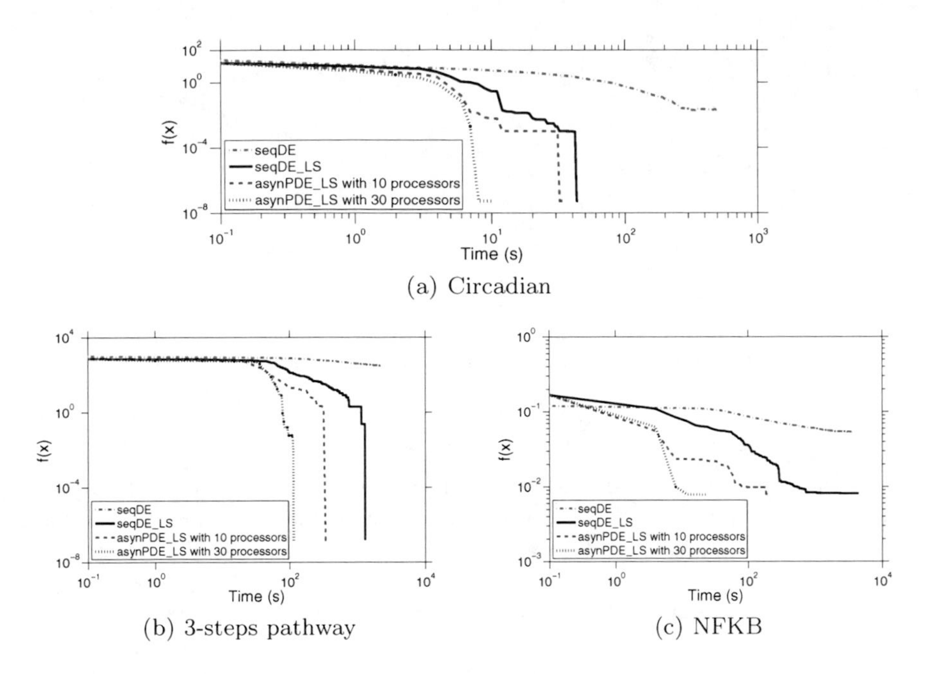

(a) Circadian

(b) 3-steps pathway

(c) NFKB

Fig. 1. Convergence curves for parameter estimation problems. Configuration: NP=10×D, μ=33,3%, LC=heterogeneous, Tp=star, SP=Best individuals, RR=Worst individuals, λ=NP, (a) DE/best/1, (b) (c) DE/best/2

problem consists of 7 ordinary differential equations with 27 parameters (13 of them were estimated) with data sets from 2 experiments.
- *Nfkb* problem: this problem is based on the model in [20] and consists of 15 ordinary differential equations with 29 parameters and data sets from 2 experiments.
- *3-step pathway* problem: problem considering a 3-step generic and highly non-linear pathway with 8 differential equations and 36 parameters, and data sets from 16 experiments, as presented in [18].

A multicore heterogeneous cluster was used to carry out these experiments. It consists of 4 nodes powered by two quadcore Intel Xeon E5420 CPUs with 16 GB of RAM and 8 nodes powered by two quadcore Intel Xeon E5520 CPUs with 24 GB of RAM. The cluster nodes are connected through a Gigabit Ethernet network.

The aim of these experiments is to demonstrate the potential of the proposed techniques in improving the convergence and execution time of very hard problems. In these benchmarks, the executions of `seqDE` can take hours or even days to complete one only test. As an example of the practical significance of our study, the 3-steps pathway problem typically requires more than 3 days of computation time using the classical version of DE in one processor, but it can be solved in less than 2 minutes using 30 processors and the asynchronous parallel method presented here.

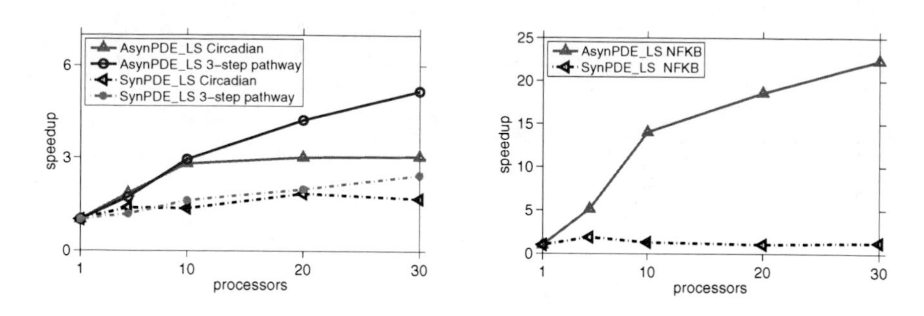

Fig. 2. Speedup with Quality Solution. Same configuration as in Figure 1

Figure 1 shows the convergence curve for these benchmarks. In order to ensure the diversity of the population, the island size is not divided among the processors, remaining to be NP. Results show that the asynPDE_LS algorithm improves quality of mean solution and reduces the variability between different iterations.

Figure 2(a) and Figure 2(b) show the speedup for a quality value test in parameter estimation problems. These speedups were calculated comparing execution times of synPDE_LS and asynPDE_LS with seqDE_LS. These figures show that the proposed asynPDE_LS reduces significantly execution time, and scales better than the synPDE_LS algorithm for a large number of processors. It is important to note that the local solver introduces a great overhead in the execution of the evaluations. Moreover, it is responsible for the lack of synchronization between processes at the migration step, thus, giving advantage to the proposed asynchronous solution. In these experiments the sequential computational load is not shared among processors, thus, the total speedup achieved over the seqDE_LS depends on the impact of the cooperation between processes. The gain compared to the synPDE_LS depends on the degree of synchronization between processes when the local solver is included. Both Circadian and 3-step pathway problems achieve more modest speedup results than the NFKB problem.

5 Conclusions

In this paper a new parallel differential evolution algorithm is proposed. The method implements an asynchronous migration strategy based in a cooperative island-model, and it also extends the original DE with several heuristics which exploit the structure of the parameter estimation problems.

The experimental results show (i) the convergence time decreases in several orders of magnitude when new search heuristics (local solver, tabu list, logarithmic search) are used in the DE algorithm (ii) the proposal not only has a competitive speedup against the synchronous solution, but it also attains a reduction in the convergence time through collaboration of the parallel processes.

References

1. Greenberg, H.J., Hart, W.E., Lancia, G.: Opportunities for combinatorial optimization in computational biology. INFORMS Journal on Computing 16(3), 211–231 (2004)
2. Larrañaga, P., Calvo, B., Santana, R., Bielza, C., Galdiano, J., Inza, I., Lozano, J.A., Armañanzas, R., Santafé, G., Pérez, A., et al.: Machine learning in bioinformatics. Briefings in Bioinformatics 7(1), 86–112 (2006)
3. Banga, J.R.: Optimization in computational systems biology. BMC Systems Biology 2(1), 47 (2008)
4. Villaverde, A.F., Banga, J.R.: Reverse engineering and identification in systems biology: strategies, perspectives and challenges. Journal of The Royal Society Interface 11(91), 20130505 (2014)
5. Crainic, T.G., Toulouse, M.: Parallel strategies for meta-heuristics. Springer (2003)
6. Alba, E.: Parallel metaheuristics: a new class of algorithms, vol. 47. Wiley-Interscience (2005)
7. Perkins, T.J., Jaeger, J., Reinitz, J., Glass, L.: Reverse engineering the gap gene network of drosophila melanogaster. PLOS Computational Biology 2(5), e51 (2006)
8. Jostins, L., Jaeger, J.: Reverse engineering a gene network using an asynchronous parallel evolution strategy. BMC Systems Biology 4(1) (2010)
9. Storn, R., Price, K.: Differential evolution–a simple and efficient heuristic for global optimization over continuous spaces. Journal of Global Optimization 11(4), 341–359 (1997)
10. Price, K., Storn, R.M., Lampinen, J.A.: Differential Evolution: A Practical Approach to Global Optimization. Natural Computing Series. Springer-Verlag New York, Inc., Secaucus (2005)
11. Chakraborty, U.K.: Advances in Differential Evolution. 1 edn. Springer Publishing Company, Incorporated (2008)
12. Das, S., Suganthan, P.N.: Differential evolution: A survey of the state-of-the-art. IEEE Transactions on Evolutionary Computation 15(1), 4–31 (2011)
13. Egea, J.A., Rodríguez-Fernández, M., Banga, J.R., Martí, R.: Scatter search for chemical and bio-process optimization. Journal of Global Optimization 37(3), 481–503 (2007)
14. Dennis Jr., J.E., Gay, D.M., Welsch, R.E.: Algorithm 573: Nl2sol an nonlinear least-squares algorithm. ACM Transactions on Mathematical Software (TOMS) 7(3), 369–383 (1981)
15. Alba, E., Luque, G., Nesmachnow, S.: Parallel metaheuristics: recent advances and new trends. International Transactions in Operational Research 20(1), 1–48 (2013)
16. Hansen, N., Auger, A., Finck, S., Ros, R.: Real-parameter black-box optimization benchmarking 2009: Experimental setup. Technical Report RR-6828, INRIA (2009)
17. CESGA: Svg specifications, `https://www.cesga.es/gl/infraestructuras/computacion/svg`
18. Moles, C.G., Mendes, P., Banga, J.R.: Parameter estimation in biochemical pathways: a comparison of global optimization methods. Genome Research 13(11), 2467–2474 (2003)
19. Locke, J., Millar, A., Turner, M.: Modelling genetic networks with noisy and varied experimental data: the circadian clock in Arabidopsis thaliana. Journal of Theoretical Biology 234(3), 383–393 (2005)
20. Lipniacki, T., Paszek, P., Brasier, A.R., Luxon, B., Kimmel, M.: Mathematical model of nf-κb regulatory module. Journal of Theoretical Biology 228(2), 195–215 (2004)

A Method to Calibrate Metabolic Network Models with Experimental Datasets

Octavio Perez-Garcia[1], Silas Villas-Boas[2], and Naresh Singhal[1]

[1] Department of Civil and Environmental Engineering, University of Auckland, New Zealand
[2] Centre for Microbial Innovation, School of Biological Sciences, University of Auckland, New Zealand

Abstract. A method to calibrate stoichiometric coefficients values related to uncharacterized or lumped reactions of metabolic network models is presented. The method finds coefficients values that produce a model version that best fits multivariable experimental data. The method was tested with a metabolic network of 44 metabolites and 49 stoichiometric reactions, with four reactions having undetermined stoichiometric coefficients values. A total of 1320 model versions with different combinations of stoichiometric coefficient values were generated. Experimental data was used to produce a calibration curve and different fitness scores were used to evaluate the accuracy of flux balance analysis (FBA) simulations of these model versions to reproduce the experimental data. The model version with highest fitness to the experimental data was found using Mean Relative Error (MRE) scores and auto-scaled transformation of estimated datasets.

Keywords: Metabolic network model, biochemical reaction stoichiometry, flux balance analysis, model calibration.

1 Introduction

Stoichiometric metabolic network (SMN) modeling and flux balance analysis (FBA) are emerging techniques in systems biology that can be used to quantify the rate of reactions within the network formed by chemical compounds and sequenced chemical reactions in cells' metabolism (Oberhardt, et al. 2009; Orth, et al. 2010). The accuracy of these techniques in estimating observation within real cells is dependent on rigorous model calibration (or model curation) (Edwards, et al. 2001). Model calibration involves contrasting the model's estimated values with experimental data (obtained from measurements in real systems) in order to refine model structure until differences between the datasets are minimal (Feist, et al. 2009). When experimental data is limited or does not contain information appropriate for model calibration, it may limit the ability of SMN models in producing relevant and accurate estimates.

The lack of experimental data for model calibration is a common problem in SMN modeling. For example, the general procedure for SMN model calibration involves comparing estimated and experimental growth rates observed under different organic carbon sources (Durot, et al. 2009; Edwards, et al. 2001). Thus, de facto this approach may not be applicable to calibrating models of organism or microbial communities

J. Sáez-Rodríguez et al. (eds.), *8th International Conference on Practical Appl. of Comput. Biol. & Bioinform. (PACBB 2014)*, Advances in Intelligent Systems and Computing 294,
DOI: 10.1007/978-3-319-07581-5_22,

that don't growth using organic molecules (i.e. autotrophs). Another scenario arises with 13C-flux analyses, which provide metabolic reaction rate measurements that can be directly compared with FBA simulations; however, experimental data is limited by expensive nature of the technique and it ability to provide rate measurements only for the central carbon pathway (Sauer 2006). Moreover, the most valuable experimental data for model calibration and validation come from transcript-, proteo- and metabol–omic analyses (Kümmel, et al. 2006; Lewis, et al. 2010). This however requires pre-existing analytical and expertise capability to obtain relevant -omic data. Because of these limitations, it is desirable to implement computational methods to calibrate SMN models using common bioprocess performance data.

In this work we introduce a method to calibrate stoichiometric coefficients values related to uncharacterized or lumped reactions of SMN models based on minimizing model estimation errors to reproduce experimental datasets composed of process variables such as culture yield, compound uptake and production rates, and other variables describing the steady state metabolism of the cell culture under specific experimental conditions. This multivariate experimental dataset is used to measure accuracy of model simulations via various goodness to fit scores.

2 Methods

2.1 Metabolic Network

A metabolic network of the nitrogen respiration pathway of *Nitrosomonas europaea* was used to develop the calibration method. The metabolic network, constructed using (Thiele and Palsson 2010) protocol, consisted of 44 metabolites, 49 stoichiometric reactions and 3 compartments. The COBRA toolbox 2.0 (Schellenberger, et al. 2011) together with the GLPK solver running in Matlab®7 R2010b software (MathWorks Inc., Natick, MA, USA) was used to convert the metabolic network into its mathematical form and perform FBA simulations. Mathematically the network was represented as a stoichiometric matrix, $S\ (m \times n)$, of m metabolites and n reactions. A non-zero entry $s_{i,j}$ in S indicates the participation of metabolite i in reaction j. All reactions within the network were mass-balanced such that $S * v = 0$, where v was the vector of reaction rates (or fluxes) (Feist, et al. 2009; Varma and Palsson 1994). The reaction rate limits (or constraints) were defined in the form $\alpha_j \leq v_j \leq \beta_j$, where α_j and β_j are the lower and upper limits placed on the reaction rate v_j (Varma and Palsson 1994).

The metabolic network of *N. europaea* respiration pathways had to be calibrated (curated) as it contained four reactions involving compounds with undetermined stoichiometric coefficients. These four reactions (presented in Table 1) corresponded to electron transport chain reactions that translocate protons across the cellular membrane and the ATP stoichiometric coefficient in the biomass reaction. The precise stoichiometry of these reactions was ambiguous or not found in the literature.

2.2 Generation of Model Versions and Simulations

Candidate stoichiometric coefficients were defined in the above four reactions so that mass and energy balance was preserved, as presented in Table 1. Although Table 1 shows that fraction values were not assigned to candidate coefficients, if needed, these can be directly assigned without method modification.

Table 1. Reactions of the SMN model calibrated in this study

Rx. ID	Stoichiometric equation	Candidate coefficients "*s*"	Number of coefficients
A	nh3[p] + o2[p] + q8h2[c] + *s* h[c] → nh2oh[p] + h2o[c] + q8[c] + *s* h[p]	0, 1, 2	3
B	q8h2[c] + 2 cyt552[p] + *s*-2 h[c] → *s* h[p] + q8[c] + 2 cyt552e[p]	2, 4	2
C	atp[c] + 10 nadh[c] + 0.25 protein[c] + *s* m[c] ←→ adp[c] + 10 nad[c] + pi[c] + 0.26 h[c] + biomass[c]	From 0 to 50 each 5	20
D	*s* atp[c] + 10 nadh[c] + 0.25 protein[c] + m[c] ←→ *s* adp[c] + 10 nad[c] + *s* pi[c] + *s* h[c] + biomass[c]	From 0 to 100 each 10	11

By systematically combining the candidate stoichiometric coefficients for each reaction with those for the remaining three reactions we obtained a total of 1320 combinations (3 x 2 x 20 x 11 = 1320) that gave 1320 model versions. These model versions and their corresponding FBA simulations were automatically generated with the following Matlab® script:

```
%% Run a FBA simulation for all model versions generated by changing the
S matrix.
%Define column vectors "coeffsA", "coeffsB", "coeffsB" and "coeffsD" of
coefficient values of each reaction. All column vectors must be of the
same length.
coeffsA=[];
coeffsB=[];
coeffsC=[];
coeffsD=[];
%Assign the values of "coeffs" to the corresponding stoichiometric
coefficient with coordinates (i,j) in model.S matrix and run a FBA
simulation. Values of column vectors are assigned row by row until the
length of "coeff1".
for j=1:length(coeff1);
    model.S(43,50)=coeff1(j);
    model.S(42,50)=coeff2(j);
    model.S(26,50)=coeff2(j);
    model.S(11,50)=coeff2(j);
    model.S(11,33)=coeff4(j);
    model.S(20,33)=coeff5(j);
    model.S(11,41)=coeff6(j);
    model.S(20,41)=coeff7(j);
    solution=optimizeCbModel(model);     %FBA simulation
    M(:,j)=solution.x            %Generation of "M" matrix
end
```

The above script generated a M matrix of n number of reactions and d number of FBA solutions (d=1320). Note that FBA simulations can be substituted by other methods to estimate network fluxes, such as random sampling or extreme pathways.

2.3 Defining Experimental Datasets for Calibration

The compound concentration curves, biomass concentration, reactor volume, inflow rate of growth medium, and other biochemical information reported in experiments previously published (Vadivelu, et al. 2006; Whittaker, et al. 2000) on *N. europaea* growing in aerobic conditions without substrate limitation (oxygen and ammonium) were used to define an experimental dataset X_l of 28 mean values of l variables that describe the metabolism of this organisms under specified conditions. Table 2 defines the categories to which the 28 variables of the experimental dataset belong.

Table 2. Definition of variables used to produce the experimental dataset and number of dataset variable values found in previously published experiments

Variable category	Formula for variable estimation with model simulation results	Number of dataset variables
Growth rate	$= v_{biomass}$	1
Specific substrate uptake rate	$= v_{substrate}$	2
Specific compound production rate	$= v_{product}$	2
Molar yield ratio of product	$= \frac{v_{product}}{v_{substrate}}$	4
Net amount of compound used in reaction	$= v_j * s_i$	2
ATP molar yield ratio	$= \frac{\sum_{j=1}^{J}(v_{ATP\ consumption} * s_{ATP})_j}{\sum_{j=1}^{J}(v_{ATP\ synthezis})_j}$	2
Proton translocation yield ratio	$= \frac{\sum_{j=1}^{J}(v_{H+\ produced} * s_{H+})_j}{v_{substrate}}$	2
Pivot compound reaction yield ratio	$= \frac{v_{consumption\ of\ i\ in\ reaction\ j}}{v_{synthezis\ of\ i}}$	6
Percentage yield ratio of element	$= \frac{v_{production\ of\ i} * s_i * a}{v_{consumption\ of\ i} * s_i * a} * 100$	4
Element mass balance	$= \sum_{j=1}^{J}(v_j * s_i * a)_j$	3
TOTAL		28

Note: v_j is the rate of reaction j; s_i is the stoichiometric coefficient of compound i; J is the total number of reactions that consume or/and produce the compound i; and a is the element's number of atoms in compound i.

The 28 variables apply to the same specific steady state condition of *N. europaea* growth. In the case of experiments with batch cultures a steady state was assumed for time periods where the change in substrate and product concentrations maintained a linear trend, therefore indicating a constant rate of consumption and production of compounds. The rate variables were normalized by the total biomass in bioreactor (expressed as grams of dry weight (gDW)).

2.4 Evaluation of Goodness to Fit

By applying the generic formulas presented in Table 2, the 28 variables were estimated using the simulation results of the M matrix to produce a x_l dataset for each of the 1320 model versions. Experimental and estimated datasets (X_l and x_l respectively) were $\log_{10}$ or auto-scale transformed because dataset values had different order of magnitude and dimensions (e.g. $v_{O2\,uptake}$ = 2.5mmol/gDW*h while the $v_{biomass}/v_{O2\,uptake}$ yield = 0.012gDW/mmol-O_2). Data transformation was necessary for capturing the deviation between the observed and estimated values in absolute terms and to minimize the effect of varying scales for different variables (Schuetz, et al. 2007; van den Berg, et al. 2006). The goodness to fit was evaluated after data transformation.

The goodness to fit of a model describes the degree to which model predictions fit experimental data (Makinia 2010). The overall fitness between experimental and estimated datasets was evaluated using the fitness scores presented in Table 3. Model version with lower fitness scores was considered to have the highest accuracy in reproducing the experimental data, and therefore considered to yield a calibrated model.

Table 3. Some formulas to evaluate goodness to fit of a model. Modified from (Makinia 2010)

Fitness score	Formula
Mean relative error (MRE)	$MRE = \frac{1}{n}\sum_{l=1}^{n}\frac{\lvert(X_l - x_l)\rvert}{X_l}$
Mean absolute error (MAE)	$MAE = \frac{1}{n}\sum_{l=1}^{n}\lvert(X_l - x_l)\rvert$
Root mean squared error (RMSE)	$RMSE = \sqrt{\frac{1}{n}\sum_{l=1}^{n}(X_l - x_l)^2}$
Root mean squared scaled error (RMSSE)	$RMSSE = \sqrt{\frac{1}{n}\sum_{i=1}^{n}\frac{m_l(X_l - x_l)^2}{(std)_l^2}}$

Note: n is the total number of variables in datasets (observed and estimated); X_l is the observed data (measured in experiments) in variable l; x_l is the estimated data (estimated in simulations) in variable l; m_l is the number of data points contributing to X_l; std is the standard deviation of the observed data (measured in experiments) in variable l.

3 Results

Fig 1 presents the fitness scores obtained for the 1320 model versions. MRE and RMSSE scores similarly ranked the fitness of model versions and identified model version number "757" as showing the best fitness to experimental data. Model version "757" had the following stoichiometric coefficient values: $s_{i,A} = 1$, $s_{i,B} = 2$, $s_{i,C} = 5$, $s_{i,D} = 80$. On other hand, according to MRE and RMSSE scores, model version "24" had the worst fitness to the experimental data. MAE and RMSE scores gave different ranks for the model versions from each other as well as from MRE and RMSSE.

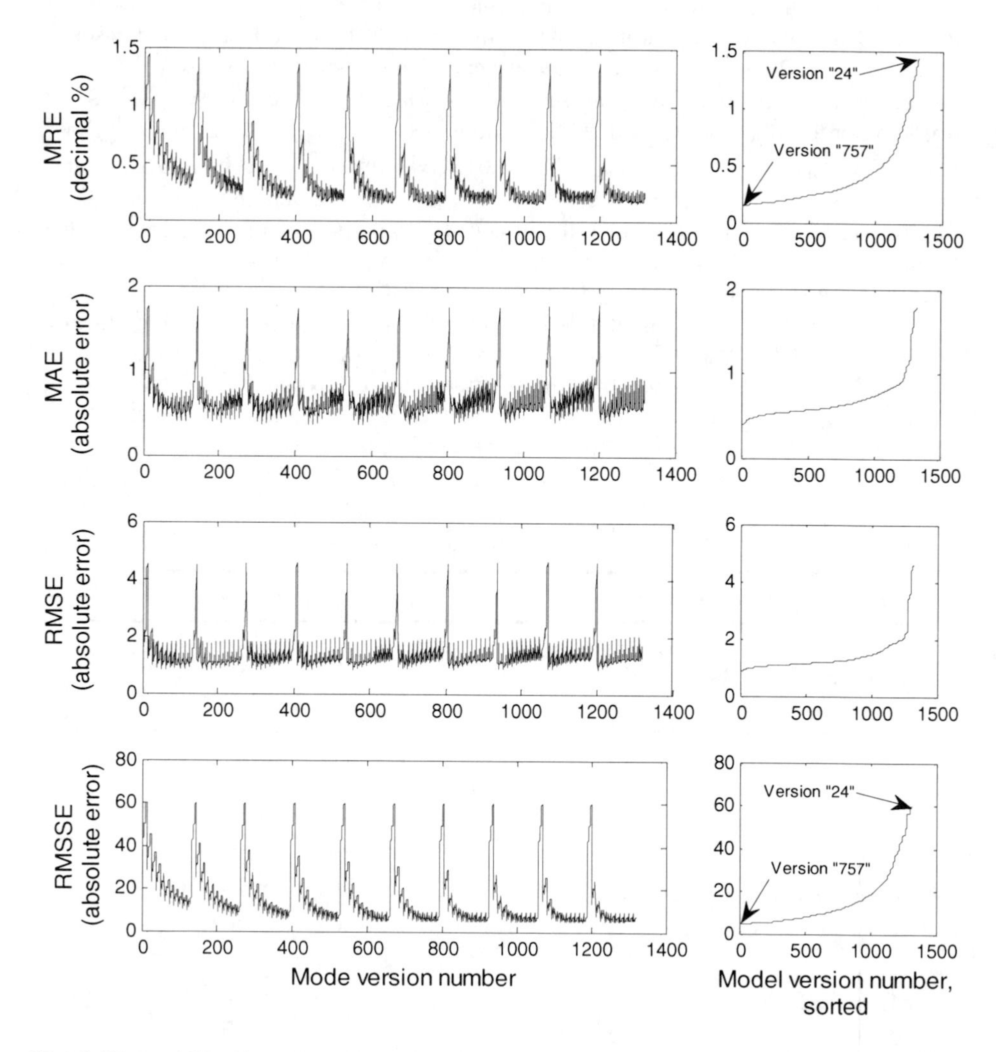

Fig. 1. Plots of fitness scores obtained for the 1320 model versions. Plots of the right column present the same scores but model versions were sorted from lower to higher fitness scores.

Fig 2 presents calibration curves generated by plotting experimental vs estimated datasets for model version "24" and "757". Fig. 2 also presents the effect of log transformed and auto-scaled datasets. Without transformation the deviation between estimated and experimental datasets cannot be visually evaluated during calibration because the large variation in the scale of different variables. The Log_{10} transformed and auto-scaled data gave less noisy calibration curves because all variables were re-scaled to the same units.

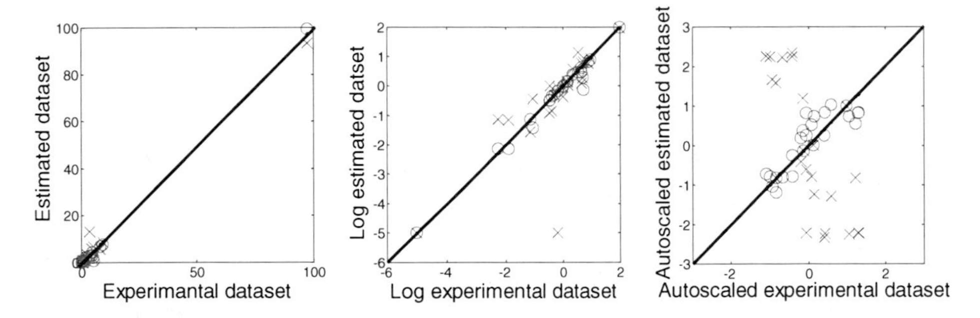

Fig. 2. Model calibration curves generated without treatment, log^{10} transformed treatment and auto-scaled treatment of datasets. Diagonal line represents a perfect fitness between experimental and estimated datasets. "X" markers represent the estimated dataset of model version "24". "O" markers represent the estimated dataset of model version "757".

The calibration curves in Fig. 2 show similar results to those obtained for fitness scores, especially with auto-scaled data. Model versions fitness can therefore be evaluated in both ways, with fitness scores or visually using calibration curves. MRE and RMSEE scores provided the most meaningful fitness scores because of two reasons: i) they corrected differences in the scales or units of variables and ii) they better reflect small deviations between experimental and estimated datasets. However MRE is a more meaning full score because error is measured on a scale from 0 to 1 (0 represents a perfect fit) which allows one to directly determine the percentage of accuracy with the formula: $accuracy\,(\%) = ((1 - MRE) * 100)$. Auto-scaled transformations provided the best way of visually representing the deviations of estimated datasets from experiments because auto-scaled datasets are more sensitive to numerical differences between the two datasets.

4 Conclusions

The method presented in this work is an easy to implement way of calibrating stoichiometric coefficients in SMN models. The essence of the method is to find those coefficients that produce a model version that best fits experimental data. In this sense, SMN model structure (coefficients of the S matrix) is defined by experimental data and this ensures realistic estimates for intracellular reaction rates. Extracting the maximum possible information on values of variables from experiments or literature is necessary to produce a robust experimental dataset with large number of variables l and thereby improve the fitness scores. This method can be applied to evaluate the fitness of multiple metabolic reactions constraints and can also be extended to fit models to transcriptomics, proteomics or metabolomics data by defining new variables (as in Table 3) using parsimonious enzyme usage FBA (pFBA) (Lewis, et al. 2010) and network embedded thermodynamic (NET) analysis (Kümmel, et al. 2006). However these applications are out of the scope of this research.

References

1. Durot, M., Bourguignon, P., Schachter, V.: Genome-scale models of bacterial metabolism: Reconstruction and applications. FEMS Microbiol. Rev. 33, 164–190 (2009)
2. Edwards, J.S., Ibarra, R.U., Palsson, B.: In silico predictions of Escherichia coli metabolic capabilities are consistent with experimental data. Nat. Biotechnol. 19, 125–130 (2001)
3. Feist, A.M., Herrgård, M.J., Thiele, I., Reed, J.L., Palsson, B.: Reconstruction of biochemical networks in microorganisms. Nature Reviews Microbiology 7, 129–143 (2009)
4. Kümmel, A., Panke, S., Heinemann, M.: Putative regulatory sites unraveled by network-embedded thermodynamic analysis of metabolome data. Molecular Systems Biology 2 (2006a)
5. Lewis, N.E., Hixson, K.K., Conrad, T.M., Lerman, J.A., Charusanti, P., Polpitiya, A.D., Adkins, J.N., Schramm, G., Purvine, S.O., Lopez-Ferrer, D., Weitz, K.K., Eils, R., König, R., Smith, R.D., Palsson, B.Ø.: Omic data from evolved E. coli are consistent with computed optimal growth from genome-scale models. Molecular Systems Biology 6 (2010)
6. Makinia, J.: Mathematical Modelling and Computer Simulation of Activated Sludge Systems. IWA Publishing, London (2010)
7. Oberhardt, M.A., Chavali, A.K., Papin, J.: Flux balance analysis: Interrogating genome-scale metabolic networks. Methods in Molecular Biology 500, 61–80 (2009)
8. Orth, J.D., Thiele, I., Palsson, B.: What is flux balance analysis? Nat. Biotechnol. 28, 245–248 (2010)
9. Sauer, U.: Metabolic networks in motion: 13C-based flux analysis. Molecular Systems Biology 2 (2006)
10. Schellenberger, J., Que, R., Fleming, R.M.T., Thiele, I., Orth, J.D., Feist, A.M., Zielinski, D.C., Bord-bar, A., Lewis, N.E., Rahmanian, S., Kang, J., Hyduke, D.R., Palsson, B.Ø.: Quantitative prediction of cellular metabolism with constraint-based models: The COBRA Toolbox v2.0. Nature Protocols 6, 1290–1307 (2011)
11. Schuetz, R., Kuepfer, L., Sauer, U.: Systematic evaluation of objective functions for predicting intracellular fluxes in Escherichia coli. Molecular Systems Biology 3 (2007)
12. Thiele, I., Palsson, B.Ø.: A protocol for generating a high-quality genome-scale metabolic reconstruction. Nature Protocols 5, 93–121 (2010)
13. Vadivelu, V.M., Keller, J., Yuan, Z.: Stoichiometric and kinetic characterisation of Nitrosomonas sp. in mixed culture by decoupling the growth and energy generation processes. J. Biotechnol. 126, 342–356 (2006)
14. van den Berg, R.A., Hoefsloot, H.C.J., Westerhuis, J.A., Smilde, A.K., van der Werf, M.: Centering, scaling, and transformations: Improving the biological information content of metabolomics data. BMC Genomics 7 (2006)
15. Varma, A., Palsson, B.: Metabolic flux balancing: Basic concepts, scientific and practical use. Bio/Technology 12, 994–998 (1994)
16. Whittaker, M., Bergmann, D., Arciero, D., Hooper, A.: Electron transfer during the oxidation of ammonia by the chemolithotrophic bacterium Nitrosomonas europaea. Biochimica et Biophysica Acta - Bioenergetics 1459, 346–355 (2000)

Metagenomic Analysis of the Saliva Microbiome with Merlin

Pedro Barbosa[1], Oscar Dias[1], Joel P. Arrais[2], and Miguel Rocha[1]

[1] CEB/IBB, School of Engineering, University of Minho, Portugal
mrocha@di.uminho.pt
[2] Dep. Informatics Engineering / CISUC, University of Coimbra, Portugal
jpa@dei.uc.pt

Abstract. In recent years, metagenomics has demonstrated to play an essential role on the study of the microorganisms that live in microbial communities, particularly those who inhabit the human body. Several bioinformatics tools and pipelines have been developed for the analysis of these data, but they usually only address one topic: to identify the taxonomic composition or to address the metabolic functional profile. This work aimed to implement a computational framework able to answer the two questions simultaneously. Merlin, a previously released software aiming at the reconstruction of genome-scale metabolic models for single organisms, was extended to deal with metagenomics data. It has an user-friendly and intuitive interface, being suitable for those with limited bioinformatics skills. The performance of the tool was evaluated with samples from the Human Microbiome Project, particularly from saliva. Overall, the results show the same patterns reported before: while the pathways needed for microbial life remain relatively stable, the community composition varies extensively among individuals.

Keywords: Metagenomics, Annotation, Human microbiome.

1 Introduction

For most of the history of life, microorganisms were the only inhabitants on Earth, and they still keep dominating the planet in many aspects. Microbial life has also an important role in human health, agriculture and ecosystem functioning. For example, the human microbiome harbors over 100 times more genes than our genome [1] and has been linked to several diseases, such as obesity and inflammatory bowel disease [2]. Such discoveries were possible with the appearance of culture-independent methods, such as the 16S ribosomal rRNA or whole-metagenome shotgun (WMS) sequencing approaches. While the former primarily focuses on identifying the organisms that compose an environmental sample and their proportions, WMS extends the potential of metagenomics by allowing gene annotation and downstream metabolic analysis of microbial communities, either from assembled contigs or unassembled reads.

Along with whole-community screenings, bioinformatics challenges have arisen and several tools have been released to analyse WMS metagenomic samples at

J. Sáez-Rodríguez et al. (eds.), *8th International Conference on Practical Appl. of Comput. Biol. & Bioinform. (PACBB 2014)*, Advances in Intelligent Systems and Computing 294,
DOI: 10.1007/978-3-319-07581-5_23,

the taxonomic and functional levels. Community profiling is usually done by using extrinsic information from genome databases, but *unsupervised* approaches also exist, featuring the binning of the sequences based on intrisic features (e.g. GC composition, k-mer distribution or codon usage). Examples of such tools are LikelyBin [3] and CompostBin [4]. Homology-based classification relies on database searches, where the major strategy for taxon assignment is the selection of the best hits. However, this type of classification needs to be interpreted carefully, since the evolutionary distance between the DNA fragments and the hit is unknown. CARMA [5], MetaPhlAn [6] or MEGAN [7] are some of the similarity based tools, showing complementary features to improve the classification.

Regarding functional annotation and metabolic reconstruction, there are also plenty of choices. First, the user can choose to perform the analysis directly from the reads or based on the assembly. While the first can be more sensitive, as a greater number of sequences is classified, the second approach is suitable to overcome the bias of higher abundance of longer genes. In both cases, the strategy is to use a search engine (e.g BLAST [8], RAPsearch2 [9]) to scan the reads or genes predicted from the contigs against protein sequence databases, such as NCBI nr [10], SwissProt [11] or KEGG Orthology [12], or protein domain databases such as NCBI Conserved Domain Database (CDD) [13]. Pathway reconstruction relies in finding the most likely set of pathways in the metagenome, usually through a gene-pathway-centric view where the biochemical functions of the community members are treated as a whole. KEGG and SEED [14] are the common resources for analysing these broader functional units. Given the described methodologies, some standalone tools (again MEGAN, HUMAnN [15]) and web services (MG-RAST [16], CAMERA [17]) have been developed.

Indeed, many efforts have been done towards a proper analysis of environmental samples, but there is still a lack of choices to perform an integrative analysis of microbial communities at taxonomic and functional levels, simultaneously. Moreover, if the user is not interested in running a web service, using some of the available standalone programs can be a hurdle since they are usually command-line based and require libraries dependencies to be run.

In this context, the main goal of this project focused on developing an user-friendly tool capable of performing a taxonomy description, as well as a robust metabolic reconstruction of a microbial community. The work was done by adapting a previously developed software, originally designed to construct genome-scale metabolic models for single organisms, Merlin. For evaluation purposes, saliva samples from the Human Microbiome Project (HMP) were used.

2 Methods and Implementation

Merlin is an open-source application implemented in $Java^{TM}$ and was built on top of the AIBench (`http://www.aibench.org`) software development framework [18]. It utilizes a relational MySQL database to locally store the data and uses different Java libraries, such as NCBI Entrez Utilities Web Service Java Application Programming Interface (API) and KEGG Representational State

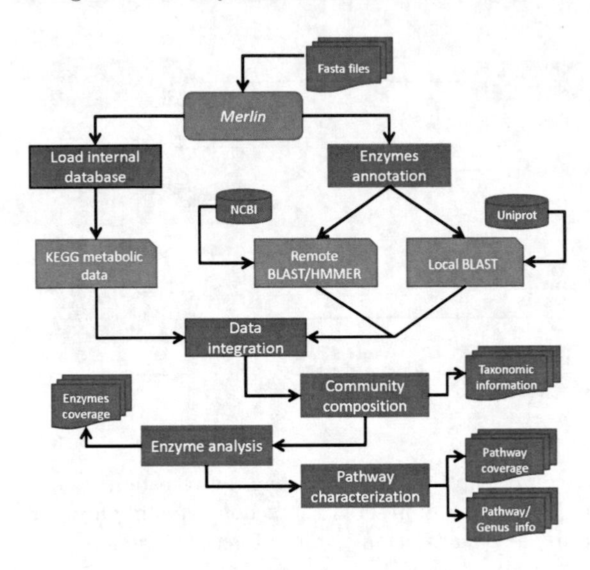

Fig. 1. Schematic representation of Merlin architecture for metagenomic analysis

Transfer (REST) API to access several web services. It requires a FASTA file of genes coding sequences as input in nucleotide or aminoacid format. The annotation of the samples is done via similarity searches using BLAST (either remote or local) and the metagenomics workflow is based on these results (Figure 1).

2.1 Taxonomic Analysis

The purpose of this operation is to assign a taxonomic label to each gene, as well as to describe the overall community composition. Thus, Merlin classifies each gene at the phylum and genus levels based on the list of homologues obtained from BLAST. Afterwards, given a classification for each gene, it calculates the proportions of each taxon in the whole set of genes.

The assignments are performed giving a weight to the number of times each phylum and genus are found within the homologues list. Merlin privileges the first five hits, since those are likely to be taxonomically more related to the target gene (Figure 2). In the end, a gene will be assigned with a taxonomic label only if it fulfills the following criteria:

- The number of homologues is higher than the minimum number required (default value is 5).
- The phylum score is higher than the defined threshold (default value is 0.5).
- The genus score is higher than the defined threshold (default value is 0.3).
- The phylum and genus are congruent.

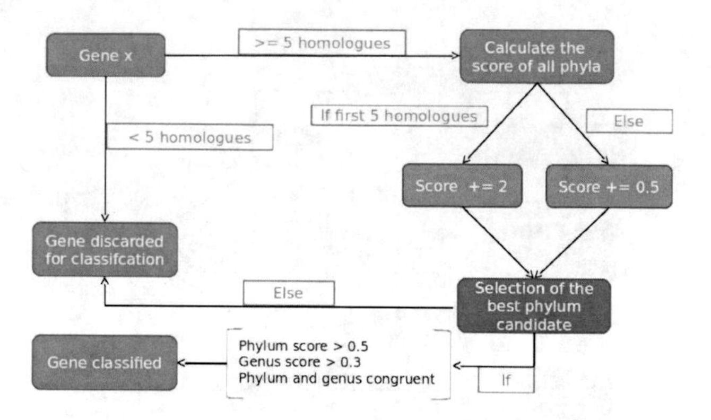

Fig. 2. Schematic representation of the taxonomic routine for gene classification employed in Merlin. The figure represents the schema for phylum classification but for genus the procedure is similar. Default values for the parameters can be changed.

2.2 Functional Analysis

The goal of this module is to identify the metabolic potential of the microbial community. Merlin uses a routine to identify enzymes from the BLAST results giving a weight to the number of times an Enzyme Commission (EC) number is found within the list of homologues for each gene [19]. For metabolic pathway inference, Merlin uses KEGG as the resource for representation and performs hypergeometric tests to find statistical significance in pathway identification. This method is commonly used in pathway analysis studies [20] and identifies enriched pathways compared with a background distribution. Merlin uses the identified enzymes from the whole set of existing enzymes in KEGG for each sample as background distribution. Then, for each pathway, it calculates the probability that the number of enzymes observed in the enzymes list that compose a pathway occurred by chance. If not, a pathway is assumed to be present.

3 Case Study

The assemblies from the saliva samples as part of the HMP were downloaded from `http://hmpdacc.org/HMASM/`. MetaGeneMark [21] was used to predict the putative genes for each sample. There were five samples from saliva available, but only three passed the quality control tests. For each sample, a remote BLAST against NCBI nr and a local one against SwissProt were performed. The results are displayed in Table 1. The differences between the two approaches are clear as the annotation against NCBI nr provides better results: a big fraction of genes present similarities. On the other hand, the local BLAST against SwissProt ran much faster, despite the large number of genes that remained unannotated ($\approx$ 60%), which was expected due to the database small size.

Table 1. Remote BLAST against NCBI nr vs Local BLAST against SwissProt for the HMP samples ran in Merlin. BLASTp set with e-value of 1^{-10}.

	NCBI nr		SwissProt	
Sample	Processed genes	With similarities	Processed genes	With similarities
SRS019210	49663	45023	49665	20231
SRS015055	46188	41073	46189	19176
SRS013942	41906	38508	41906	18677

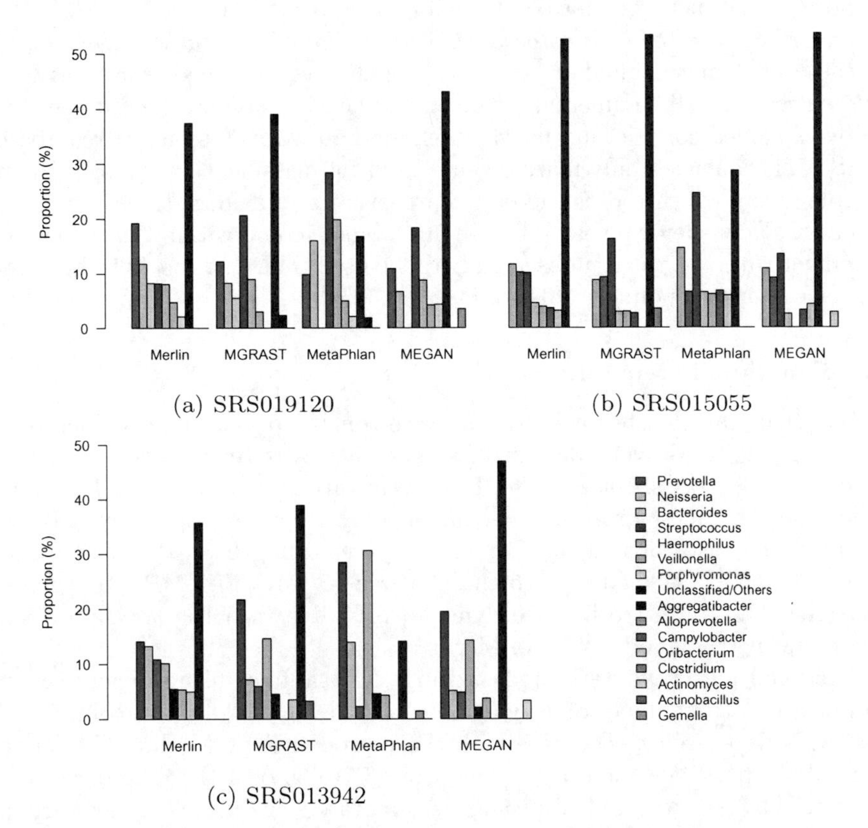

Fig. 3. Genus distribution of the most abundant taxa in each sample by different tools

3.1 Taxonomic Composition of Saliva Microbiome

Regarding the samples annotated using SwissProt, and given the low number of homologies found (Table 1), it becomes clear that this is not the best approach to taxonomically characterize metagenomes. Since the Merlin routine is highly dependent of the BLAST results, poor outputs on this step compromised the performance of the algorithm. Furthermore, using SwissProt as the reference database creates biased results because few organisms are well represented there, inducing the taxonomic assignments towards these organisms.

Using annotations against NCBI nr, Merlin was able to assign a taxonomic label in more than half of genes in each sample. The proportion could be increased if the default value for the minimum number of homologues required was changed, but it was decided to keep a conservative approach. Concerning the phylum analysis, three clearly stand out: *Bacteroidetes*, *Firmicutes* and *Proteobacteria*, despite none dominates the microbiome. The genus composition was also assessed for a comparison between samples and different tools (Figure 3). The high percentage of unclassified sequences in all cases is evident. Although the darkblue bars also represent organisms with residual abundance, results show the potential of metagenomics on unveiling new forms of life.

While the *Prevotella*, *Streptococcus*, *Veillonella*, *Neisseria* and *Haemophilus* are the overall most abundant genera in all samples, no consistency was found between the tools. The different proportions of each taxon on the different tools can be explained considering the way each method works (assembly/read based, BLAST all sequences/only marker genes used for classification). Thus, it is not possible to say which tool is the best. Furthermore, previous studies of the oral flora at the genus level reveal a diverse microbiome composition [22], which is in agreement with the pattern observed here. Overall, Merlin appears to be a good alternative for taxonomic studies of metagenomes.

3.2 Functional Capabilities

The enzymes encoded by each method were compared to those obtained in the IMG/M-HMP web server [23]. Table 2 shows a large discrepancy between assignments based on SwissProt and NCBI nr as the latter presents a smaller number of identified enzymes. Propagated errors on enzymes annotation in NCBI might be the main reason for these. The numbers regarding SwissProt annotations seem to agree in cardinality with those stored in IMG/M-HMP. Further tests confirmed that the majority of enzymes overlap between the two approaches, demonstrating the good results of Merlin.

Functional pathways were predicted in saliva samples using hypergeometric tests based on the number of enzymes encoded in each. The results obtained by HUMAnN were used to compare with those produced by Merlin. The number of metabolic pathways identifed ranged from 37 to 56 over the different samples and methods. As expected, the samples annotated against NCBI-nr harbored less pathways, since the number of encoded enzymes was smaller too (Table 2).

Table 2. Comparison of the complete EC numbers annotated by IMG/M and Merlin in each sample

	IMG/M-HMP		Merlin SwissProt		Merlin NCBI nr	
Sample	Encoded	Unique	Encoded	Unique	Encoded	Unique
SRS019210	12143 (24.34%)	957	10058 (20.25%)	977	4871 (9.81%)	605
SRS015055	11988 (25.81%)	997	9776 (21.17%)	977	2287 (4.95%)	506
SRS013942	10642 (25.46%)	954	8922 (21.29%)	957	2739 (6.54%)	507

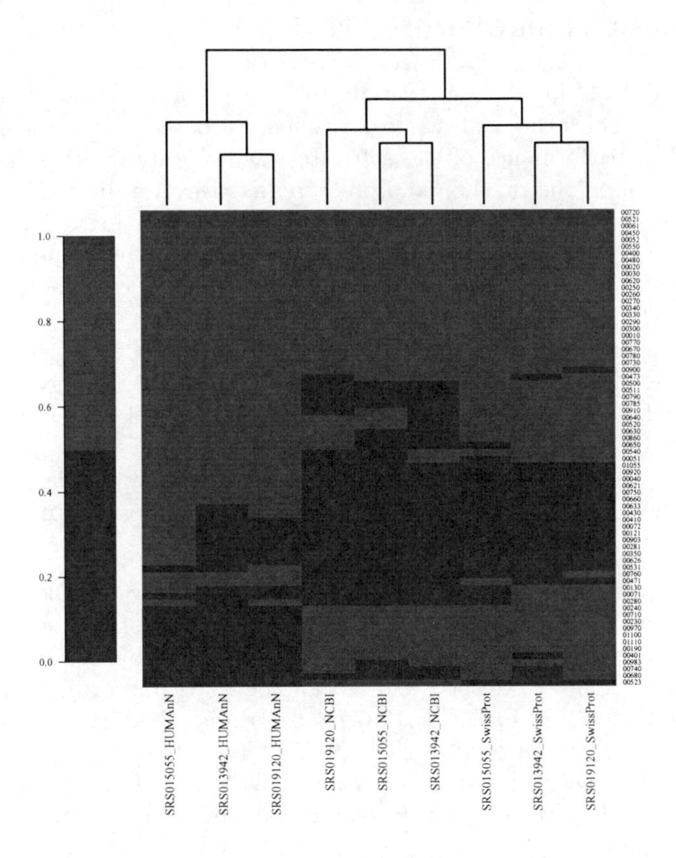

Fig. 4. Presence of metabolic pathways in the samples from saliva across different annotations. Columns represent the samples. Rows represent the binary value for pathway coverage. Green colors stands for present pathways whilst red cells account for the absent ones. Pathways with no enzymes in their constituents were filtered out from HUMAnN results for an easier comparison. Furthermore, as HUMAnN calculates coverage as a likelihood that ranges from 0 to 1, values higher than 0.5 were treated as present (binary value 1) and those with values lower than 0.5 were handled as absent(binary value 0).

To inspect if the inferred pathways were concordant across different methods, an heatmap was constructed (Figure 4).This data representation enabled to cluster the samples according to their similarity in terms of coverage. It is clear that the method used for pathways assignment influences more the clustering of the samples than the sample itself. Furthermore, only 70 different pathways out of 154 possible (with complete EC numbers) were found over all methods and samples, which also confirms the stable functional capability of the human microbiome described before [15] [24].

4 Conclusions and Future Work

An extension of Merlin, an user friendly tool for metabolic reconstruction, was presented. It enables the analysis of metagenomes based on an assembly-based approach. The performance of the software was evaluated with saliva samples from the HMP and the taxonomic profile predicted in Merlin was in agreement with other tools, despite some differences in the proportions. The functional characterization showed a conserved pool of pathways through different samples, although Merlin sometimes presented less pathways than expected because the routine is highly dependent on the enzymes annotation.

There are also some aspects that should be improved in the future. The most relevant one is to implement annotations against KEGG Orthology, or any other catalog of orthologs. This feature would increase the speed of the process maintaining high sensitivity for the taxonomic analysis.

Merlin is freely available from `http://www.merlin-sysbio.org` where a tutorial with more detailed information about the methods is also provided.

Acknowledgments. The work is partially funded by ERDF - European Regional Development Fund through the COMPETE Programme (operational programme for competitiveness) and by National Funds through the FCT (Portuguese Foundation for Science and Technology) within projects ref. COMPETE FCOMP-01-0124-FEDER-015079 and Strategic Project PEst-OE/EQB/LA0023/2013, and also by Project 23060, PEM - Technological Support Platform for Metabolic Engineering, co-funded by FEDER through Portuguese QREN under the scope of the Technological Research and Development Incentive system, North Operational.

References

1. Qin, J., Li, R., Raes, J., et al.: A human gut microbial gene catalogue established by metagenomic sequencing. Nature 464(7285), 59–65 (2010)
2. Greenblum, S., Turnbaugh, P.J., Borenstein, E.: Metagenomic systems biology of the human gut microbiome reveals topological shifts associated with obesity and inflammatory bowel disease. Proceedings of the National Academy of Sciences of the United States of America 109(2), 594–599 (2012)
3. Kislyuk, A., Bhatnagar, S., Dushoff, J., et al.: Unsupervised statistical clustering of environmental shotgun sequences. BMC Bioinformatics 10(1), 316 (2009)
4. Chatterji, S., Yamazaki, I., Bai, Z., Eisen, J.A.: CompostBin: A DNA composition-based algorithm for binning environmental shotgun reads. In: Vingron, M., Wong, L. (eds.) RECOMB 2008. LNCS (LNBI), vol. 4955, pp. 17–28. Springer, Heidelberg (2008)
5. Gerlach, W., Stoye, J.: Taxonomic classification of metagenomic shotgun sequences with CARMA3. Nucleic Acids Research 39(14), e91 (2011)
6. Segata, N., Waldron, L., Ballarini, A., et al.: Metagenomic microbial community profiling using unique clade-specific marker genes. Nature Methods 9(8), 811–814 (2012)

7. Huson, D.H., Mitra, S., Ruscheweyh, H.J., et al.: Integrative analysis of environmental sequences using MEGAN4. Genome Research 21(9), 1552–1560 (2011)
8. Altschul, S., Gish, W., et al.: Basic Local Alignment Search Tool. J. Mol. Biol. 215(3), 403–410 (1990)
9. Zhao, Y., Tang, H., Ye, Y.: RAPSearch2: a fast and memory-efficient protein similarity search tool for next-generation sequencing data. Bioinformatics 28(1), 125–126 (2012)
10. Pruitt, K.D., Tatusova, T., Brown, G.R., et al.: NCBI Reference Sequences (RefSeq): current status, new features and genome annotation policy. Nucleic Acids Research 40(Database issue), D130–D135 (2012)
11. Consortium, U.: Update on activities at the Universal Protein Resource (UniProt) in 2013. Nucleic Acids Research 41(Database issue), D43–D47 (2013)
12. Kanehisa, M., Goto, S., Furumichi, M., et al.: KEGG for representation and analysis of molecular networks involving diseases and drugs. Nucleic Acids Research 38(Database issue), D355–D360 (2010)
13. Marchler-Bauer, A., Lu, S., Anderson, J.B., et al.: CDD: a Conserved Domain Database for the functional annotation of proteins. Nucleic Acids Research 39(Database issue), D225–D229 (2011)
14. Overbeek, R., Begley, T., Butler, R.M., et al.: The subsystems approach to genome annotation and its use in the project to annotate 1000 genomes. Nucleic Acids Research 33(17), 5691–5702 (2005)
15. Abubucker, S., Segata, N., Goll, J., et al.: Metabolic reconstruction for metagenomic data and its application to the human microbiome. PLoS Computational Biology 8(6), e1002358 (2012)
16. Meyer, F., Paarmann, D., D'Souza, M., et al.: The metagenomics RAST server - a public resource for the automatic phylogenetic and functional analysis of metagenomes. BMC Bioinformatics 9(1), 386 (2008)
17. Sun, S., Chen, J., Li, W., et al.: Community cyberinfrastructure for Advanced Microbial Ecology Research and Analysis: the CAMERA resource. Nucleic Acids Research 39(Database issue), D546–D551 (2011)
18. Glez-Peña, D., Reboiro-Jato, M., Maia, P., et al.: AIBench: a rapid application development framework for translational research in biomedicine. Computer Methods and Programs in Biomedicine 98(2), 191–203 (2010)
19. Dias, O., Rocha, M., Eugenio, F., et al.: Merlin: Metabolic Models Reconstruction using Genome-Scale Information. Computer Applications in Biotechnology 11(1), 120–125 (2010)
20. Evangelou, M., Rendon, A., Ouwehand, W.H., et al.: Comparison of methods for competitive tests of pathway analysis. PloS One 7(7), e41018 (2012)
21. Zhu, W., Lomsadze, A., Borodovsky, M.: Ab initio gene identification in metagenomic sequences. Nucleic Acids Research 38(12), e132 (2010)
22. Keijser, B., Zaura, E., Huse, S., et al.: Pyrosequencing analysis of the Oral Microflora of healthy adults. Journal of Dental Research 87(11), 1016–1020 (2008)
23. Markowitz, V.M., Chen, I.M.A., Chu, K., et al.: IMG / M-HMP: A Metagenome Comparative Analysis System for the Human Microbiome Project. PLoS One 7(7), 1–7 (2012)
24. The Human Microbiome Project Consortium: Structure, function and diversity of the healthy human microbiome. Nature 486(7402), 207–14 (June 2012)

Networking the Way towards Antimicrobial Combination Therapies

Paula Jorge[1], Maria Olívia Pereira[1], and Anália Lourenço[1,2]

[1] CEB - Centre of Biological Engineering, University of Minho,
Campus de Gualtar, 4710-057 Braga, Portugal
paulajorge@ceb.uminho.pt, mopereira@deb.uminho.pt
[2] ESEI - Escuela Superior de Ingeniería Informática, Edificio Politécnico, Campus Universitario As Lagoas s/n, Universidad de Vigo, 32004 Ourense, Spain
analia@ceb.uminho.pt, analia@uvigo.es

Abstract. The exploration of new antimicrobial combinations is a pressing concern for Clinical Microbiology due to the growing number of resistant strains emerging in healthcare settings and in the general community. Researchers are screening agents with alternative modes of action and interest is rising for the potential of antimicrobial peptides (AMPs). This work presents the first ever network reconstruction of AMP combinations reported in the literature fighting *Pseudomonas aeruginosa* infections. The network, containing 193 combinations of AMPs with 39 AMPs and 154 traditional antibiotics, is expected to help in the design of new studies, notably by unveiling different mechanisms of action and helping in the prediction of new combinations and synergisms. The challenges faced in the attempted text-mining approaches and other considerations regarding the manual curation of the data are pointed out, reflecting about the future automation of this type of reconstruction as means to widen the scope of analysis.

Keywords: Antimicrobial peptides, drug synergism, interaction network, *Pseudomonas aeruginosa* infections.

1 Introduction

Clinical Microbiology is currently facing major challenges regarding the discovery and/or design of new antimicrobial agents and the development of novel antimicrobial strategies. Drug and even multi-drug resistant (MDR) strains are emerging with increasing frequency, and rendering ineffective many conventional antibiotic treatments. Therefore, research is focused on finding alternatives to keep new resistance from developing and to prevent the resistance that already exists from spreading, either by discovering biomolecules with antimicrobial potential and different mode of action, or by combining agents and potentiating their efficacy. Notably, there is a growing interest in the use of antimicrobial combinations as a strategy to increase the antimicrobial spectrum, prevent the emergence of resistance, reduce toxicity and side

J. Sáez-Rodríguez et al. (eds.), *8th International Conference on Practical Appl. of Comput. Biol. & Bioinform. (PACBB 2014)*, Advances in Intelligent Systems and Computing 294,
DOI: 10.1007/978-3-319-07581-5_24,

effects and provide synergistic activity. In fact, synergy testing has been encouraged to guide clinical treatments for MDR strains, namely *Pseudomonas aeruginosa* associated pulmonary exacerbation [1].

Although combinations may be accomplished by using traditional antimicrobials, the most promising strategy at the moment is the use of novel antimicrobials with new mechanisms of actions, either combined with each other or with traditional compounds. Notably, antimicrobial peptides (AMPs) are short-length peptides (between 15 and 30 amino acids) that exert activity against a broad spectrum of microorganisms, such as Gram-negative and Gram-positive bacteria (including drug-resistant strains), and are effective both in planktonic and biofilm scenarios [2]. These peptides have been recognized as promising candidates to replace classical antibiotics due to their multiple mechanisms of action and low specificity in terms of molecular targets, which reduces the chance of acquired resistance [3]. Besides, AMPs can influence processes which support antimicrobial action, like cytokine release, chemotaxis, antigen presentation, angiogenesis and wound healing [4].

Recent advances in large-scale experimental technologies have resulted in an accumulation of data that reflect the interplay between biomolecules on a global scale. Bioinformatics approaches, such as network reconstruction, can help in profiling and interpreting the activity of AMPs and thus, in exploiting their potential as antimicrobial drugs. Networks can be used to map the interaction data outputted by combination studies, can be explored to unveil new interactions at the global scale, and also to classify new drugs by their mechanism of action [5]. In particular, the investigation of antimicrobial combinations has been supported by network models that have demonstrated that the partial inhibition of few cell targets can be more efficient than the complete inhibition of a single target [6].

Pharmacological networks can be constructed and integrated from heterogeneous and complementary sources of chemical, biomolecular and clinical information, but most of the information related to drug combinations is scattered over scientific literature. Manual curation is effort and time consuming, and virtually unfeasible if a systematic and up-to-date screening is desired. In this regard, some works have introduced text mining approaches to mine drug information. Most of these works focus on drug-drug interactions (DDIs), which are related to adverse events of combinations of available drugs, usually targeting the same gene or pathway [7].

The aim of this work was to reconstruct the first ever AMP interaction network. Currently, several databases collect AMP related data, namely sources, targets and minimum inhibitory concentrations [4]. However, it is difficult to find information about AMP interactions in these databases. Scientific literature is the primary source of data and thus, the use of text mining tools was investigated to alleviate manual curation. The application of AMP-based therapies to the treatment of infections caused by the bacterium *P. aeruginosa*, one of the most studied pathogenic microorganisms, was chosen as proof-of-concept to the development of a more systematic reconstruction framework.

2 Materials and Methods

PubMed was searched for papers on synergistic interactions including AMPs. Specifically, we required the presence of any variant of the term synergism, using synergis* (where the * is a wildcard), following the query "antimicrobial peptide *Pseudomonas aeruginosa*". Then, we manually curated the interactions described in the papers yielded by the search, with at least on AMP as one of the combined antimicrobials. These interactions were represented in a network where nodes identify antimicrobial agents and edges encode the interactions among agents. Edge labels encode information on experimental evidence, such as *P. aeruginosa* strain(s), mode of growth (planktonic, biofilm and in vivo), method of combination analysis (determination of the fractional inhibitory concentration (FIC), time-kill assay, among others), and the PMID and URL of the publication.

Besides producing a high-quality reconstruction of known AMP interactions, this manual curation provided a "gold standard" for text mining. Notably, authors were able to identify the main elements of information to be collected as well as challenges in the interpretation of texts. The outputs of two public text mining tools were then evaluated – Chilibot [8] and PubTator [9].

3 Results and Discussion

3.1 AMP Literature Curation

From a total of 203 papers resulting from our PubMed search, 132 papers were manually curated. Some papers were excluded (37 %) on the basis of their relevance to the topic. Notably, these papers do not mention AMPs, *P. aeruginosa* as target and/or cover for antitumor and food preservation areas. Review works were also excluded (9.1 %), since most would represent repeated data from the other curated papers whilst they do not provide necessary details on the experiments. Then, some interesting evidences arose from the analysis of the information retrieved from the literature (Table 1).

Regarding the mode of bacterial growth, 91 % of the studies focused on the use of combinations on planktonic cells, and, surprisingly, only 3 % of the studies covered the biofilm mode of growth. Biofilms are recognized as one of the main causes of several infections in humans [10], being highly related to nosocomial and chronic infections, and more resistant to treatment [2]. As such, more studies should be devoted to AMP combinations towards biofilm treatment.

Most of the combinations (80 %) involved one AMP combined with another compound, mainly traditional antibiotics. This is linked to the rational that AMPs can enhance the activity of antibiotics, which act upon intracellular targets, by disrupting the cell membrane and facilitating the access within the cell. In terms of the bacterial strains most used, these studies focused on the reference strain *P. aeruginosa* ATCC 27853, *P. aeruginosa* PAO1, and clinical isolates. Finally, the determination of the fractional inhibitory concentration (FIC) and the time-kill assays stand out as the most common methods for synergy assessment, which is in concordance with recent reviews of the field [1].

Table 1. Statistics on the manual curation process. Legend: * - not in the network; ** - info retrieved from abstract; could be on the network.

		No. of papers (%)	*No. of combinations (%)*
Total curated		132 (100)	-
Off topic*		49 (37)	-
Reviews*		12 (9.1)	-
Manuscript not available/different language**		16 (12)	-
Total in the network		**56 (100)**	**193 (100)**
Mode of growth	Biofilm	2 (3.6)	3 (1.6)
	Planktonic	51 (91)	187 (97)
	In vivo	7 (13)	8 (4.1)
	Unknown	2 (3.6)	3 (1.6)
Combination with	AMPs	18 (32)	39 (20)
	Antibiotics and others	48 (86)	154 (80)

Regarding the text mining approaches, the success of the tools tested in automatically curate AMP knowledge was poor. Chilibot limits the search to a maximum of 50 terms at a time. So, it is not possible to execute a systematic screening of interactions among all known AMPs (using AMP database records as input, for example). In turn, PubTator does not recognize most AMPs as chemical entities. Moreover, authors became aware that full-text curation should be a requirement for this line of reconstruction. Most abstracts do not cover all combinations tested, focusing only on the best outcomes, and a great part do not give information about strains and methodology.

3.2 Network Topology

The constructed network (Figure 1) is represented as a non-oriented graph, containing 121 nodes, representing AMPs and other antimicrobial compounds, and 193 edges, each correspondent to a combination. The network is non-homogeneous, with an interior containing highly connected drugs and an exterior comprised of some drugs with low interactions. Each node is linked to an average of 3 nodes, which means that each AMP was combined with an average of 3 antimicrobials.

The network is dominated by a small number of highly connected nodes. The colors on the nodes on Figure 1 correspond to their degree of connectivity, i.e. the number of nodes to which they are directly connected, ranging from the highly connected nodes (red) to nodes with only one connection (green). The most highly connected node (degree = 49), hence the most used compound in studies concerning AMP combinations, is colistin. Colistin (polymyxin E) is an AMP currently used as a last resource treatment for *P. aeruginosa* infections in the respiratory tract of cystic fibrosis patients, as well as other MDR Gram negative bacterial infections [11]. The use of colistin combined with other antimicrobial could lower its dosage and, thus, the associated toxicity. Colistin is followed in connectivity by polymyxin B (degree = 19), which belongs to the same family of AMPs, which indicates a preference for polymyxins in combination studies towards *P. aeruginosa*.

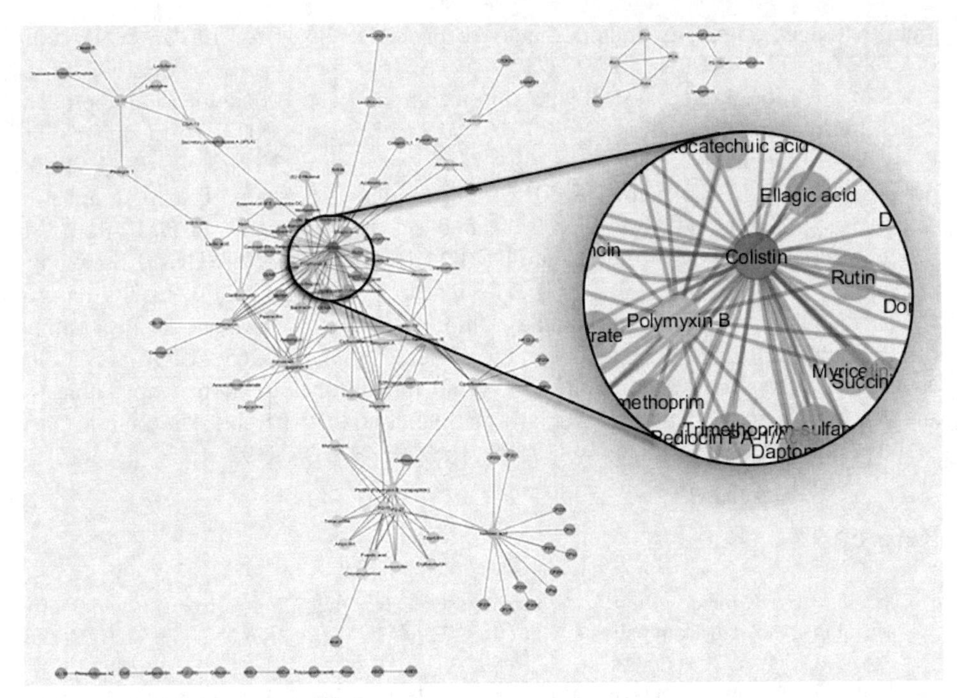

Fig. 1. General view of the AMP interaction network for *P. aeruginosa*, highlighting the two most connected AMPs – colistin and polymyxin B. Colors range from red (higher degree of connectivity) to green (lower degree of connectivity).

4 Conclusions

The increasing number of publications on the antimicrobial potential of drug combinations offers a wealth of information that can support interpretation of experimentally derived data and greatly enhance hypothesis generation. Drug interaction and functional networks are not simply new renditions of existing data: they provide the opportunity to gain insights into the impact of antimicrobial strategies in pathogenic systems.

We presented in this work the first reconstruction of an AMP combination network, specifically for the bacterial pathogen *P. aeruginosa*. Text-mining approaches that are currently available online came short in helping in the construction of the network. By recurring to the manual curation of PubMed articles, the authors were able to identify far more information that the one provided in the abstracts, which brings attention to the lack of systematization in data presentation. The resulting network currently contains 193 annotated combinations, with AMPs combined with 39 AMPs and 154 traditional antibiotics. The network is dominated by few highly connected AMPs, such as colistin and polymyxin B, characterizing them as current favorites in these studies. Future work will soon lead to a more complete network, encompassing also studies regarding antagonism. The annotation of this information is of interest, since antagonism is becoming a hot topic due to its possible role on the evolution and spread of drug resistance [5].

Acknowledgements. The authors thank the project PTDC/SAU-ESA/646091/2006/ FCOMP-01-0124-FEDER-007480FCT, the Strategic Project PEst-OE/EQB/ LA0023/2013, the Project "BioHealth - Biotechnology and Bioengineering approaches to improve health quality", NORTE-07-0124-FEDER-000027, co-funded by the Programa Operacional Regional do Norte (ON.2 – O Novo Norte), QREN, FEDER, the project "RECI/BBB-EBI/0179/2012 - Consolidating Research Expertise and Resources on Cellular and Molecular Biotechnology at CEB/IBB", FCOMP-01-0124-FEDER-027462, and the Agrupamento INBIOMED from DXPCTSUG-FEDER unha maneira de facer Europa (2012/273). The research leading to these results has received funding from the European Union's Seventh Framework Programme FP7/REGPOT-2012-2013.1 under grant agreement n° 316265, BIOCAPS. This document reflects only the author's views and the European Union is not liable for any use that may be made of the information contained herein. The authors also acknowledge the PhD Grant of Paula Jorge, Ref. SFRH/BD/88192/2012.

References

1. Saiman, L.: Clinical utility of synergy testing for multidrug-resistant Pseudomonas aeruginosa isolated from patients with cystic fibrosis: "the motion for.". Paediatr. Respir. Rev. 8, 249–255 (2007)
2. Jorge, P., Lourenço, A., Pereira, M.O.: New trends in peptide-based anti-biofilm strategies: a review of recent achievements and bioinformatic approaches. Biofouling 28, 1033–1061 (2012)
3. Wimley, W.C., Hristova, K.: Antimicrobial peptides: successes, challenges and unanswered questions. J. Membr. Biol. 239, 27–34 (2011)
4. Lai, Y., Gallo, R.L.: AMPed up immunity: how antimicrobial peptides have multiple roles in immune defense. Trends Immunol. 30, 131–141 (2009)
5. Yeh, P.J., Hegreness, M.J., Aiden, A.P., Kishony, R.: Drug interactions and the evolution of antibiotic resistance. Nat. Rev. Microbiol. 7, 460–466 (2009)
6. Csermely, P., Agoston, V., Pongor, S.: The efficiency of multi-target drugs: the network approach might help drug design. Trends Pharmacol. Sci. 26, 178–182 (2005)
7. Percha, B., Altman, R.B.: Informatics confronts drug-drug interactions. Trends Pharmacol. Sci. 34, 178–184 (2013)
8. Chen, H., Sharp, B.M.: Content-rich biological network constructed by mining PubMed abstracts. BMC Bioinformatics 5, 147 (2004)
9. Wei, C.-H., Kao, H.-Y., Lu, Z.: PubTator: a web-based text mining tool for assisting biocuration. Nucleic Acids Res 41, W518–W522 (2013)
10. Fey, P.D.: Modality of bacterial growth presents unique targets: how do we treat biofilm-mediated infections? Curr. Opin. Microbiol. 13, 610–615 (2010)
11. Li, J., Nation, R.L., Turnidge, J.D., Milne, R.W., Coulthard, K., Rayner, C.R., Paterson, D.L.: Colistin: the re-emerging antibiotic for multidrug-resistant Gram-negative bacterial infections. Lancet Infect Dis. 6, 589–601 (2006)

A Logic Computational Framework to Query Dynamics on Complex Biological Pathways*

Gustavo Santos-García[1], Javier De Las Rivas[2], and Carolyn Talcott[3]

[1] Computing Center, Universidad de Salamanca, Salamanca, Spain
santos@usal.es

[2] Bioinformatics and Functional Genomics Research Group,
Cancer Research Center (CiC-IBMCC, CSIC/USAL), Salamanca, Spain
jrivas@usal.es

[3] Computer Science Laboratory, SRI International, 333 Ravenswood Ave,
Menlo Park, CA 94025, USA
clt@csl.sri.com

Abstract. Biological pathways define complex interaction networks where multiple molecular elements work in a series of reactions to produce a response to different biomolecular signals. These biological systems are dynamic and we need mathematical methods that can analyze symbolic elements and complex interactions between them to produce adequate readouts of such systems. Rewriting logic procedures are adequate tools to handle dynamic systems which are applied to the study of specific biological pathways behaviour. Pathway Logic is a rewriting logic development applied to symbolic systems biology. Rewriting logic language Maude allows us to define transition rules and to set up queries about the flow in the biological system. In this paper we describe the use of Pathway Logic to model and analyze the dynamics in a well-known signaling transduction pathway: epidermal growth factor (EGF) pathway. We also use Pathway Logic Assistant (PLA) tool to browse and query this system.

Keywords: signal transduction, symbolic systems biology, epidermal growth factor signaling, Pathway Logic, rewriting logic, Maude, Petri net, executable model.

1 Introduction: Rewriting Logic

Rewriting logic [1,2] is a logic of concurrent change that can naturally deal with states and with highly nondeterministic concurrent computations. It has good properties as a flexible and general semantic framework in order to give semantics to a wide range of languages and models of concurrency. Moreover, it

* Pathway Logic development has been funded in part by NIH BISTI R21/R33 grant (GM068146-01), NIH/NCI P50 grant (CA112970-01), and NSF grant IIS-0513857. This work was partially supported by NSF grant IIS-0513857. Research was supported by Spanish projects Strongsoft TIN2012-39391-C04-04 and PI12/00624 (MINECO, Instituto de Salud Carlos III).

J. Sáez-Rodríguez et al. (eds.), *8th International Conference on Practical Appl. of Comput. Biol. & Bioinform. (PACBB 2014)*, Advances in Intelligent Systems and Computing 294,
DOI: 10.1007/978-3-319-07581-5_25,

allows user-definable syntax with complete freedom to choose the operators and structural properties appropriate for each problem.

The naturalness of rewriting logic for modeling and experimenting with mathematical and biological problems has been illustrated in a number of works [3]. The basic idea is that we can model a cell as a concurrent system whose concurrent transitions are precisely its biochemical reactions. In this way we can develop symbolic models of biological systems which can be analyzed like any other rewrite theory, as, for example, the use of search and model checking.

2 Symbolic Models for Biological Signaling Pathways

The growth of genomic sequence information combined with technological advances in the analysis of global gene expression has revolutionized research in biology and biomedicine [4]. Investigation of mammalian signaling processes, the molecular pathways by which cells detect, convert, and internally transmit information from their environment to intracellular targets such as the genome, would greatly benefit from the availability of predictive models.

Various models for the computational analysis of cellular signaling networks have been proposed for approaches that incorporate rate and/or concentration information to simulate responses to specific stimuli [5,6]. Alternatively, networks consisting of signaling pathway targets (genes) are modeled using differential equations to represent changes in the concentrations of both the targets and their pathway components [7,8]. The objectives here include predictions of transcription signal sequencing, gene expression and understanding of regulation mechanisms. Symbolic models are based on formalisms that provide a language to represent the states of the system, mechanisms to model their changes, such as reactions, and tools for analysis based on computational or logical inference.

A variety of formalisms have been used to develop symbolic models of biological systems [9], including Petri nets [10,11]; ambient/membrane calculi [12]; statecharts [13]; live sequence charts; and rule-based systems [14,15]. Each of these formalisms was initially developed to model and analyze computer systems with multiple processes executing concurrently.

Simulations using in silico models founded on kinetic measurements of signaling pathways or networks are important in order to achieve a detailed understanding of the biochemistry of signal transduction [16]. However, the development of such models is impeded by the great difficulty in obtaining experimental data. Our approach focuses on developing abstract qualitative models of metabolic and signaling processes that can be used as the basis of further analyses by powerful tools, such as those developed in the formal methods community.

Rewriting Logic and Maude. Rewriting logic was first proposed by Meseguer in 1990 as a unifying framework for concurrency [1]. Since then a large number of researchers have contributed to the development of several aspects of the logic and its applications in different areas of computer science [2].

Rewriting logic is a logic of change in which the distributed states of a system are understood as algebraically axiomatized data structures, and the basic local changes that can concurrently occur in a system are axiomatized so as to rewrite rules that correspond to local patterns that, when present in the state of a system, can change into other patterns.

A rewrite theory consists of a signature (which is taken to be an equational theory) and a set of labelled (conditional) rewrite rules. The signature of a rewrite theory describes a particular structure for the states of a system (e.g., multiset, binary tree, etc.) so that its states can be distributed according to the laws of such a structure. The rewrite rules in the theory describe those elementary local transitions which are possible in the distributed state by concurrent local transformations. The deduction rules of rewriting logic allow us to reason formally which general concurrent transitions are possible in a system satisfying such a description. Computationally, each rewriting step is a parallel local transition in a concurrent system.

Maude [3,17] is a high performance language and system supporting both equational and rewriting logic computation. A key novelty of Maude is the efficient support for rewriting, narrowing, and unification modulo equational theories such as those used to model lists or multisets. Maude modules are theories in rewriting logic. The Maude system, its documentation, a collection of examples, some case studies, and related papers are available on the Maude web page at `http://maude.csl.sri.com`.

3 Pathway Logic

Pathway Logic [18,19,20] is an approach to the modeling and analysis of molecular and cellular processes based on rewriting logic. Pathway Logic models of biological processes are developed using the Maude system. A Pathway Logic knowledge base includes data types representing cellular components such as proteins, small molecules, or complexes; compartments/locations; and post-translational modifications. Rewrite rules describe the behavior of proteins and other components depending on modification state and biological context. Each rule represents a step in a biological process such as metabolism or intra/inter- cellular signaling. A collection of such facts forms a formal knowledge base. A model is then a specification of an initial state (cell components and locations) interpreted in the context of a knowledge base. Such models are executable and can be understood as specifying possible ways in which a system can evolve. Logical inference and analysis techniques are used for simulation of possible ways in which a system could evolve, for the assemblage of pathways as answers to queries, and for the reasoning of the dynamic assembly of complexes, cascading transmission of signals, feedback-loops, cross talk between subsystems, and larger pathways. Logical and computational reflection can be used to transform and further analyze models.

Given an executable model such as the one described above, there are many kinds of computation that can be carried out, including: static analysis, forward simulation, forward search, backward search, explicit state model checking, and meta analysis.

Pathway Logic models are structured in four layers: sorts and operations, components, rules, and queries. The *sorts* and *operations* layer declares the main sorts and subsort relations, the logical analog to ontology. The sorts of entities include `Chemical`, `Protein`, `Complex`, and `Location` (cellular compartments), and `Cell`. These are all subsorts of the `Soup` sort that represents unordered multisets of entities. The sort `Modification` is used to represent post-translational protein modifications (e.g., activation, binding, phosphorylating). Modifications are applied using the operator `[ - ]`. For example, the term `[EGFR - act]` represents the activation of the epidermal growth factor receptor EGFR.

A cell state is represented by a term of the form `[cellType | locs]` where `cellType` specifies the type of cell (e.g., Fibroblast) while `locs` represents the contents of a cell organized by cellular location. Each location is represented by a term of the form `{locName | components}` where `locName` identifies the location (for example `CLm` for cell membrane, `CLc` for cell cytoplasm, `CLo` for the outside of the cell membrane, `CLi` for the inside of the cell membrane) while `components` stands for the mixture of proteins and other compounds in that location.

The *components* layer specifies particular entities (proteins, chemicals) and introduces additional sorts for grouping proteins in families. The *rules* layer contains rewrite rules specifying individual steps of a process. These correspond to reactions in traditional metabolic and interaction databases. The *queries* layer specifies initial states and properties of interest.

The Pathway Logic Assistant (PLA) provides an interactive visual representation of Pathway Logic models and facilitates the following tasks: it displays the network of signaling reactions for a given dish; it formulates and submits queries to find pathways; it visualizes gene expression data in the context of a network; or it computes and displays the downstream subnet of one or more proteins. Given an initial dish, the PLA selects the relevant rules from the rule set and represents the resulting reaction network as a Petri net. This provides a natural graphical representation that is similar to the hand drawn pictures used by biologists, as well as very efficient algorithms for answering queries.

The Pathway Logic and PLA system, its documentation, a collection of examples, some case studies, and related papers are available at
`http://pl.csl.sri.com`.

4 Understanding Dynamics on a Biological Pathway

In this section we explain how an experimental biologist might use the Pathway Logic knowledge bases and PLA in their research. We will focus on the Pathway Logic model of response to the Epidermal growth factor (EGF) stimulation. This is an important model for the study of cancer and many other phenomena since the signaling Epidermal growth factor receptor (EGFR) regulates growth, survival, proliferation, and differentiation in mammalian cells.

To model biochemical events such as signaling processes, we use the dynamic part of a rewrite theory (rewrite rules) to express biochemical processes or reactions involving single or multiple subcellular compartments. For example, consider a rule (Rule 757, from MedLine database article with ID 11964154 [21]) that establishes: *In the presence of PIP3, activated Pdk1 recruits PKCe from the*

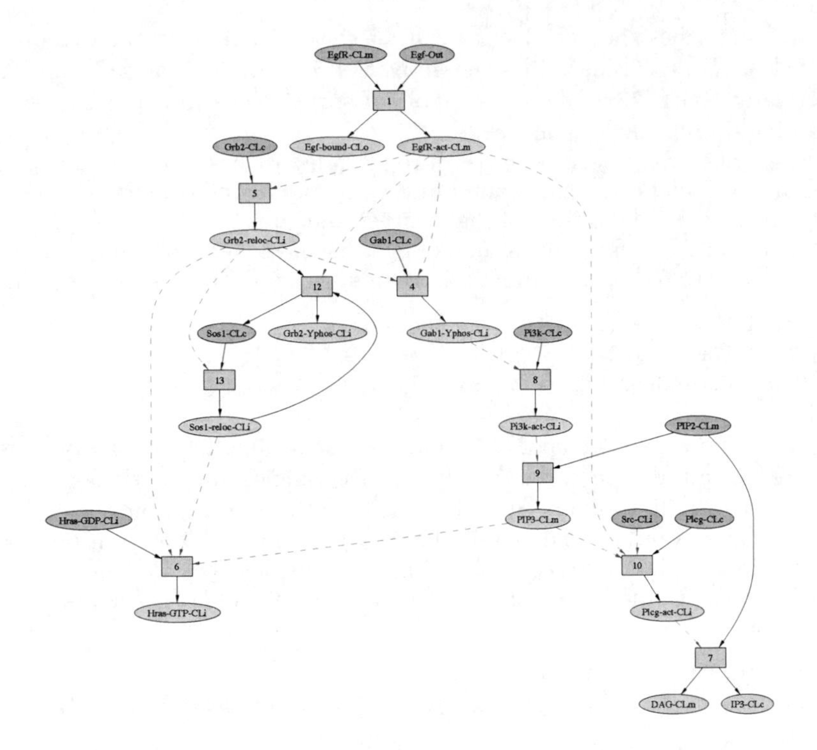

Fig. 1. RasDish as a Petri net using Pathway Logic

cytoplasm to the cell membrane and activates it. In Maude syntax, this signaling process is described by the following rewrite rule:

```
rl[757.PIP3.Pdk1.act.PKCe]:
  {CM | cm:Soup PIP3 [Pdk1 - act] {cyto:Soup PKCe}}
  =>
  {CM | cm:Soup PIP3 [Pdk1 - act] [PKCe-act] {cyto:Soup}}
  [metadata "cite = 11964154"] .
```

Now we consider the binding of EGF to the EGFR (Rules 410 and 438) [22]: *Activated Erk1 is rapidly translocated to the nucleus where it is functionally sequestered and can regulate the activity of nuclear proteins including transcription factors.* In Maude syntax, this signaling process is described by the following rewrite rules:

```
rl[410.Erk1/2.to.nuc]:
  {CM | cm:Soup {cyto:Soup [Erk1 - act] {NM | nm:Soup {nuc:Soup}}}}
  =>
  {CM | cm:Soup {cyto:Soup
  {NM | nm:Soup {nuc:Soup [Erk1 - act]}}}} .
rl[438.Erk.act.Elk]:
  [?Erk1/2 - act] Elk1 => [?Erk1/2 - act] [Elk1 - act] .
```

Rule 410 describes the translocation of activated Erk1 from the cytoplasm to the nucleus. Rule 438 describes the activation of Elk1 (a transcription factor) by activated Erk1. ?Erk1/2 is a variable of sort Erk1/2 (a subsort of Protein containing proteins Erk1 and Erk2).

The queries layer specifies initial states (called dishes) to be studied. Initial states are in silico Petri dishes containing a cell and ligands of interest. An initial state is represented by a term of the form PD(out cell), where cell represents a cell state and out represents a soup of ligands and other molecular components in the cells surroundings. Our analysis begins with a initial dish state defined by

```
eq rasDish =     PD(EGF [HMEC | {CLo | empty }
  {CLm | EGFR  PIP2 }   {CLi | [Hras - GDP]  Src }
  {CLc | Gab1 Grb2 Pi3k Plcg Sos1 }]) .
```

Figure 1 shows the Petri net representation of rasDish. Ovals are occurrences in which the initial occurrences are darker. Rectangles are transitions. Two way dashed arrows indicate an occurrence that is both input and output.

Suppose we want to find out if there is a pathway (computation) leading to activation of Hras (loaded with GTP). In this case one can use the search command with a suitable search pattern and parameters ([1]: the first solution; =>+: at least one step).

```
Maude> search [1] rasDish =>+
  PD(out:Soup [HMEC | cyto:Soup {CLi | cli:Soup [Hras - GTP]}]) .
```

The solution to this query given by Maude is:

```
Solution 1 (state 15)
  out:Soup --> empty
  cyto:Soup --> {CLo |[EGF - bound]}
                {CLm | PIP3 [EGFR - act]}  {CLc | Plcg}
  cli:Soup --> Src[Gab1 - Yphos][Grb2 - reloc]
               [Pi3k - act][Sos1 - reloc]
```

Then we can ask Maude for the rule labels:

```
Maude> show path labels 15 .
  1.EGFR.act       9.PIP3.from.PIP2.by.Pi3k        5.Grb2.reloc
  13.Sos1.reloc    4.Gab1.Yphosed    6.Hras.act.1    8.Pi3k.act
```

Now we consider a new dish d1 and want to find out if a state with cJun and cFos activated is reachable. Besides we want to know what happens if PI3Ka is removed from d1 (call it d1x). The Maude model-checker can be used to find a counter-example to the assertion that such a state is never reachable. The interface to the Maude model checker is embodied in the MODEL-CHECKER module [17]. This module defines syntax for Linear Temporal Logic (LTL) formulas built over a sort Prop. It also introduces a sort State and a satisfaction relation |= on states and LTL formulas.

To use the model checker, the user defines particular states and propositions and axiomatizes satisfaction on these states and propositions. A proposition prop1 that is satisfied by dishes with cJun and cFos activated is defined as follows.

```
eq d1 = PD(EGF {CM | EGFR Pak1 PIP2 nWasp [H-Ras - GDP]
  {Akt1 Gab1 Grb2 Gsk3 Eps8 Erk1 Mek1 Mekk1 Mekk4 Mkk4 Mkk3
  Mlk2 Jnk1 p38a p70s6k Pdk1 PI3Ka PKCb1 Raf1 Rsk1 Shc Sos
  [Cdc42 - GDP] {NM | empty {cJun cFos }}}}) .
eq PD(out:Soup {CM | cm:Soup {cyto:Soup  {NM | nm:Soup
  {nuc:Soup [cJun - act] [cFos - act] }}}})   |= prop1 = true .
```

The formula `~<> prop1` says that `prop1` never holds. Executing the command `red d1 |= ~<>prop1.` results in a counter example (a reachable state satisfying `prop1` together with a path leading to that state). A more biologist friendly notation for such uses of the model-checker is obtained by defining

```
eq findPath(S:State,P:Prop) = getPath(P:Prop, S:State |= ~<> P:Prop) .
```

where `getPath` abstracts the counter example returned by Maude to a list of rule labels followed by a state satisfying the given proposition. If we execute `findPath(d1,prop1)` and `findPath(d1x,prop1)` we discover that, in both cases, we find a state in which both `cJun` and `cFos` are activated. However, `Akt1` and `Eps8` are both activated when `PI3Ka` is present, but not when it is removed.

Models of cellular response to many different stimuli, including a much more complete model of EGF signaling, as well as a tutorial guide for using PLA to query the models, can be found at `pl.csl.sri.com`.

5 Conclusions

Rewriting logic procedures are powerful symbolic methods that can be applied to understand the dynamics of complex biological systems. It provides many benefits, including the ability to build and analyze models with multiple levels of detail; represent general rules; define new kinds of data and properties; and execute queries using logical inference.

We are interested in formalizing models that molecular biologists can use to think about signaling pathways and other processes in familiar terms while allowing them to computationally ask questions about possible outcomes and dynamics. Here we have described the use of Pathway Logic as a rewriting logic tool to model signal transduction processes, and also the use Pathway Logic Assistant in order to browse and analyze these models. The models are useful for clarifying and organizing experimental data and knowledge from the literature. The eventual goal is to reach a level of global understanding of complex systems that supports prediction of new and possibly unexpected results.

References

1. Meseguer, J.: Conditional rewriting logic as a unified model of concurrency. Theor. Comput. Sci. 96, 73–155 (1992)
2. Meseguer, J.: Twenty years of rewriting logic. J. Log. Algebr. Program. 81(7-8), 721–781 (2012)

3. Clavel, M., Durán, F., Eker, S., Lincoln, P., Martí-Oliet, N., Meseguer, J., Talcott, C.: All About Maude - A High-Performance Logical Framework. LNCS, vol. 4350. Springer, Heidelberg (2007)
4. Vukmirovic, O.G., Tilghman, S.M.: Exploring genome space. Nature 405, 820–822 (2000)
5. Weng, G., Bhalla, U.S., Iyengar, R.: Complexity in biological signaling systems. Science 284, 92–96 (1999)
6. Asthagiri, A.R., Lauffenburger, D.A.: A computational study of feedback effects on signal dynamics in a mitogen-activated protein kinase (mapk) pathway model. Biotechnol. Prog. 17, 227–239 (2001)
7. Smolen, P., Baxter, D.A., Byrne, J.H.: Mathematical modeling of gene networks. Neuron 26, 567–580 (2000)
8. Saadatpour, A., Albert, R.: Discrete dynamic modeling of signal transduction networks. In: Liu, X., Betterton, M.D. (eds.) Computational Modeling of Signaling Networks, pp. 255–272 (2012)
9. Fisher, J., Henzinger, T.A.: Executable cell biology. Nat. Biotechnol. 25(11), 1239–1249 (2007)
10. Hardy, S., Robillard, P.N.: Petri net-based method for the analysis of the dynamics of signal propagation in signaling pathways. Bioinformatics 24(2), 209–217 (2008)
11. Li, C., Ge, Q.W., Nakata, M., Matsuno, H., Miyano, S.: Modelling and simulation of signal transductions in an apoptosis pathway by using timed Petri nets. J. Biosci. 32, 113–127 (2006)
12. Regev, A., Panina, E.M., Silverman, W., et al.: Bioambients: An abstraction for biological compartments. Theor. Comput. Sci. 325(1), 141–167 (2004)
13. Efroni, S., Harel, D., Cohen, I.: Towards rigorous comprehension of biological complexity: modeling, execution and visualization of thymic T-cell maturation. Genome Res. 13(11), 2485–2497 (2003)
14. Faeder, J.R., Blinov, M.L., Hlavacek, W.S.: Rule-Based Modeling of Biochemical Systems with BioNetGen. Systems Biology. Methods Mol. Biol. 500, 113–167 (2009)
15. Hwang, W., Hwang, Y., Lee, S., Lee, D.: Rule-based multi-scale simulation for drug effect pathway analysis. BMC Med. Inform. Decis. Mak 13(suppl. 1), S4 (2013)
16. Eduati, F., De Las Rivas, J., Di Camillo, B., Toffolo, G., Saez-Rodríguez, J.: Integrating literature-constrained and data-driven inference of signalling networks. Bioinformatics 28(18), 2311–2317 (2012)
17. Clavel, M., Durán, F., Eker, S., Lincoln, P., Martí-Oliet, N., Meseguer, J., Talcott, C.: Maude Manual, version 2.6 (2011), `http://maude.cs.uiuc.edu`
18. Talcott, C., Eker, S., Knapp, M., Lincoln, P., Laderoute, K.: Pathway logic modeling of protein functional domains in signal transduction. In: Proceedings of the Pacific Symposium on Biocomputing (2004)
19. Talcott, C., Dill, D.L.: The pathway logic assistant. In: Plotkin, G. (ed.) Proc. of the Third Intl Conf. CMSB 2005, pp. 228–239 (2005)
20. Talcott, C.L.: Pathway logic. In: Bernardo, M., Degano, P., Zavattaro, G. (eds.) SFM 2008. LNCS, vol. 5016, pp. 21–53. Springer, Heidelberg (2008)
21. Cenni, V., Döppler, H., Sonnenburg, E.D., et al.: Regulation of novel protein kinase C epsilon by phosphorylation. Biochem. J. 363(Pt.3), 537–545 (2002)
22. Schlessinger, J.: Cell signaling by receptor tyrosine kinases. Cell 103, 211–225 (2000)

Evaluating Pathway Enumeration Algorithms in Metabolic Engineering Case Studies

F. Liu, P. Vilaça, I. Rocha, and Migael Rocha

CEB/IBB, Universidade do Minho, Portugal

Abstract. The design of cell factories for the production of compounds involves the search for suitable heterologous pathways. Different strategies have been proposed to infer such pathways, but most are optimization approaches with specific objective functions, not suited to enumerate multiple pathways. In this work, we analyze two pathway enumeration algorithms based on graph representations: the Solution Structure Generation and the Find Path algorithms. Both are capable of enumerating exhaustively multiple pathways using network topology. We study their capabilities and limitations when designing novel heterologous pathways, by applying these methods on two case studies of synthetic metabolic engineering related to the production of butanol and vanillin.

1 Introduction

The quest for sustainable industries lead to an increased interest in Biotechnology. One of its key features is to re-engineer microbes to produce valuable compounds [5]. The development of cell factories is an iterative process involving steps as the search for suitable hosts and viable synthetic pathways. Heterologous pathways augment their capabilities to produce non native compounds. The definition of pathways allows to organize chemical reactions into set providing a coherent function, such as transforming a substrate to a target compound.

The constraint based modeling (CBM) approach is often adopted for *in silico* analysis of genome scale metabolic models (GSMM) not requiring kinetic information. The system is subjected to constraints such as reaction stoichiometry, reversibility and assumption of a pseudo-steady state, allowing the computation of a feasible flux space that characterizes the system. Flux Balance Analysis (FBA) is a popular method to determine the flux distribution that maximizes an objective (e.g. related to cellular growth) using linear programming [16].

Pathway optimization has been approached using different strategies. Regarding CBM, FBA was used to determine producible non native compounds [3] by merging GSMMs with large databases as KEGG, allowing to infer heterologous reactions. A limitation of FBA is the fact that it determines a single solution, while multiple optimal solutions exist. Furthermore, sub-optimal solutions may offer valuable information on alternative routes. On the other hand, Elementary Flux Modes (EFM) are defined as the minimal subsets of reactions to maintain steady state. However, their computation is restricted to small networks [14].

J. Sáez-Rodríguez et al. (eds.), *8th International Conference on Practical Appl. of Comput. Biol. & Bioinform. (PACBB 2014)*, Advances in Intelligent Systems and Computing 294,
DOI: 10.1007/978-3-319-07581-5_26,

Figueiredo *et al* [6] propose an enumeration strategy to compute the k shortest EFMs expanding the size of partially computable problems. Nonetheless, database size networks (e.g. KEGG or MetaCyc) still offer a great challenge for full EFM computation. The OptStrain algorithm [17] uses mixed integer linear optimization to obtain the pathway with the smallest number of heterologous reactions, but does not enumerate alternatives.

Other methods have applied standard graph methods, taking advantage of shortest path algorithms to infer the shortest pathway between two compounds [8,18]. This strategy can also be augmented by using shortest path enumerating methods, such as the k-shortest path algorithm [4]. A major problem with this strategy is that graph paths return linear routes between compounds, while in reality these may involve more compounds. Additionally, compounds represented as hubs in the network mislead the algorithms by shortening the paths since they connect many reactions. To circumvent this problem, weighting [8] or filtering methods [7] have been proposed to reroute the solutions.

An alternative is to use more complex representations. Hypergraphs or process graphs (which are directed bipartite graphs) are capable to model chemical reactions with higher detail. This allows to address the problem of multiple products and reactants, since edges connect to vertex sets instead of a single vertex. Process graphs were used by Friedler *et al* [9–11] in an exhaustive approach for decision mapping in synthesis processes, being later adapted for pathway identification [13]. The work of Carbonell *et al* [2] introduced an enumeration strategy to extract pathways using hypergraphs.

In this work, we analyze two existing algorithms for multiple pathway enumeration, the Solution Structure Generation (SSG) and the Find Path (FP), both based on set systems representations. These algorithms are implemented and tested with two case studies, regarding the production of butanol and vanillin, using the bacterium *Escherichia coli* and the yeast *Saccharomyces cerevisiae*, two model organisms for which there are available GSMMs. The results obtained by both are provided and discussed, being clear the need to introduce some improvements to allow the scalability of the methods.

2 Problem Definition

In a topological approach, a pathway extraction problem can be defined as a dependency problem. Thus, a reaction needs to be satisfied and satisfies metabolites (that are dependencies of other reactions), that correspond to reactants and products, respectively. Here, the notation used in the following is defined. Mostly, it is based on the axioms and algorithms presented in [9–11].

Networks will be composed only by metabolites and reactions. In this system, metabolites are the vertex entities, while reactions are represented by an ordered pair $\langle M_1, M_2 \rangle$, that connects two disjoint sets of metabolites.

Definition 1. *(Reaction) A reaction is an ordered pair* $\langle M_1, M_2 \rangle$ *of two disjoint sets of metabolites (i.e.,* $M_1 \cap M_2 = \emptyset$*). The first set represents the reactants, while the second represents the products.*

Definition 2. *(Metabolic Network) A metabolic network Σ is a pair composed by a set of metabolites Π and a set of reactions Υ.*

A reversible reaction r is represented by including another entity r', such that the metabolite sets are swapped. Additionally, a network $\Sigma' = \langle \Pi', \Upsilon' \rangle$ is defined as a subnetwork of $\Sigma\langle \Pi, \Upsilon \rangle$ if every element of Σ' is contained in Σ (i.e., $\Pi' \subseteq \Pi$ and $\Upsilon' \subseteq \Upsilon$), then $\Sigma' \subseteq \Sigma$.

A retrosynthetic metabolic problem can be defined as follows:

Definition 3. *(Retrosyntehtic Metabolic Problem) A retrosynthetic metabolic problem Γ is defined by a triplet $\langle \Sigma, S, T \rangle$, where Σ is a metabolic network that represents the search space, while S and T are two disjoint sets of metabolites (i.e, $S \cap T = \emptyset$) which are the constraints of the heterologous pathways. The set S keeps the initial substrates (e.g., supplies or raw materials), while the set T defines the target compounds of interest.*

An heterologous pathway is a set of reactions, in most cases a subnetwork of a larger network (defined as the search space), if it satisfies the following:

Definition 4. *(Heterologous Pathway) An heterologous pathway σ of a synthetic problem Γ is any network (or subnetwork), such that: a) the product set T is included in $\langle M, R \rangle$, i.e., $T \subset M$ and b) for every metabolite m in the subnetwork that is not included in the substrate sets of Γ (i.e., $M - S$) there is a reaction r in R such that m is a product of R.*

The heterologous pathway definition is not sufficient to guarantee that the solution is feasible, because it omits the stoichiometry of the reactions. Both algorithms addressed in this work do not take account this property for the computation of heterologous solutions. This eventually will lead to the computation of infeasible solutions that later can be verified by applying FBA.

3 Algorithms

3.1 Solution Structure Generation

The Solution Structure Generation (SSG) algorithm enumerates solutions of Γ by recursively branching all possible combinations. This technique, denoted as decision mapping, can be described as follows: let Σ' be a subnetwork such that condition a) verifies. Then, in order to fulfill condition b), the sub-problem Γ' is solved producing the unsatisfied metabolites in Σ'. Let $\Sigma = \langle T, \emptyset \rangle$ be a network containing T and no reactions, then a) trivially verifies. Then, $\wp$(producers of t), $t \in T$ (where $\wp(X)$ denotes the power set of X) are candidates for partial solutions of Γ, since if solutions of Γ exists, then at least one element of $\wp$ eventually must be present in one or more solutions of Γ. Recursively, we solve the sub-problem Γ', with the new target set $T' = R - S - M$, where R is the set of reactants of the newly introduced reactions (minus the initial set S and producible metabolites in the partial solution), until eventually either there are no possible reactions to add, and this implies that we have reached a dead end that happens when we pick a producer of T that does not belong to any solution, or $T = \emptyset$ which implies that we achieved a solution.

There are several limitations of the SSG method. The first is the high amount of memory that is required to compute power sets which grows exponentially with the number of elements (2^n). Additionally, this generates an extensive amount of possible combinations. If the network is not pruned, meaning that the network contains reactions that do not belong to any solution, then the algorithm may contain branches that return no solutions and, depending the depth of these branches, this increases severely the computation time to obtain solutions. Friedler *et al* [10] proposed a polynomial algorithm to prune process graphs to remove all reactions that might exhibit this behavior. Because of these limitations, in this work, some modifications were implemented to the original algorithm. Given space constraints, the full algorithms including these changes are fully given and explained in supplementary material that is available in `http://darwin.di.uminho.pt/pacbb14-liu`.

3.2 Find Path

The Find Path (FP) algorithm proposed by Carbonell *et al* [2] enumerates pathways by using hypergraphs. In a metabolic context, both hypergraphs and process graphs are much similar. A solution of the FP algorithm is defined as a *hyperpath* P, which is an hypergraph (usually a subgraph) where the hyperarcs (reactions) can be ordered as $r_1, r_2, \ldots, r_m$ such that r_i is dependent only on the substrates in S and the products of the previous reactions. This is computed with a subroutine, Find All [2], that sorts the entire network satisfying this condition. Additionally, reactions that cannot be satisfied are removed.

Not all pathways can be expressed by the definition of an hyperpath [2]. Lets consider for instance co-factor metabolites m_a and m_b. Usually, these metabolites are both present in a single reaction $r = \langle M_1, M_2 \rangle$ where $m_a \in M_1$ and $m_b \in M_2$ or vice versa. These reactions can be satisfied by each other in a way where there is an $r' = \langle M_1', M_2' \rangle$ where $m_b \in M_1'$ and $m_a \in M_2'$. Therefore, it is impossible to sort an hyperpath if neither m_a or m_b are included in S. Examples of these metabolites are ATP/ADP, NADH/NAD, etc. Fortunately, if assuming S to be an organism chassis, these metabolites are usually include in S. However, this does not guarantee that other more complex cycles do not exist.

This issue enables the generation of redundant solutions. Let $\Gamma = \langle \Sigma, \{s_0\}, \{t_0\} \rangle$ be a retrosynthetic problem, assuming that a) an heterologous pathway $\Sigma' \subset \Sigma$ exists from s_0 to t_0, such that b) $r, r' \in \Sigma'$ where $r = \langle \{m_0, p_0\}, \{m_1, p_1\} \rangle$ and $r' = \langle \{m_1, p_1\}, \{m_2, p_0\} \rangle$. The FP algorithm can only identify such pathway if $\Gamma' = \langle \Sigma, \{s_0\}, \{p_0, m_0\} \rangle$ is feasible. Since r, r' satisfy the metabolites p_0, p_1 of each other (i.e., $r + r' = \langle \{m_0\}, \{m_2\} \rangle$) this implies that any effort to produce p_0 in Γ' is unnecessary and every solution that b) verifies may contain multiple redundant solutions (the reactions included in the solutions are unique but in steady state they are redundant).

In this work, to extend the capabilities of the FP algorithm a modification was implemented in the Minimize subroutine (see supplementary material in `http://darwin.di.uminho.pt/pacbb14-liu`. The redundancy problem still remains an open topic for further improvement.

4 Experiments and Results

4.1 Case Studies

The algorithms were tested by applying two case studies of synthetic metabolic engineering. The first example is the production of 1-butanol using *E. coli* [1], while the second concerns vanillin synthesis using *S. cerevisiae* [12]. Both algorithms (i.e., SSG and FP) are applied using the set of compounds in the KEGG Ligand and MetaCyc databases as the chemical search space. Additionally, to integrate and test the obtained solutions *in silico*, a GSMM is required: the *i*JO1366 GSMM for *E. coli* and *i*MM904 [15] for *S. cerevisiae* were used. Therefore, a total of 8 result sets were generated for two algorithms, two case studies and two search spaces (databases).

4.2 Data Preprocessing

Before running the algorithms, several pre-processing tasks needed to be performed. The first was to select and define the constraints of the problem, selecting the search space Σ, the initial set S and the target compounds T. For both case studies, the target set is a singleton containing only the compound of interest, 1-butanol in the first case and vanillin in the second. For the substrate set, all metabolites included in the GSMMs were selected. This later will allow to integrate the obtained solutions with these models and evaluate their performance. The BiGG database [19] aided in the transformation of the species identifiers of the model to those in the databases. The species that did not match any cross-referencing were discarded.

The reference pathway of the 1-butanol synthesis was mostly present in the *i*JO1366 GSMM. So, to obtain alternative pathways we removed the following species: M_btcoa_c (Butanyl-CoA), M_btal_c (Butanal), M_b2coa_c (Crotonyl-CoA), M_3hbcoa_c (3-hydroxybuty), M_aacoa_c (Acetoacetyl-Coa). Additionally, every reaction connected to these compounds was also removed. The impact in the biomass value calculated using the FBA was minimal reducing to 0.977 (from 0.986). Removing these species will allow to find alternative paths from other internal metabolites of *i*JO1366 to 1-butanol, since an alternative solution to the identified in [1] is desired which may or may not be optimal against existing pathway. Note that the algorithms do not generate solutions including reactions to produce the initial substrate set since these are defined as supplied. Regarding the other case study, no modifications were made in the *i*MM904 GSMM.

A minor modification was done in the MetaCyc database, since it contains reactions with the metabolite pairs `NAD-P-OR-NOP/NADH-P-OR-NOP` which are instances of either NAD/NADH or NADP/NADHP. These reactions were unfolded to their correct instances. This is essential to infer the 1-butanol pathway, as several reactions of this pathway were expressed in this format. The KEGG Ligand database did not require pre-processing.

4.3 Algorithm Setup

Because of the combinatorial explosion of possible pathways, it is impossible to obtain every solution existing in a database size network using any of the algorithms. To compare the algorithms, the search space was split into subsets by *radius*. The *radius* is an integer that defines the minimum number of links (i.e., reactions) required to reach that reaction from an initial set of metabolites. This implies that a reaction that belongs to radius i also belongs to $i + 1$, and therefore a sub-network Σ_i of *radius* i always complies to $\Sigma_i \subseteq \Sigma_{i+1}$.

With these reduced search spaces, solutions were computed using each of the algorithms. An attempt was made to obtain the entire set of candidate solutions for each *radius*, until either the proces crashes due to lack of memory or exceeds computational time allotted ($>$ 24 hours). To validate the solutions, FBA was used to maximize the yield of the target product and validate its feasibility integrating the solution into the respective GSMM.

4.4 Results

Figure 1 shows the number of solutions computed and their feasibility. SSG is more limited than FP by the size of the search space. A major problem of the SSG algorithm is the high memory demand because of the power set computation. With the reduction of the power set size (only partial sets are computed), it still presents high memory demand to branch all the possible combinations. Moreover, the SSG computes every solution that satisfies Definition 4 which eventually leads to the computation of infeasible pathways.

Still, in general, the SSG shows better performance in the computation of solutions (Figure 2) mainly because of the branching technique which gives a major advantage to the computation time per solution because of the backtracking. As the algorithm moves to a candidate solution, the next solution reuses the previous partial solution. This results in a neglectable impact on the computation time per solution as the search space increases (i.e., increasing size of the radius). However, since the number of solutions exponentially grows with the increasing size of the search space, the total computation time increases.

The FP is capable to compute larger search spaces, being the major bottleneck the computation time per solution, since the internal Minimize routine has quadratic complexity to the number of reactions [2]. A scenario was also found where FP computes multiple distinct redundant solutions.

For every solution that satisfies the feasibility test, its performance was evaluated by integrating into the corresponding GSMM. The farthest *radius* that either algorithm was able to compute was selected for this process. For the

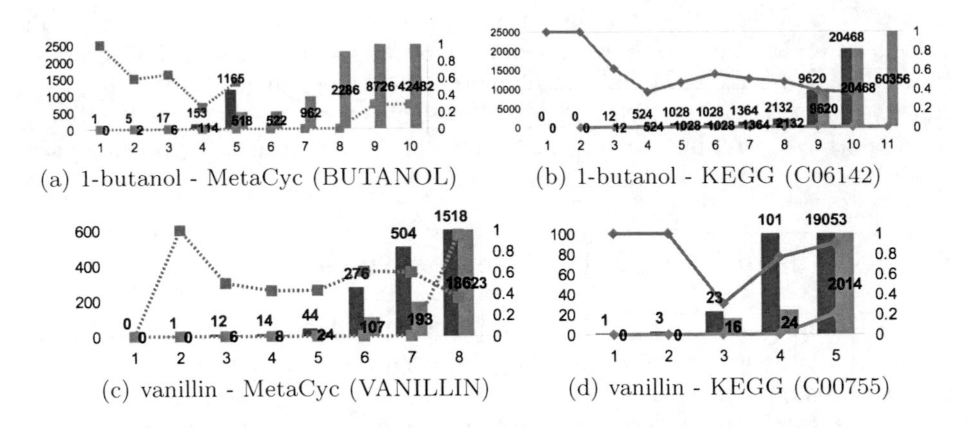
(a) 1-butanol - MetaCyc (BUTANOL) (b) 1-butanol - KEGG (C06142)

(c) vanillin - MetaCyc (VANILLIN) (d) vanillin - KEGG (C00755)

Fig. 1. Pathways computed for each of the problems by radius. The number of solutions on the left. The percentage of infeasible or redundant solutions on the right. Blue - SSG. Orange - FP.

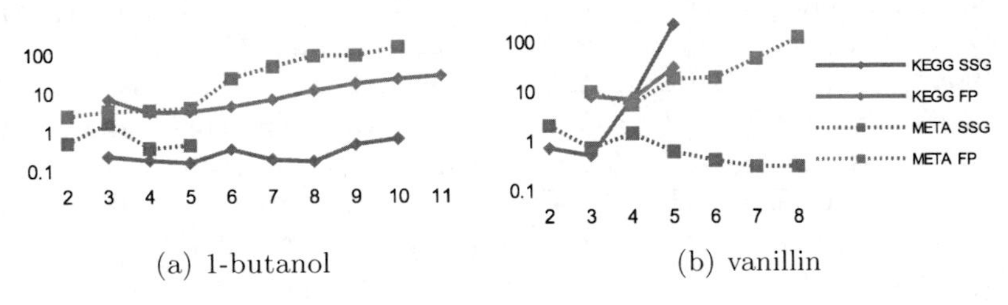

(a) 1-butanol (b) vanillin

Fig. 2. Time cost (milliseconds) per each solution

1-butanol case, from the 42482 and 60356 solutions obtained from the FP algorithm, a total of 32692 and 22968 were compatible with the *i*JO1366 GSMM for search spaces of MetaCyc and KEGG, respectively. In the vanillin case, 944 of 974 computed solutions are valid (MetaCyc), being the numbers for KEGG of 1600 out of 1852. The 1-butanol case shown a massive amount of solutions mostly because of the NAD/NADH alternatives for many reactions.

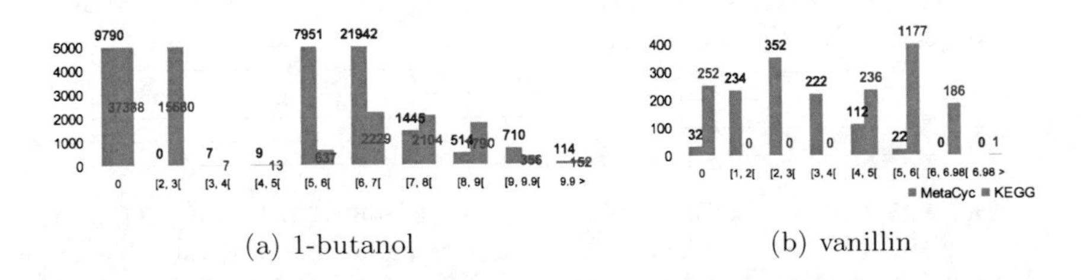

(a) 1-butanol (b) vanillin

Fig. 3. Histogram of theoretical yield values of 1-butanol in *i*JO1366 and vanillin in *i*MM904 . On the y-axis - number of solutions, x-axis yield range. Last value is the optimal solution (for better yield).

KEGG provided the solutions with highest yield, with 6.98 of vanillin in *i*MM904 , and 152 pathways with 9.99 for 1-butanol compared to 114 pathways from MetaCyc. There is a noticeable difference in the stoichiometry of the reactions between KEGG and MetaCyc in the 1-butanol pathways. A detailed view of the pathways obtained in this case study can be found in the supplementary material (http://darwin.di.uminho.pt/pacbb14-liu).

5 Conclusions and Future Perspectives

The algorithms analyzed both present errors in the computation of heterologous pathways. Although topologically they are correct, as they have the common goal which is to infer heterologous pathways (subnetworks) that satisfy the rules of initial substrates and target product, in a steady state point of view several examples may be infeasible. However, by using post-processing methods such as FBA, the correct solutions can be identified, which allows to correctly enumerate multiple pathways. The case study of 1-butanol shows that there are many viable routes for 1-butanol production in *i*JO1366 all with the same optimal yield. Moreover, even if a problem contains only a single optimal solution (e.g., vanillin in *i*MM904), examples of sub-optimal pathways also show a broad range of yield values many near the optimal. Other methods hardly can achieve such a range of feasible steady state heterologous pathways.

Thus, it is shown that although neither of the algorithms is readily suitable to compute steady state heterologous pathways for large databases, they are still able extract potential pathways, after targeted improvements in scalability. Additionally, they offer a generic method to infer pathways for multiple purposes, since they to not follow any strict objective function (e.g., yield or size).

Acknowledgments. The work is partially funded by ERDF - European Regional Development Fund through the COMPETE Programme (operational programme for competitiveness) and by National Funds through the FCT (Portuguese Foundation for Science and Technology) within projects ref. COMPETE FCOMP-01-0124-FEDER-015079 and Strategic Project PEst-OE/EQB/LA0023/2013, and also by Project 23060, PEM - Technological Support Platform for Metabolic Engineering, co-funded by FEDER through Portuguese QREN under the scope of the Technological Research and Development Incentive system, North Operational.

References

1. Atsumi, S., Cann, A.F., Connor, M.R., Shen, C.R., Smith, K.M., Brynildsen, M.P., Chou, K.J.Y., Hanai, T., Liao, J.C.: Metabolic engineering of Escherichia coli for 1-butanol production. Metabolic Engineering 10(6), 305–311 (2008)
2. Carbonell, P., Fichera, D., Pandit, S.B., Faulon, J.-L.: Enumerating metabolic pathways for the production of heterologous target chemicals in chassis organisms. BMC Systems Biology 6(1), 10 (2012)

3. Chatsurachai, S., Furusawa, C., Shimizu, H.: An in silico platform for the design of heterologous pathways in nonnative metabolite production. BMC Bioinformatics 13(1), 93 (2012)
4. Croes, D., Couche, F., Wodak, S.J., van Helden, J.: Metabolic PathFinding: inferring relevant pathways in biochemical networks. Nucleic Acids Research 33(Web Server issue), W326–W330 (2005)
5. Curran, K.A., Alper, H.S.: Expanding the chemical palate of cells by combining systems biology and metabolic engineering. Metabolic Engineering 14(4), 289–297 (2012)
6. de Figueiredo, L.F., Podhorski, A., Rubio, A., Kaleta, C., Beasley, J.E., Schuster, S., Planes, F.J.: Computing the shortest elementary flux modes in genome-scale metabolic networks. Bioinformatics 25(23), 3158–3165 (2009)
7. Faust, K., Croes, D., van Helden, J.: Metabolic pathfinding using RPAIR annotation. Journal of Molecular Biology 388(2), 390–414 (2009)
8. Faust, K., Dupont, P., Callut, J., Helden, J.V.: Pathway discovery in metabolic networks by subgraph extraction. Bioinformatics 26(9), 1211–1218 (2010)
9. Friedler, F., Tarján, K., Huang, Y., Fan, L.: Graph-theoretic approach to process synthesis: axioms and theorems. Chemical Engineering Science 47(8), 1973–1988 (1992)
10. Friedler, F., Tarjan, K., Huang, Y.W., Fan, L.: Graph-theoretic approach to process synthesis: Polynomial algorithm for maximal structure generation. Computers & Chemical Engineering 17(9), 929–942 (1993)
11. Friedler, F., Varga, J., Fan, L.: Decision-mapping: A tool for consistent and complete decisions in process synthesis. Chemical Engineering Science 50(11), 1755–1768 (1995)
12. Hansen, E.H., Møller, B.L., Kock, G.R., Bünner, C.M., Kristensen, C., Jensen, O.R., Okkels, F.T., Olsen, C.E., Motawia, M.S., Hansen, J.R.: De Novo Biosynthesis of Vanillin in Fission Yeast (Schizosaccharomyces pombe) and Baker's Yeast (Saccharomyces cerevisiae). Applied and Environmental Microbiology 75(9), 2765–2774 (2009)
13. Lee, D.-Y., Fan, L.T., Park, S., Lee, S.Y., Shafie, S., Bertók, B., Friedler, F.: Complementary identification of multiple flux distributions and multiple metabolic pathways. Metabolic Engineering 7(3), 182–200 (2005)
14. Machado, D., Soons, Z., Patil, K.R., Ferreira, E.C., Rocha, I.: Random sampling of elementary flux modes in large-scale metabolic networks. Bioinformatics (Oxford, England) 28(18), i515–i521 (2012)
15. Mo, M.L., Palsson, B.O., Herrgård, M.J.: Connecting extracellular metabolomic measurements to intracellular flux states in yeast. BMC Systems Biology 3, 37 (2009)
16. Orth, J.D., Thiele, I., Palsson, B.O.: What is flux balance analysis? Nature Biotechnology 28(3), 245–248 (2010)
17. Pharkya, P., Burgard, A.P., Maranas, C.D.: OptStrain: A computational framework for redesign of microbial production systems. Genome Research 814, 2367–2376 (2004)
18. Rahman, S.A., Advani, P., Schunk, R., Schrader, R., Schomburg, D.: Metabolic pathway analysis web service (Pathway Hunter Tool at CUBIC). Bioinformatics (Oxford, England) 21(7), 1189–1193 (2005)
19. Schellenberger, J., Park, J.O., Conrad, T.M., Palsson, B.O.: BiGG: a Biochemical Genetic and Genomic knowledgebase of large scale metabolic reconstructions. BMC Bioinformatics 11, 213 (2010)

T-HMM: A Novel Biomedical Text Classifier Based on Hidden Markov Models

A. Seara Vieira, E.L. Iglesias, and L. Borrajo

Computer Science Dept., Univ. of Vigo, Escola Superior de Enxeñería Informática, Ourense, Spain
{adrseara,eva,lborrajo}@uvigo.es

Abstract. In this paper, we propose an original model for the classification of biomedical texts stored in large document corpora. The model classifies scientific documents according to their content using information retrieval techniques and Hidden Markov Models.

To demonstrate the efficiency of the model, we present a set of experiments which have been performed on OHSUMED biomedical corpus, a subset of the MEDLINE database, and the Allele and GO TREC corpora. Our classifier is also compared with Naive Bayes, k-NN and SVM techniques.

Experiments illustrate the effectiveness of the proposed approach. Results show that the model is comparable to the SVM technique in the classification of biomedical texts.

Keywords: Hidden Markov Model, Text classification, Bioinformatics.

1 Introduction

The number of scientific documents stored in public corpora as MedLine is prohibitively large to handle manually. Therefore there is great interest to automate the classification process and produce relevant documents to a topic. A large number of techniques have been developed for text classification, including Naive Bayes, k-Nearest Neighbours (kNN), Neural Networks, and Support Vector Machines (SVMs). Among them SVM has been recognized as one of the most effective text classification methods.

In other studies, Hidden Markov Models (HMM) have been used to describe the statistical properties of a sequential random process. They are known for their application in language problems like speech recognition and pattern matching [1]. However, their application has been extended to fields of text processing such as information extraction [2, 3], information retrieval [4], text categorization [5, 6], and text classification [7].

Miller et al. [4] use HMM in an information retrieval model. Given a set of documents and a query Q, the system searches a document D relevant to the query Q. It computes the probability that D is the relevant document in the user's mind, given query Q, *i.e* $P(D \text{ is } R|Q)$, and ranks the documents based on this measure. The proposed model generates a user query with a discrete Hidden

J. Sáez-Rodríguez et al. (eds.), *8th International Conference on Practical Appl. of Comput. Biol. & Bioinform. (PACBB 2014)*, Advances in Intelligent Systems and Computing 294,
DOI: 10.1007/978-3-319-07581-5_27, © Springer International Publishing Switzerland 2014

Markov process depending on the document the user has in mind. The HMM is viewed as a generator of the query, and is used to estimate the probability that each document will be produced in the corpus.

Kwan Yi et al. [7] use the previous idea in a similar approach. They describe the text classification as the process of finding a relevant category c for a given document d. They implement a Hidden Markov Model to represent each category. Thus, given a document d, the probability that a document d belongs to category c is computed on the specific HMM model c. In their system, a document is treated as a wordlist, and the HMM for each category is viewed as a generator of a word sequence.

Kairong Li et al. [6] research the text categorization process based on Hidden Markov Model. The main idea of the article lies in setting up an HMM classifier, combining χ^2 and an improved TF-IDF method and reflecting the semantic relationship in the different categories. The process shows the semantic character in different documents to make the text categorization process more stable and accurate.

Following this line, we propose a novel and original HMM for biomedical text classification based on document content. The main goal is to develop a simpler classifier than existing methods. The model is focused on distinguishing relevant and non-relevant documents from a dataset, and deals with the problem in common search systems where a document may be relevant or not given the specific user query. The new classifier is evaluated and compared with the results obtained by the Naive Bayes and SVM techniques.

The remainder of this paper is organized as follows: Section 2 explains the text classification process using HMM; Section 3 details the proposed solution; Section 4 discusses the experiments and results obtained in the case study; Section 5 presents the conclusions and future work.

2 Text Classification Process with HMM

Text classification is the task of automatically assigning a document set to a predefined set of classes or topics [8].

In our context, given a training set $T = \{(d_1, c_1), (d_2, c_2)...(d_n, c_n)\}$, which consists of a set of preclassified documents in categories, we want to build a classifier using HMM to model the implicit relation between the characteristics of the document and its class, in order to be able to accurately classify new unknown documents.

Each document d_i has a binary class attribute c_i which can have a value of Relevant or Non-relevant. Our work is therefore focused on building a classifier based on the training set that can classify new documents as relevant or non-relevant without previously knowing their class information.

Following the idea proposed by Kwan Yi et al. [7], we use HMM as a document generator. Fig. 1 shows the proposed framework. According to the structure, an HMM is implemented for each category: Relevant and Non-Relevant. Each model is then trained with documents belonging to the class that it represents. When a

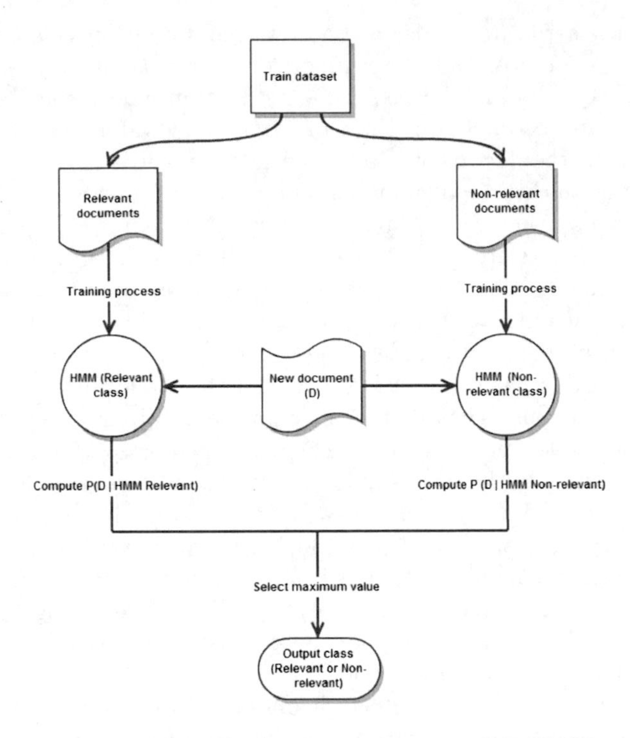

Fig. 1. General classifying process with T-HMM

new document needs to be classified, the system evaluates the probability of this document being generated by each of the Hidden Markov models. As a result, the class with the maximum probability value is selected and considered as the output class for the document.

For further information and details on the principles and algorithms of HMM, see [1].

3 T-HMM Classifier Proposal

The proposed model (T-HMM) aims to classify biomedical documents according to their content. To achieve that, input data needs to be expressed in a format that HMM algorithms can handle.

The most common approach in document classification tasks is the bag-of-words approach [9]. In this case, every document is represented by a vector where elements describe the word frequency (number of occurrences). Words with a higher number of occurrences are more relevant because they are considered the best representation of the document semantic. For training purposes, words are placed in descending order according to their ranking to represent each document.

The complexity of text classification in terms of time and space depends on the size of these vectors. In order to reduce their dimension, a text preprocessing step is required, where rare words and those which do not provide any useful information (such as prepositions, determiners or conjunctions) are removed. This step is further explained in the Experiments Section where some adjustments such as TFIDF are also made in the word frequency value. The final selected words to represent the documents are called feature words.

When building an HMM model, it is important to reflect what "hidden states" stand for. In Kwan Yi et al. [7], states are designed to represent different sources of information; other related works use hidden states to represent internal sections of the document [10, 5]. However, as stated before, our purpose is focused on classifying documents by their content rather than their structure.

In our model, hidden states reflect the difference in relevance (ranking) between words in a document. Each state represents a relevance level for words appearing in the corpus. That is, the most probable observations for the first state are the most relevant words in the corpus. The most probable observations for the second state are the words holding the second level of relevance in the corpus, and so on. The number of states N is a modifiable parameter that depends on the training corpus and how much flexibility we want to add to the model. Specifically, the number of states is a cut-off parameter. It should be high enough to represent multiple relevance levels without overfitting the model to the training data. A good starting point is to consider n the average number of words with non-zero value in a document.

Considering that each document is ultimately represented by a vector or a wordlist ranked in decreasing order, and ignoring words with zero value, a Hidden Markov model is proposed to represent a predefined category c as follows:

1. The union of words from the training corpus is taken as the set of observation symbols V. For each word, there is a symbol v_k. The set of possible observations is the same for every HMM, taking into account all words in the corpus, regardless of their category.
2. As mentioned above, states represent ranking positions. Therefore, states are ordered from the first rank to the last one. The state transitions are ordered sequentially in the same way, forming a left-right HMM [1] without self-state loops, in which the probability of state S_{i+1} behind state S_i is 1. The transition probability matrix A is then defined as:

$$a_{ij} = \begin{cases} 1 & \text{if } j = i + 1 \\ 0 & \text{in other case} \end{cases}$$

3. The observation output probability distribution of each state is defined according to the training corpus and category c. A word/observation v_k will have a higher output probability at a given state s_i if the word appears frequently with the same ranking position that s_i represents. In addition, all states, regardless of the rank they represent, will also have a probability of emitting words appearing in documents with the c category that HMM was built for. The weight (importance) of these two separate probabilities is controlled by a f parameter.

Given a category c and a dataset D_c of documents that belong to that category, the output probability matrix B for an HMM that represents category c is defined as follows:

$$b_i(v_k) = f \cdot \frac{\sum_{d \in D_c} R_d(v_k, i)}{\sum_{d \in D_c} E_d(i)} + (1 - f) \cdot \frac{\sum_{d \in D_c} A_d(v_k)}{\sum_{j=0}^{|V|} \left(\sum_{d \in D_c} A_d(v_j) \right)}$$

(a) $b_i(v_k)$ stands for the probability of the word/observation v_k being emitted at state s_i

(b) $f \in [0, 1]$

(c) $R_d(v_k, i) = \begin{cases} 1 & \text{if word } v_k \text{ appears at } i\text{th rank position in document } d \\ 0 & \text{in other case} \end{cases}$

(d) $E_d(i) = \begin{cases} 1 & \text{if there is any word with } i\text{th rank position in document } d \\ 0 & \text{in other case} \end{cases}$
This factor is necessary because documents have a different number of feature words. If the number of states is too high, some documents may not have enough feature words to complete all position ranks.

(e) $A_d(v_k) = \begin{cases} 1 & \text{if word } v_k \text{ appears at least one time in document } d \\ 0 & \text{in other case} \end{cases}$

(f) $|V|$ is the number of feature words.

4. The initial probability distribution π is defined by giving probability 1 to the first state s_0.

The first part of the formula represents the relevance of the ranking order. The more weight this part has, the more restrictive the model is when classifying a new document, as it takes into account the exact order of words in relevance from the document training set. Although it can increase the precision of the categorization process, this can lead to an overfit if the f-value is too high.

The second part maintains the same value for all states and provides the model with a better generalization to classify documents.

Finally, taking all the different possibilities for word ranks into account, if a specific document has the same relevance for two or more feature words (which is very improbable after applying TFIDF), then these words will appear in every rank they belong to.

3.1 Classifying New Documents

Once the two Hidden Markov models are created and trained (one for each category), a new document d can be classified by, first of all, formatting it into an ordered wordlist L_d in the same way as in the training process. Then, as words are considered observations in our HMM, we calculate the probability of the word sequence L_d being produced by the two HMMs. That is, $P(L_d|\lambda_R)$ and $P(L_d|\lambda_N)$ need to be computed, where λ_R is the model for relevant documents and λ_N the model for non-relevant documents. The final output class for document d will be the class represented by the HMM with the highest calculated probability.

The calculation of these probabilities is made by applying the forward-backward algorithm explained in the Rabiner article [1].

4 Experiments

4.1 Corpus Preprocessing

To demonstrate the efficiency of the model, we present a set of experiments which have been performed on the OHSUMED biomedical corpus [11], a subset of the MEDLINE database, and the Allele and GO TREC Genomics 2005 corpora [12].

In both of the original TREC datasets, documents can be classified as relevant (R) or non-relevant (N). For the OHSUMED dataset, each document in the set has one or more associated categories (from 23 disease categories). In order to adapt them to a scheme similar to the TREC corpus, we select one of these categories as relevant and consider the others as non-relevant. If a document has been assigned two or more categories and one of them is considered relevant, then the document itself will also be considered relevant and will be excluded from the set of non-relevant documents.

Five categories are chosen as relevant: Neoplasms(C04), Digestive (C06), Cardio (C14), Immunology (C20) and Pathology (C23), since they are by far the most frequent categories of the OHSUMED corpus. The other 18 categories are considered as the common bag of non-relevant documents. For each one of the five relevant categories, a different corpus is created in the way mentioned above, ending up with five distinct matrices.

Additionally, the corpora need to be pre-processed. Following the bag-of-words approach, we format every document into a vector of feature words in which elements describe the word occurrence frequencies. All the different words that appear in the training corpus are candidates for feature words. In order to reduce the initial feature size, standard text pre-processing techniques are used. A predefined list of stopwords (common English words) is removed from the text, and a stemmer based on the Lovins stemmer [13] is applied. Then, words occurring in fewer than ten documents of the entire training corpus are also removed.

Once the initial feature set is determined, a dataset matrix is created where rows correspond to documents and columns to feature words. The value of an element in a matrix is determined by the number of occurrences of that feature word (column) in the document (row). This value is adjusted using the tf-idf statistic in order to measure the word relevance. The application of tf-idf decreases the weight of terms that occur very frequently in the collection, and increases the weight of terms that occur rarely [13].

In addition, a feature reduction algorithm is applied to the resultant feature word set in order to further reduce its dimensionality. The feature selection method based in Information Gain that is implemented in WEKA [14] is used as the feature reduction algorithm for this study, since it was previously employed and proved its effectiveness in similar text classification tasks [15]. This algorithm uses a threshold to determine the size of the reduced feature set. In this case, the threshold is set to the minimum value, so that every feature word with a non-zero value of information gain is included in the resultant feature set.

4.2 Comparison with Other Related Methods on Text Classification

Commonly used text classification models such as Naive Bayes, *k*-NN and SVM are among the best performers, particularly the latter. In this research, tests were made with these classifiers using the same corpus in order to compare them with the novel T-HMM. Tests consist of a supervised classification using a

Table 1. Comparative between models (T-HMM, Naive Bayes, *k*-NN and SVM). The values for each measure correspond to the average value obtained in the 10 fold-cross evaluation.

Corpus	Measure	T-HMM	SVM	Bayes	*k*-NN
Allele	Kappa	0,530	0,268	0,349	0,149
	F-measure (N)	0,963	0,975	0,924	0,971
	F-measure (R)	0,564	0,282	0,403	0,166
GO	Kappa	0,314	0,000	0,241	0,061
	F-measure (N)	0,914	0,958	0,887	0,956
	F-measure (R)	0,388	0,000	0,329	0,081
Ohsumed (C04)	Kappa	0,754	0,750	0,641	0,423
	F-measure (N)	0,935	0,942	0,899	0,882
	F-measure (R)	0,819	0,807	0,741	0,531
Ohsumed (C06)	Kappa	0,772	0,667	0,616	0,447
	F-measure (N)	0,968	0,965	0,940	0,950
	F-measure (R)	0,803	0,701	0,673	0,486
Ohsumed (C14)	Kappa	0,768	0,793	0,616	0,380
	F-measure (N)	0,943	0,953	0,893	0,892
	F-measure (R)	0,824	0,839	0,720	0,466
Ohsumed (C20)	Kappa	0,603	0,583	0,503	0,401
	F-measure (N)	0,932	0,958	0,905	0,946
	F-measure (R)	0,666	0,621	0,586	0,443
Ohsumed (C23)	Kappa	0,420	0,472	0,337	0,251
	F-measure (N)	0,780	0,829	0,771	0,773
	F-measure (R)	0,640	0,637	0,564	0,458

10 fold-cross validation for all the mentioned corpus. Evaluation measures used are F-measure [13] and Kappa Statistic [16].

Table 1 shows the results of these evaluations. The configuration for Naive Bayes, k-NN and SVM models used for the tests are those utilized by default in WEKA environment [17], taking $k = 3$ and applying a RBF kernel for SVM. These parameters have proven to be effective in a general classifying process, and in the case of the k-NN algorithm, using a higher number of neighbours hinders the classification performance.

In the case of the T-HMM parameters, they are also established with a general approach. The number of states n is set to the average number of feature words with a non zero value in the traning set in each fold-cross validation step, while the f-factor is set to 0.5.

According to the results, T-HMM outperforms Naive Bayes and k-NN in relevant class F-measure and Kappa measures for each tested corpus. It is important to note that in this case, the F-measure for the relevant category has a greater importance in the evaluation, since the focus of this kind of text classification is to retrieve the relevant documents to a topic. On the other hand, the Kappa statistic provides a better comparison in terms of overall performance.

In the case of the SVM, T-HMM provides better overall results in 4 of 5 evaluations. This is particulary prominent in the GO corpus, where the SVM cannot classify correctly any of the relevant documents due to the class distribution of the corpus, being the number of relevant documents much inferior to the number of non-relevant documents.

5 Conclusions and Future Work

This paper has introduced a novel method in biomedical text classification using Hidden Markov Models. It offers a simple and parametrizable structure, based on relevance of words, which allows the model to be adjusted to future documents. The system was made for distinguishing relevant documents from a dataset, giving a start point to common search systems, where a document may be relevant or not according to a user query. Its application can be extended to other domains related to text processing. As an example, in [18] a variant of this T-HMM model is proposed as an over-sampling technique to improve the efficiency of other classifiers, focusing solely on the stage prior to the classification process.

The experimental results of this study show that the application of the proposed HMM-based classifier appears to be promising. In automatic classification of the OHSUMED medical corpus, our model outperforms commonly used text classification techniques like Naive Bayes, and achieves comparable results to the SVM approach, which is recognized as one of the most effective text classification methods. In addition, T-HMM is more efficient in terms of running time than the other approaches.

In the whole process, there are still some areas that could be improved. Firstly, our model can represent the relevance of words by their ranking, but there is no representation for how large the difference of this relevance is. This may help distinguish between general and technical documents, since a big gap between word relevance in a document is probably the result of a few words with high frequency value, which would indicate a more specialized document.

Another line of research, following the proposed structure, is to adapt the model into a ranking system. The current information retrieval (IR) methods have severe difficulties to generate a pertinent rank of the documents retrieved as relevant. Thus, a new scheme based on our HMM approach could be useful to re-rank the list of documents returned by another IR system.

Acknowledgements. This work has been funded from the European Union Seventh Framework Programme [FP7/REGPOT-2012-2013.1] under grant agreement n 316265, BIOCAPS.

References

1. Rabiner, L.R.: A tutorial on hidden Markov models and selected applications in speech recognition. In: Waibel, A., Lee, K.-F. (eds.) Readings in Speech Recognition, pp. 267–296. Morgan Kaufmann Publishers Inc., San Francisco (1990)
2. Freitag, D., Mccallum, A.K.: Information extraction with hmms and shrinkage. In: Proceedings of the AAAI 1999 Workshop on Machine Learning for Information Extraction, pp. 31–36 (1999)
3. Leek, T.R.: Information extraction using hidden markov models. Master's thesis, UC San Diego (1997)
4. Miller, D.R.H., Leek, T., Schwartz, R.M.: A hidden markov model information retrieval system. In: Proceedings of the 22nd Annual International ACM SIGIR Conference on Research and Development in Information Retrieval, SIGIR 1999, pp. 214–221. ACM, New York (1999)
5. Frasconi, P., Soda, G., Vullo, A.: Hidden markov models for text categorization in multi-page documents. Journal of Intelligent Information Systems 18, 195–217 (2002)
6. Li, K., Chen, G., Cheng, J.: Research on hidden markov model-based text categorization process. International Journal of Digital Content Technology and its Application 5(6), 244–251 (2011)
7. Yi, K., Beheshti, J.: A hidden markov model-based text classification of medical documents. Journal of Information Science 35(1), 67–81 (2009)
8. Sebastiani, F.: Text categorization. In: Text Mining and its Applications to Intelligence, CRM and Knowledge Management, pp. 109–129. WIT Press (2005)
9. Nikolaos, T., George, T.: Document classification system based on hmm word map. In: Proceedings of the 5th International Conference on Soft Computing as Transdisciplinary Science and Technology, CSTST 2008, pp. 7–12. ACM, New York (2008)
10. Barros, F.A., Silva, E.F.A., Cavalcante Prudêncio, R.B., Filho, V.M., Nascimento, A.C.A.: Combining text classifiers and hidden markov models for information extraction. International Journal on Artificial Intelligence Tools 18(2), 311–329 (2009)
11. Hersh, W.R., Buckley, C., Leone, T.J., Hickam, D.H.: Ohsumed: An interactive retrieval evaluation and new large test collection for research. In: SIGIR, pp. 192–201 (1994)

12. Hersh, W., Cohen, A., Yang, J., Bhupatiraju, R.T., Roberts, P., Hearst, M.: Trec 2005 genomics track overview. In: TREC 2005 Notebook, pp. 14–25 (2005)
13. Baeza-Yates, R.A., Ribeiro-Neto, B.: Modern Information Retrieval. Addison-Wesley Longman (1999)
14. Witten, I.H., Frank, E.: Data Mining: Practical Machine Learning Tools and Techniques. Morgan Kaufmann Series in Data Management Sys. Morgan Kaufmann (June 2005)
15. Janecek, A.G., Gansterer, W.N., Demel, M.A., Ecker, G.F.: On the relationship between feature selection and classification accuracy. In: JMLR: Workshop and Conference Proceedings, vol. 4, pp. 90–105 (2008)
16. Viera, A.J., Garrett, J.M.: Understanding interobserver agreement: the kappa statistic. Family Medicine 37(5), 360–363 (2005)
17. Sierra Araujo, B.: Aprendizaje automático: conceptos básicos y avanzados: aspectos prácticos utilizando el software Weka. Pearson Prentice Hall (2006)
18. Iglesias, E.L., Seara Vieira, A., Borrajo, L.: An hmm-based over-sampling technique to improve text classification. Expert Systems with Applications 40(18), 7184–7192 (2013)

TIDA: A Spanish EHR Semantic Search Engine

Roberto Costumero, Consuelo Gonzalo, and Ernestina Menasalvas

Universidad Politécnica de Madrid - Centro de Tecnología Biomedica, Madrid, Spain
{roberto.costumero,consuelo.Gonzalo,ernestina.menasalvas}@upm.es

Abstract. Electronic Health Records (EHR) and the constant adoption of Information Technologies in healthcare have dramatically increased the amount of unstructured data stored. The extraction of key information from this data will bring better caregivers decisions and an improvement in patients' treatments. With more than 495 million people talking Spanish, the need to adapt algorithms and technologies used in EHR knowledge extraction in English speaking countries, leads to the development of different frameworks. Thus, we present TIDA, a Spanish EHR semantic search engine, to give support to Spanish speaking medical centers and hospitals to convert pure raw data into information understandable for cognitive systems. This paper presents the results of TIDA's Spanish EHR free-text treatment component with the adaptation of negation and context detection algorithms applied in a semantic search engine with a database with more than 30,000 clinical notes.

Keywords: Natural Language Processing, Electronic Health Records, Negation Detection.

1 Introduction

The adoption and effective use of information technologies in healthcare (in particularly the Electronic Health Record - EHR) has been hindered for decades. The leading English speaking countries have brought new technologies to build health decision-making systems in the last years such as Natural Language Processing (NLP), Negation and context detection algorithms, frameworks like Mayo's cTAKES [1], terminology systems as the Unified Medical Language System (UMLS) [2] or actual systems such as IBM Watson^{TM}[3]. Despite the interest shown in the adoption of such technologies in healthcare and having more than 495 million people talking Spanish throughout the world [4] there has not been great efforts in the translation and adoption in the Spanish environment.

In this paper we present the design of TIDA (Texts and Images Data Analyzer), the architecture of a Spanish EHR semantic search engine that makes it possible to: I) process clinical notes and radiological information identifying negation and context of medical terms; II) process images and other media to identify patterns of abnormality and key factors physicians need in order to complete a patient's study and III) extract knowledge using Big Data Analytics using the sources aforementioned and patient's structured data to build intelligent end-user applications. TIDA's components and their functionality are presented over the different sections, though, in this paper, we concentrate on TIDA's component responsible for medical natural text processing.

J. Sáez-Rodríguez et al. (eds.), *8th International Conference on Practical Appl. of Comput. Biol. & Bioinform. (PACBB 2014)*, Advances in Intelligent Systems and Computing 294,
DOI: 10.1007/978-3-319-07581-5_28,

The main contributions of the paper are: I) to present the architecture to homogenize the structure of the data as the basis for the analytic system; II) to introduce the problems regarding the health related texts in Spanish (clinical notes and imaging reporting) and III) to adapt NegEx and ConText algorithms to be able to deal with Spanish texts, integrating with the output of existing indexing tools so they can be used as the input for the adapted ConText algorithm for Spanish towards semantic enrichment of queries.

The rest of the paper has been organized as follows: Section 2 presents the related works with emphasis on natural language processing . In Section 3 we present the architecture of the system, the algorithms and technologies used. Section 4 presents preliminary results. To end with, conclusion and future works are presented in section 5.

2 Related Work

In Natural Language Processing (NLP) the input of natural language is transformed in several steps of a pipeline to get computers to understand it. In the treatment of free-text NLP, the text serves as an input to the system which will lead to several structured components with semantic meaning so the information can be manipulated knowing the importance of the different parts of the speech. To do the proper training for NLP algorithms to learn, a properly annotated corpus is needed.

These components learn from the training data included in the corpus to analyze future input sentences. Although there are several studies introducing different English corpora to train models to get NLP process working, we focus here on the ones which have been used in the healthcare domain. As cited in Savova et al. [1], there are no community resources such as annotated medical corpus in the clinical domain, so in the evaluation on tools like cTAKES, own corpus has been developed. Using the gold standard linguistic annotations of Penn TreeBank (PTB) [10] and GENIA corpus [11] together with their own Mayo Clinic EMR corpus, cTAKES models were trained. The lack of corpus in Spanish language for the healthcare domain makes such training difficult these days.

The usual NLP pipeline components are a *Sentence Detector*, a *Tokenizer*, a *Part of Speech Tagger*, a *Chunker or Shallow Parser* and a *Named Entity Recognition*. In Figure 2 the process we have implemented is illustrated. Another module that is useful once terms are found is the negation detector that identifies when a particular term or expression appears negated in a text. Depending on the domain context and hypothetical flavor of an expression may also be required.

Natural Language Processing is being extensively used in fields like medicine with many different framework approaches such as cTAKES [1]. It is an open source framework for information extraction from EHR's free-text. It has been developed by the Mayo clinic and open sourced by the Apache Foundation. It is built upon IBM's UIMA (Unstructured Information Management Architecture) [7] and Apache's OpenNLP [8]. cTAKES also includes several components that can be reused independently or executed together in a pipelined process whose output is a structured representation of the unstructured free-text.

cTAKES processes clinical notes written in English identifying different medical concepts from several dictionaries, included own developed ones, but also Unified Medical Language System (UMLS)[2] For this purpose, the algorithm NegEx [9] is used.

Once preprocessed, clinical notes can be searched to find relevant information for clinical and research purposes. In this process two important aspects require special attention: I) negation of terms and II) context of terms. Negation can invert the sense of a term consequently yielding numerous false-positive matches in a search. On the other hand, the context of a term identifies the subject it refers to (also known as the experiencer). This is specially important in clinical notes so to be able to identify the subject of a symptom or disease.

NegEx [9] is a simple algorithm for identifying negated findings and diseases in clinical notes based on the appearance of key negation phrases on the text, limiting the scope of the negation trigger. It is a well known algorithm for negation detection in the healthcare domain and the terms and the algorithm have already been translated and tested in other languages such as Swedish, German and French. In the original paper [9] the algorithm was tested with 1235 findings and diseases in 1000 sentences from discharge summaries. It gave an specificity of 94.5% and a positive predictive value of 84.5% with a sensitivity of 77.8%.

ConText [13] is an extension of the NegEx algorithm that not only identifies negation but also identifies an hypothetical status. It is intended to work moderately at finding the person experiencing the symptoms or whether the condition occurred historically. ConText algorithm [13] was tested with 4654 annotations combining the development and test set of 240 reports (2377 annotated conditions in the development set and 2277 annotated conditions in the test set). Overall negation detection has a F-Measure of 0.93. Overall historical detection has a F-Measure of 0.76 while the overall hypothetical detection has a F-Measure of 0.86.

3 TIDA Spanish EHR Semantic Search Engine

3.1 Introduction

The complexity of healthcare information management is not only due to the amount of data generated but also by its diversity and the challenges of extracting knowledge from unstructured data. Solutions proposed until now have been focused on different aspects of the information process, ranging from unstructured text analysis from discharge summaries and radiology reports, to analysis of PET imaging or Rx, but none of them giving an integrated solution to process and mine obtaining information from all sources.

3.2 TIDA's Architecture Design

We present TIDA (Text, Image and Data Analytics) a Spanish EHR semantic search engine. TIDA makes it possible to address the problem of Spanish text indexing in the healthcare domain by adapting different techniques and technologies which are explained in this section.

TIDA is designed to build flexible applications over a typical data storage system. TIDA's architecture mainly relies on the information obtained from healthcare databases, which has previously been gathered together into a common storage, so different components get the information from a common warehouse. In order to fulfill this requirements, the architecture (see Figure 1) presents the following components: a **DB** as common data storage system with all hospital's data, from reports to images and patient's structured information which will serve information to the immediate upper level of components; the **Mayo/Apache cTAKES** as free-text analysis system built upon Mayo's cTAKES framework. This framework relies on information gathered through UMLS from different dictionaries for diseases, drugs or laboratory test classifications such as ICD, RxNorm, LOINC or SNOMED CT; an **Image transformation framework** including a set of own developed applications to determine anomalies and automatically annotate medical images using the IBM UIMA architecture which cTAKES is built upon; the **IBM/Apache UIMA** component to gather the two previous components' output in order to get a structured view of the unstructured data; the **Patient's structured data**; a **Structured data, images and text annotator** in charge of annotating text and images supported by UIMA and the structured information; An instance of **Apache Lucene** which indexes all the previously annotated data to serve different kinds of applications; **Apache Solr** to bring quick, reliable semantic search into the picture; An **API** powered by Solr's output, to bring more functionality and link end-user web applications to the whole set of the architecture; and finally, the **End-user web application** to serve end-user applications to give different functionalities on the same data structures.

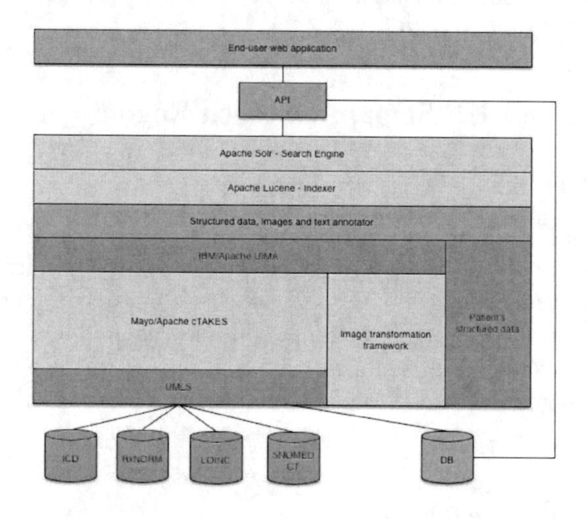

Fig. 1. Principal components of TIDA's architecture

TIDA's architecture is designed to be language-independent, so the platform principles stay the same no matter which language texts are written in. Language-dependent components UMLS and cTAKES can be adapted so the architecture works as expected. In this paper we present the Spanish adapted version of TIDA, which has been built incorporating the components in Spanish.

3.3 Text Analysis in TIDA

It should be noted that, though this architecture is presented to get the whole picture of the work that needs to be done, this paper concentrates in the advances obtained in the free-text analysis. Figure 2 shows the process followed in the free-text analysis done in this paper.

Due to the lack of health related corpora in Spanish, we decided to use a general domain annotated corpus, so at least the models can be trained to be used with Spanish words. AnCora [14] is one of the fully annotated Spanish corpus containing more than 500,000 words.

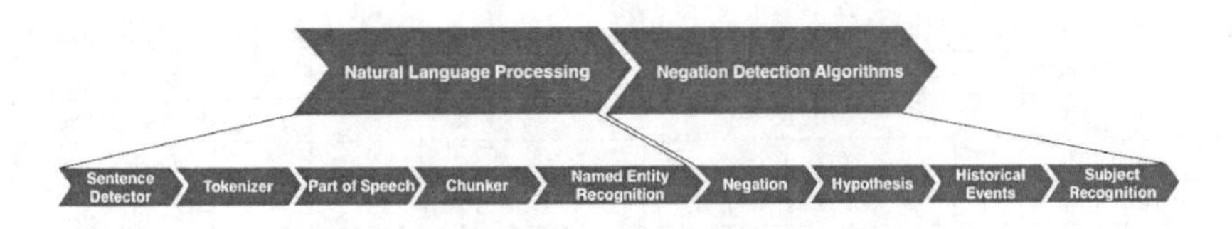

Fig. 2. TIDA's text analysis process

There should be considered two main processes in the text analysis:

1. **Indexation**, which is a heavy process due to a complete text analysis and which runs in a pipeline the different processes presented in Figure 2, starting with Natural Language Processing and following with the Negation detection algorithms.
2. **Searching**, which thanks to indexation is a very lightweight process.

TIDA's text analysis relies on cTAKES [1], which has been briefly introduced in Section 2. cTAKES is an English-centric development, which means that, although it has a very good design and is very modular, it has been developed towards an English comprehensive system. This introduces a challenge to make it suitable to use in other languages. It is not on the scope of this paper to introduce cTAKES architecture, but we will introduce key components to develop our work.

cTAKES uses a common NLP pipeline including a Sentence Detector, a Tokenizer, a Part of Speech Tagger, a Chunker and a Dictionary Lookup for Named Entity Recognition. Although this components should be modified, using the newly trained models to work in Spanish, we are focusing on this paper in the Assertion module, which has been developed outside the scope of cTAKES in a prototype to prove its functionality prior its integration in cTAKES pipeline. cTAKES assertion module, the one involved in negation detection, is the one replicating the functionality on the analysis of the negation, hypothesis and the experiencer on a particular condition.

We are working in line with the NegEx algorithm, which relies on several keywords acting as "negation triggers". The translation of the triggering terms helps us to detect in a moderated way the detection of the negation in Spanish. An example of such terms can be seen in Table 1.

When analyzing the experiencer for a certain condition in ConText, triggering terms must be used. The results obtained when querying for the condition must have the actual condition but associated to an experiencer distinct from the patient. Examples of different historical and hypothetical triggering terms can be found in Table 2.

Table 1. Example of NegEx triggers' translation

Trigger in English	Trigger in Spanish
can be ruled out can rule him out	se puede descartar
no, not	no
no evidence	sin evidencia, no evidencia
no suspicion of	ninguna sospecha de

Table 2. Example of ConText triggers' translation

Historical in English	Historical in Spanish	Hypothesis in English	Hypothesis in Spanish
again noted	observado de nuevo	could	podría
change in	cambio en	likely	probablemente
chronic	crónico	looks like	parece
clinical history	historia clínica de	not certain	sin certeza

4 Results

TIDA's text analysis component has been trained with AnCora [14] corpus, which has more than 500,000 words. Around 500 words have been translated for the negation detection algorithms. In order to test the system clinical notes and reports (> 30,000) from different services of several public hospitals in Spain have been provided to test the system doing the proper indexation and then querying the system with some medical terms. In this section, we show some results of the application of the process both indexation and searching through the different processes of free-text analysis. The software has been executed in a machine with a Quad-core 2.83GHz CPU with 4GB of RAM.

Figure 3 shows a particular case used to test the NLP component before doing the indexation of the clinical notes with Apache Lucene [15]. This sentence has been extracted from a clinical note. This same figure serves to see what the actual output of the Sentence Detector would be. Tokenization is shown below the Sentence Detection, where 11 are returned, one for each of the words in the sentence. Afterwards, the process analyzes the part of speech for each token. Note that *NN* stands for a Noun, *JJ* stands for an Adjective, *IN* stands for a preposition and *NNS* stands for a Noun in plural, following Penn Treebank's PoS tags [10], which are widely understood in English speaking environment. Finally, the chunker output is found. PoS tokens are grouped into: Noun Phrases (NP), Verb Phrases (VP), Prepositional Phrases (PP) or Others (O).

Testing of the searching system has been done with several queries using Apache Solr [16] to get the user input and the results displayed in a user-friendly environment. A particular case is the use of the metastasis medical term. Figure 4 shows an example of the query results for "metástasis" (Spanish spelling of metastasis) in which the patient actually has metastasis and below the result where the patient does not. The system returned a total of 6,835 results found in 7ms. In the latter query, the system returned a total of 6,296 results found in 8ms. As indexation is done separately and before the process of the actual search, the performance of the searching process is very good,

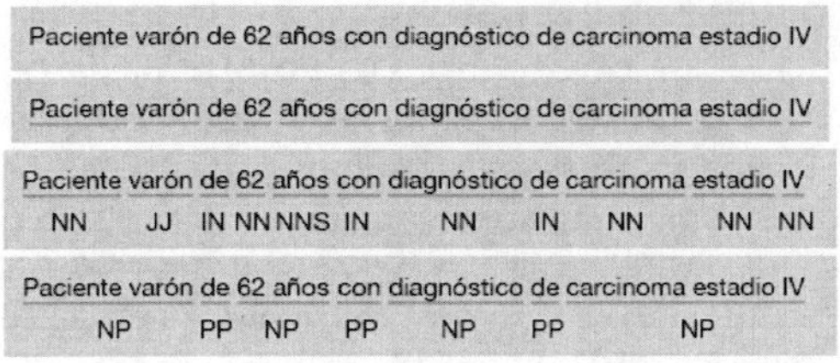

Paciente varón de 62 años con diagnóstico de carcinoma estadio IV

Paciente varón de 62 años con diagnóstico de carcinoma estadio IV

Paciente varón de 62 años con diagnóstico de carcinoma estadio IV
NN JJ IN NN NNS IN NN IN NN NN NN

Paciente varón de 62 años con diagnóstico de carcinoma estadio IV
NP PP NP PP NP PP NP

Fig. 3. NLP process example for a given sentence

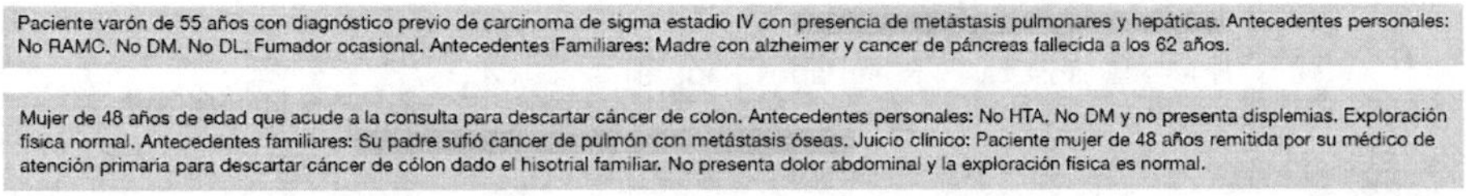

Paciente varón de 55 años con diagnóstico previo de carcinoma de sigma estadio IV con presencia de metástasis pulmonares y hepáticas. Antecedentes personales: No RAMC. No DM. No DL. Fumador ocasional. Antecedentes Familiares: Madre con alzheimer y cancer de páncreas fallecida a los 62 años.

Mujer de 48 años de edad que acude a la consulta para descartar cáncer de colon. Antecedentes personales: No HTA. No DM y no presenta displemias. Exploración física normal. Antecedentes familiares: Su padre sufió cancer de pulmón con metástasis óseas. Juicio clínico: Paciente mujer de 48 años remitida por su médico de atención primaria para descartar cáncer de cólon dado el hisotrial familiar. No presenta dolor abdominal y la exploración física es normal.

Fig. 4. Metastasis positive and negative query result

with response in the range of milliseconds. However, the analysis of the performance of the system is left for future work.

5 Conclusions and Future Work

The increase of data generated and the adoption of IT in healthcare have motivated the development of systems such as IBM Watson and frameworks like cTAKES (integrated exclusively in an English speaking environment). The huge amount of Spanish speaking people leads to the development of new solutions adapted to Spanish. Thus, in this paper we have presented TIDA, an approach which focuses into the generation of a complete EHR semantic search engine which brings an integrated solution to medical experts to analyze and identify markers in a system that brings together text, images and structured data analysis.

Results presented demonstrate that the adaptation of existing algorithms and technologies to the Spanish environment is currently working and that cognitive systems can be built to work in the health domain. The proper treatment of Spanish texts together with the correct adaptation of ConText algorithm lead to the correct indexation for better queries.

Future work leads to the process on integrating the newly trained models and the Spanish adapted algorithms into the cTAKES framework as well as validating the whole system's performance is left to future work. Also, we are leaving for future work the application of data mining techniques so the indexing component boosts medical related terms and the search engine returns better results.

References

1. Savova, G.K., et al.: Mayo clinical Text Analysis and Knowledge Extraction System (cTAKES): architecture, component evaluation and applications. Journal of the American Medical Informatics Association 17(5), 507–513 (2010)

2. Bodenreider, O.: The unified medical language system (UMLS): integrating biomedical terminology. Nucleic Acids Research 32(suppl. 1), D267–D270 (2004)
3. Watson, I.B.M.: (December 13, 2013), http://www.ibm.com/watson
4. Cervantes Institute. El espaol: una lengua viva. Informe (2012), http://cvc.cervantes.es/lengua/anuario/anuario_12/i_cervantes/p01.htm (January 8, 2014)
5. Logica and Nordic Healthcare group. Results from a survey conducted by Logica and Nordic Healthcare Group (January 2012), http://www.logica-group.com/we-are-logica/media-centre/thought-pieces/2012/market-study-of-electronic-medical-record-emr-systems-in-europe/~/media/Global%20site/Media%20Centre%20Items/Thought%20pieces/2012/WPPSEMRJLv16LR.ashx (January 7, 2014)
6. Hospitales, H.M.: Estadsticas y Resultados Sanitarios (2012), http://www.hmhospitales.com/grupohm/Estadisticas/Paginas/Estadisticas-Generales.aspx (January 7, 2014)
7. Ferrucci, D., Lally, A.: UIMA: an architectural approach to unstructured information processing in the corporate research environment. Natural Language Engineering 10(3-4), 327–348 (2004)
8. Apache OpenNLP, http://opennlp.apache.org (November 21, 2013)
9. Chapman, W.W., et al.: A simple algorithm for identifying negated findings and diseases in discharge summaries. Journal of Biomedical Informatics 34(5), 301–310 (2001)
10. Marcus, M.P., Marcinkiewicz, M.A., Santorini, B.: Building a large annotated corpus of English: The Penn Treebank. Computational Linguistics 19(2), 313–330 (1993)
11. Kim, J.-D., et al.: GENIA corpusa semantically annotated corpus for bio-textmining. Bioinformatics 19(suppl. 1), i180–i182 (2003)
12. ILSP. Hellenic National Corpus (January 9, 2014), http://hnc.ilsp.gr/en/
13. Harkema, H., et al.: ConText: An algorithm for determining negation, experiencer, and temporal status from clinical reports. Journal of Biomedical Informatics 42(5), 839–851 (2009)
14. Taul, M., Mart, M.A., Recasens, M.: AnCora: Multilevel Annotated Corpora for Catalan and Spanish. In: LREC (2008)
15. Lucene, A.: (November 21, 2013), https://lucene.apache.org/core/
16. Solr, A.: (November 21, 2013), http://lucene.apache.org/solr/

BioClass: A Tool for Biomedical Text Classification

R. Romero, A. Seara Vieira, E.L. Iglesias, and L. Borrajo

Univ. of Vigo, Computer Science Dept., Escola Superior de Enxeñería Informática, Ourense, Spain
{rrgonzalez,adrseara,eva,lborrajo}@uvigo.es

Abstract. Traditional search engines are not efficient enough to extract useful information from scientific text databases. Therefore, it is necessary to develop advanced information retrieval software tools that allow for further classification of the scientific texts. The aim of this work is to present BioClass, a freely available graphic tool for biomedical text classification. With BioClass an user can parameterize, train and test different text classifiers to determine which technique performs better according to the document corpus. The framework includes data balancing and attribute reduction techniques to prepare the input data and improve the classification efficiency. Classification methods analyze documents by content and differentiate those that are best suited to the user requeriments. BioClass also offers graphical interfaces to get conclusions simply and easily.

Keywords: Biomedical text mining tool, Text classification, Bioinformatics, Computer-based software development.

1 Introduction

In recent years, the area of Information Retrieval (IR) has increased its primary goals of indexing, searching and categorizing documents of a collection. Today, research in IR includes modeling, document classification and categorization, system architecture, user interfaces, data visualization, filtering, languages, etc.

The constant growth of information services and available resources has decreased the effectiveness in obtaining relevant information for a specific research interest. Without an efficient strategy for extracting relevant information for the researcher, existing tools may become unusable.

In order to extract information from scientific texts stored in public data sources and to determine if a collection is relevant to a specific topic, several challenges must be met. First of all, documents must be transformed into an intermediate representation before their content can be analyzed. The typical representation of a document is a set of *keywords*, which can either be extracted directly from the text of the document or specified manually by an expert. The keyword relevance is different depending on the document. This effect is captured by assigning numerical weights for the keyword in each document [1]. In this way

J. Sáez-Rodríguez et al. (eds.), *8th International Conference on Practical Appl. of Comput. Biol. & Bioinform. (PACBB 2014)*, Advances in Intelligent Systems and Computing 294,
DOI: 10.1007/978-3-319-07581-5_29,

the corpus becomes a matrix, referred to as a *sparse matrix*, in which rows are the documents (instances) and columns are the keywords.

After corpus preprocessing, reasoning models are usually used to analyze and classify documents. However, two problems may arise. First, the sparse matrix is often very large and bulky. Second, the number of documents in the majority class (non-relevant documents) is usually much higher than that of the minority class (relevant documents) [2].

Using attribute reduction algorithms, the most relevant characteristics are selected, producing a much more specific dataset containing fewer noisy attributes. On the other hand, by applying balancing techniques it is possible to change the number of documents per class in order to solve problems related to overfitting [3,4].

Finally, once the results from the classification process are obtained, the researcher needs graphical interfaces, which provide scenarios that allow simply and easily make conclusions.

The aim of this work is to present BioClass, a new tool for biomedical text classification, which includes balancing and attribute reduction techniques, classification algorithms and data visualization. The framework is based on a well-defined workflow to guide the user through the classification process.

2 BioClass Architecture

As shown in Figure 1, BioClass is composed of six software modules.

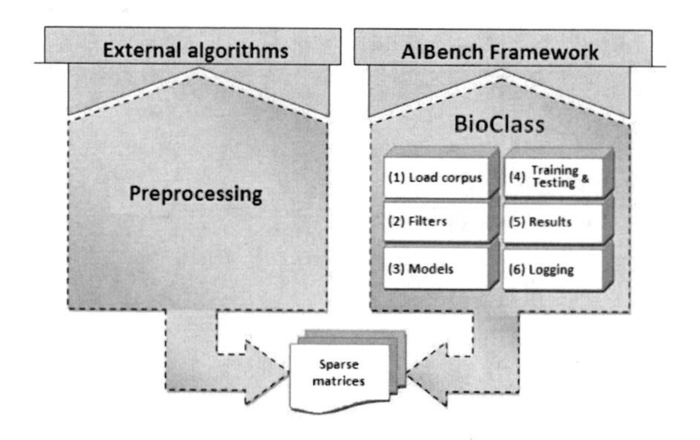

Fig. 1. BioClass architecture module

The *Load Corpus module* is responsible for obtaining the preprocessed dataset (sparse matrix) for the classification. It was designed as a highly scalable abstraction layer, which allows different data access connectors to be added. It currently supports CSV and ARFF [5] file types.

The *Filtering module* carries out operations on datasets in order to improve the classification processes. It is divided into two parts. The first is to filter instances or documents, which makes it possible to handle the number of documents that belong to each class (relevant or not relevant). In order to support these operations, *subsampling* and *oversampling* algorithms are implemented. The subsampling algorithm artificially decreases the number of samples that belong to the majority class. The oversampling algorithm redistributes the number of samples that belong to the minority class, taking the majority class into account [3,6].

The second part reduces the number of attributes of the corpus. This approach is intended to decrease the dimensionality without losing expressive power. The following parameters are implemented:

- *Operation set* applies union, intersect and minus operations between the test, and trains matrices to increase the number of correlated attributes.
- *Cfs subset eval* evaluates the worth of an attribute subset by considering the individual predictive ability of each feature along with the degree of redundancy between them [7].
- *Chi squared* evaluates the worth of an attribute by computing the value of the chi-squared statistic with respect to the class [8].
- *Consistency subset eval* determines the most consistent attribute subset with respect to the original attribute set [9].
- *Info gain attribute eval* evaluates the worth of an attribute by measuring the information gain with respect to the class [10].
- *Gain ratio attribute eval* evaluates the worth of an attribute by measuring the gain ratio with respect to the class [10].
- *One R attribute Eval* uses a ruled-bases system to select the most relevant attributes [11].
- *Principal Components* generates a new attribute subset merging attributes with a greater independence of values [12].

The *Models module* supports different reasoning algorithm implementations and provides methods for training and testing. In order to select the algorithms implemented in BioClass, an exhaustive analysis using BioCreative [13], TREC [14] and Semantic Mining in Biomedicine [15] conferences was performed. As a result of the analysis, two Support Vector Machine types (Cost-SVM and Nu-SVM) [16], Naive Bayes [17] and K-Nearest Neighbour classifiers [18] were implemented.

The *Train and test module* allows a reasoning model to be chosen, parameterized, trained with a specific dataset, and tested. It is an abstraction layer with respect to the classification process through which the developer can implement new reasoning models and integrate them with BioClass.

The *Results module* provides a graphical user interface to interact with BioClass. It allows the user to select between reasoning models or filters for data processing, and to edit or manage datasets and classifiers. Additionally it provides an user-friendly interface to analyze the obtained results from a classification process.

Finally, the *Log module* provides additional information about processes carried out. It can be useful for solving problems that may arise during beta testing or start up. This module also provides an abstraction layer for persistence, and offers different formats to store the data log.

Bioclass was developed with the Java version 6 programming language and is based on the AIBench framework (www.aibench.org) [19]. It is distributed under GNU LGPLv3 license. AIBench allows the research effort to be centered on the algorithm development process, leaving the design and implementation of input-output data interfaces in the background. Reasoning model APIs are those implemented by Weka [5], except for the SVMs for which LibSVM [20] was used.

3 BioClass User Interaction

The main BioClass interface is composed of a Menu bar, a Clipboard Panel, a Viewer panel and a Log Panel (see Figure 2).

- Menu bar provides access to each operation available in BioClass, e.g., load corpus or data, apply instance and attribute filters, or create and use reasoning models to perform the classification process.
- Clipboard panel contains those elements which have been generated by previous operations. These objects can be used as input data by new operations, or can launch their associated views.
- Viewer panel shows the appropriate view attached to a specific Clipboard object. Visualization is a very simple process to accomplish; the user needs only select an object from the Clipboard and click on it.
- Log panel offers acknowledgement over a process or action trace.

Operations in BioClass are grouped by categories: Corpus, Classification and Filtering. Available operations in the Corpus menu allow datasets to be loaded onto the system and added to the Clipboard. The researcher can use the Classification menu to create, train and test reasoning models in different ways. Finally, Filtering menu contains instance algorithms to reduce data dimensionality or reorganize instances.

In order to understand how BioClass works, we have developed a general scenario to guide the user through the full classification process. Figure 3 shows a cyclic process covering all classification tasks and the correspondence of each task with the operations in BioClass. It was drawn as cyclic to highlight the fact that obtained results can be used as feedback in a new classification process.

As shown in Figure 3, the Load Corpus task is directly connected to the Load Corpus module, and enclosed in the Corpus category. The researcher can use it to load their preprocessed datasets into BioClass.

Once the dataset has been loaded in BioClass, the researcher can select it in the Clipboard and obtain some relevant information from its viewers. Figure 4 show information about the selected dataset grouped by attributes. Instance and attribute tabs can obtain information about documents or attributes used to filter or to delete them. Some relevant information is represented by statistical

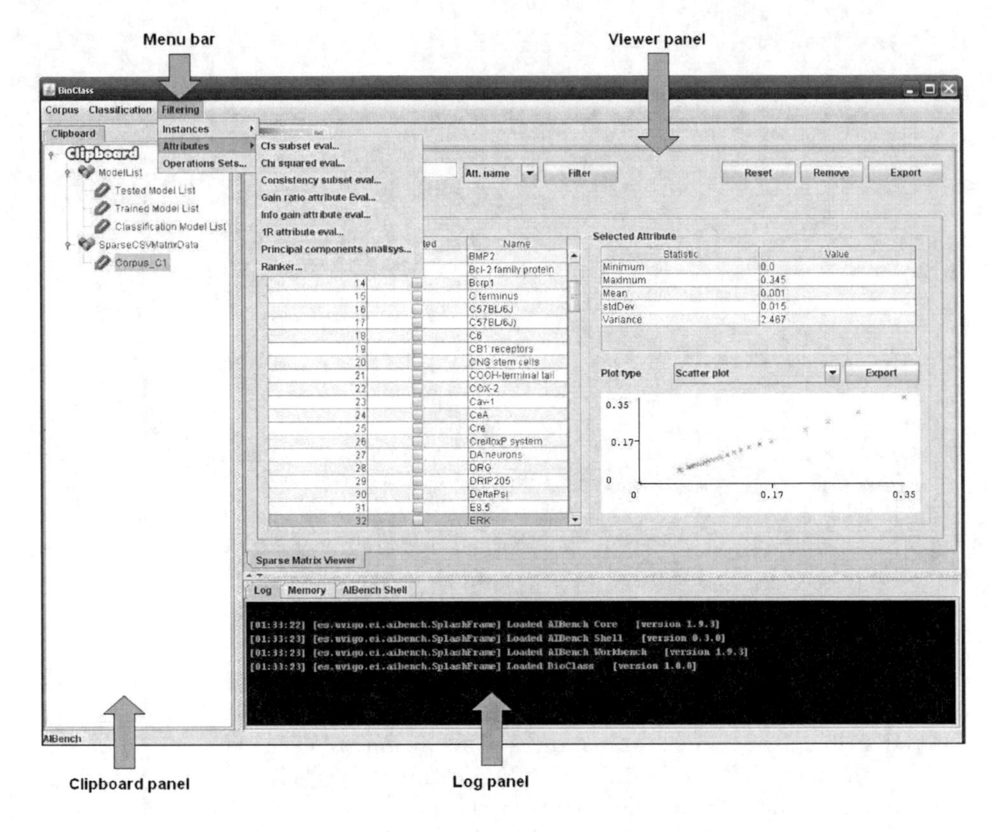

Fig. 2. BioClass main interface

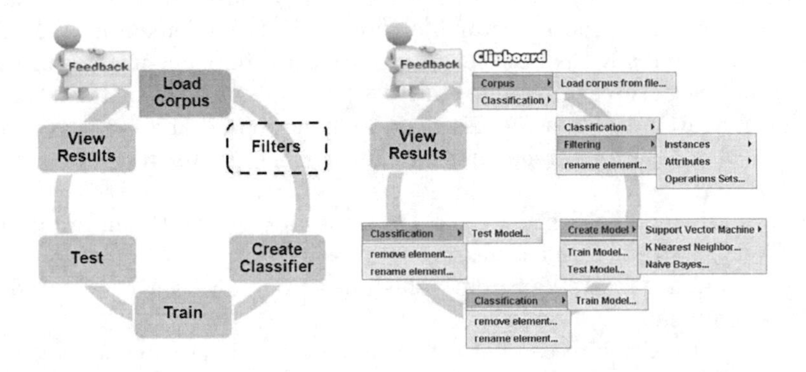

Fig. 3. General classification process scenario

measures and plots in order to extract valuable conclusions. Additionally, modified data can be exported as a new dataset or reset. Plots can also be exported as JPEG or GIF files.

The second task, Filters, refers to those algorithms that can increase the accuracy of the datasets according to a specific classifier, or help to improve the

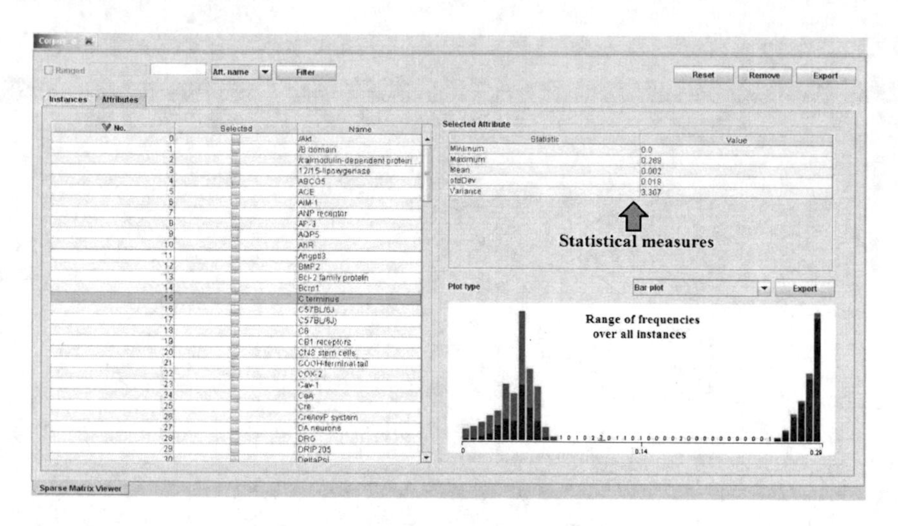

Fig. 4. Sparse matrix viewer. It shows information about each attribute contained in the dataset.

obtained results. The Filtering Menu is divided into two sub-menus: Instances and Attributes (see Figure 2). The first one corresponds to those algorithms that can redistribute the number of instances which belong to each class (subsampling or oversampling). The second refers to algorithms that reduce data dimensionality, as mentioned above). In Figure 3 the task is encircled by a dotted line because these algorithms are optional in the classification process.

In the next task, the researcher must create a new classifier. The Classification Menu includes two Support Vector Machines (SVM) implementations, a Naive Bayes classifier and a K Nearest Neighbour. BioClass implementations are highly configurable because each classifier has its own specific parameters. Once the classifier is created, it can be accessed through Clipboard and launch an edition viewer in which the researcher can get its current parameter values or even modify them.

Following the general scenario, the Train task focuses on the internal process of training a selected classifier according to a specific dataset (modified or not), and generating a new trained item. This operation can be accessed under the Classification menu.

In the Test task, the researcher can test the trained classifier with either the same or a different dataset. Test operation, located again in the Classification Menu, needs a trained classifier and a loaded dataset as input data. This process can take some time depending on the computer capabilities. Once it has been concluded, it produces a result object which can be accessed by the corresponding viewer.

Finally, in the Results task a result object must be selected in the Clipboard in order to appear in the Results viewer (Figure 5). The interface is divided into two sections, a prediction table and the summary. The prediction table shows

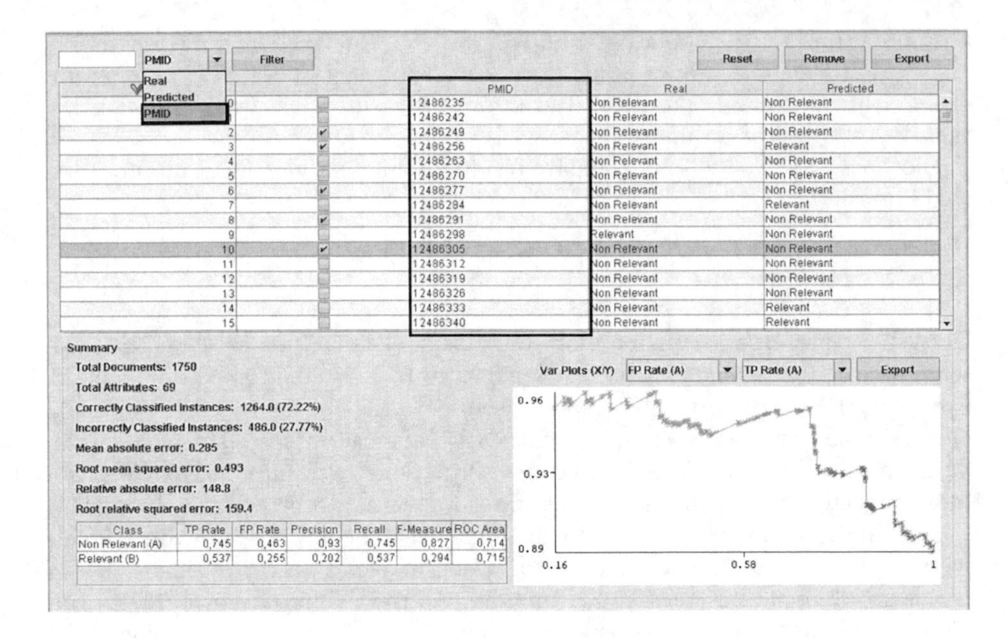

Fig. 5. Results viewer interface of a classification process

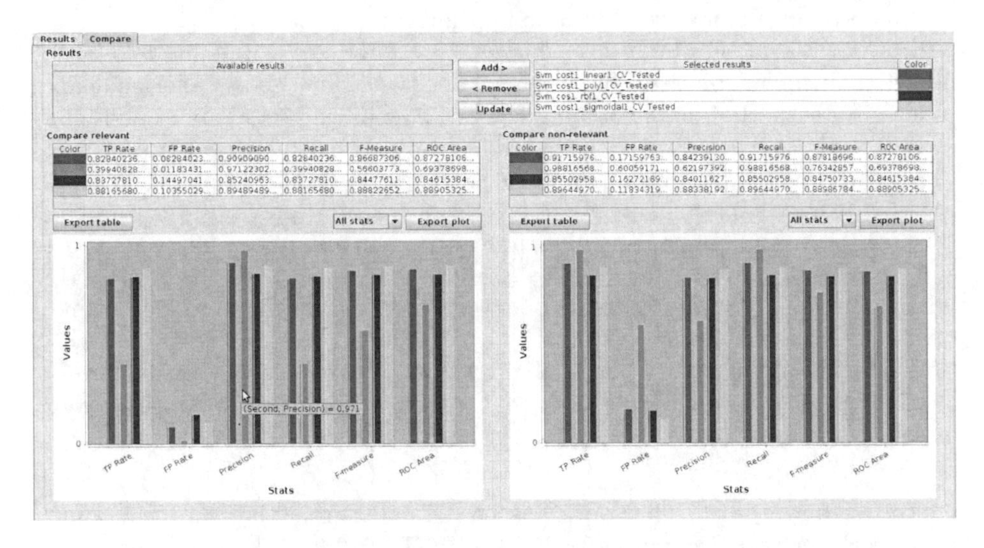

Fig. 6. Result comparative viewer interface

both the real and the predicted values per each document. With the Toolbar it is possible to filter, remove, reset or export current results.

The Summary section contains statistic results associated with the classification process. Information about the following statistical measures is shown on the left: total documents, total attributes, correctly classified instances, incorrectly classified instances, mean absolute error, root mean squared error, relative absolute error, and root relative squared error. On the right side, a configurable plot can provide a helpful graphic. The user can select between a combination of TP Rate, FP Rate, Precision, Recall or F-Measure measures. The plot can then be exported as an image file.

BioClass also allows easily compare results of different classification processes (see Figure 6).

4 Conclusion

This study presents a tool for biomedical text classification with which a researcher can distinguish a document set related to a specific topic as relevant or irrelevant. BioClass also supports algorithms in order to increase the efficiency of the classification process, and provides a set of powerful interfaces to analyze, filter and compare classification results. All actions performed in BioClass are interconnected, which allows the classification process to be completely guided.

The BioClass design is based on the AIBench Model View Controller architecture and is divided into modules in order to facilitate future inclusions of new algorithms and increase its scalability. As the magnitude of datasets can be quite different, and computers may not have enough memory to allocate all these data, a great deal of effort must be placed on developing algorithms that can manage huge amounts of data and optimize memory costs.

BioClass is a good solution to evaluate the performance of different text classifiers. Other similar solutions, like Weka [5] or Orange [21], do not offer guidelines to classify texts. BioClass is based on a workflow having its modules and interfaces directly connected with it, so users do not need to know the order of steps to be performed. In addition, BioClass supports several additional functionalities: (i) it is possible reuse any model which has been previously trained, (ii) through a rich attribute perspective, users can compare, plot or filter attributes and extract analysis and conclusions, (iii) filters do not modify the original corpora, so users may study differences between them (iv) it allows to create a new corpus with the classification results.

Finally, a full documentation, guidelines and example files about this project can be accessed through `code.google.com/p/bioclass` and `sing.ei.uvigo.es/bioclass/`.

Acknowledgements. This work has been funded from the European Union Seventh Framework Programme [FP7/REGPOT-2012-2013.1] under grant agreement n 316265, BIOCAPS.

References

1. Baeza-Yates, R.A., Ribeiro-Neto, B.: Modern Information Retrieval. Addison-Wesley Longman (1999)
2. Tan, S.: Neighbor-weighted k-nearest neighbor for unbalanced text corpus. Expert Systems with Applications 28(4), 667–671 (2005)

3. Anand, A., Pugalenthi, G., Fogel, G.B., Suganthan, P.N.: An approach for classification of highly imbalanced data using weighting and undersampling. Amino Acids 39, 1385–1391 (2010)
4. Chawla, N.V., Japkowicz, N., Kotcz, A.: Editorial: special issue on learning from imbalanced data sets. SIGKDD Explorations 6(1), 1–6 (2004)
5. Garner, S.R.: Weka: The waikato environment for knowledge analysis. In: Proc. of the New Zealand Computer Science Research Students Conference, pp. 57–64 (1995)
6. Zhang, J., Mani, I.: knn approach to unbalanced data distributions: A case study involving information extraction. In: Proceedings of the ICML 2003 Workshop on Learning from Imbalanced Datasets (2003)
7. Hall, M.A.: Correlation-based Feature Subset Selection for Machine Learning. PhD thesis, Department of Computer Science, University of Waikato, Hamilton, New Zealand (April 1999)
8. Duda, R.O., Hart, P.E., Stork, D.G.: Pattern Classification, 2nd edn. Wiley, New York (2001)
9. Liu, H., Setiono, R.: A probabilistic approach to feature selection - a filter solution. In: 13th International Conference on Machine Learning, pp. 319–327 (1996)
10. Lorenzo, J.: Selección de Atributos en Aprendizaje Automático basado en la Teoría de la Información. PhD thesis, Faculty of Computer Science, Univ. of Las Palmas (2002)
11. Holte, R.C.: Very Simple Classification Rules Perform Well on Most Commonly Used Datasets. Machine Learning 11(1), 63–90 (1993)
12. Jolliffe, I.T.: Principal Component Analysis, 2nd edn. Springer (2002)
13. Hirschman, L., Yeh, A., Blaschke, C., Valencia, A.: Overview of biocreative: critical assessment of information extraction for biology. BMC Bioinformatics 6(suppl.1), s1 (2005)
14. Ando, R.K., Dredze, M., Zhang, T.: Trec 2005 genomics track experiments at ibm watson. In: Proceedings of TREC 2005. NIST Special Publication (2005)
15. Collier, N., Hahn, U., Rebholz-Schuhmann, D., Rinaldi, F., Pyysalo, S. (eds.): Proceedings of the Fourth International Symposium for Semantic Mining in Biomedicine, Cambridge, United Kingdom. CEUR Workshop Proceedings, vol. 714. CEUR-WS.org (October 2010)
16. Osuna, E., Freund, R., Girosi, F.: Support vector machines: Training and applications. Technical report, Cambridge, MA, USA (1997)
17. Domingos, P., Pazzani, M.: On the optimality of the simple bayesian classifier under zero-one loss. Machine Learning 29, 103–130 (1997)
18. Dasarathy, B.V.: Nearest Neighbor (NN) Norms: NN Pattern Classification Techniques. IEEE Computer Society Press, Los Alamitos (1991)
19. Glez-Peña, D., Reboiro-Jato, M., Maia, P., Rocha, M., Díaz, F., Fdez-Riverola, F.: Aibench: A rapid application development framework for translational research in biomedicine. Comput. Methods Prog. Biomed. 98, 191–203 (2010)
20. Chang, C., Lin, C.: LIBSVM: a library for support vector machines (2001)
21. Demšar, J., Zupan, B., Leban, G., Curk, T.: Orange: From experimental machine learning to interactive data mining. In: Boulicaut, J.-F., Esposito, F., Giannotti, F., Pedreschi, D. (eds.) PKDD 2004. LNCS (LNAI), vol. 3202, pp. 537–539. Springer, Heidelberg (2004)

Chemical Named Entity Recognition: Improving Recall Using a Comprehensive List of Lexical Features

Andre Lamurias, João Ferreira, and Francisco M. Couto

Dep. de Informática, Faculdade de Ciências, Universidade de Lisboa, Portugal
{alamurias,joao.ferreira}@lasige.di.fc.ul.pt, fcouto@di.fc.ul.pt

Abstract. As the number of published scientific papers grows everyday, there is also an increasing necessity for automated named entity recognition (NER) systems capable of identifying relevant entities mentioned in a given text, such as chemical entities. Since high precision values are crucial to deliver useful results, we developed a NER method, Identifying Chemical Entities (ICE), which was tuned for precision. Thus, ICE achieved the second highest precision value in the BioCreative IV CHEMDNER task, but with significant low recall values. However, this paper shows how the use of simple lexical features was able to improve the recall of ICE while maintaining high levels of precision. Using a selection of the best features tested, ICE obtained a best recall of 27.2% for a precision of 92.4%.

Keywords: Text mining, Conditional Random Fields, Named Entity Recognition, Chemical Compounds, ChEBI.

1 Introduction

As the number of published scientific papers grows everyday, there is also an increasing necessity for automated named entity recognition (NER) systems capable of identifying relevant entities mentioned in a given text. The BioCreative challenge is a community effort to evaluate text mining and information extraction systems applied to the biological domain. One of the tasks proposed for the fourth edition of this competition consisted in the detection of mentions of chemical compounds and drugs in MEDLINE titles and abstracts (CHEMDNER task) [1]. The chemical entities to identify were those that can be linked to a chemical structure. This task was divided in two subtasks: The first, Chemical Document Indexing (CDI), expected an ordered list of unique chemical entities referenced in given text. The second subtask, Chemical Entity Mention recognition (CEM), expected the exact position of each chemical entity mentioned in the text. The task organizers also provided a training corpus composed by 10,000 MEDLINE abstracts that were annotated manually by domain experts. The specific rules used by the annotators and the criteria used for choosing the MEDLINE entries were defined by the organization and released with the corpus.

J. Sáez-Rodríguez et al. (eds.), *8th International Conference on Practical Appl. of Comput. Biol. & Bioinform. (PACBB 2014)*, Advances in Intelligent Systems and Computing 294,
DOI: 10.1007/978-3-319-07581-5_30,

To participate in BioCreative IV, we started by adapting our method [2] based on Conditional Random Fields (CRF) classifiers trained with the CHEMDNER corpus. Our system trained one classifier for each type of entity annotated in the corpus. The final confidence score of each token classified as a chemical entity by at least one classifier was calculated by averaging the three best classifier scores. The identified entities were validated by resolving each chemical entity recognized to the ChEBI ontology and by calculating the semantic similarity between other ChEBI entities detected on the same text. The resolution and similarity method enabled us to filter most false positives and achieve the second highest precision value on both subtasks [3]. However, the set of features used to train the classifiers was relatively small comparing with other CRF based approaches that participated in BioCreative also, which use more general and domain specific features [4, 5].

In this paper, we present a new version of our system (ICE) that achieved significantly better results using more lexical features derived from the word tokens. The rest of this paper is organized as follows: Section 2 presents an overview on the BioCreative 2013 competition which we used to evaluate our system, Section 3 (Methods) describes the approach we used for the competition and how we improved it since then, Section 4 (Results) compares the effect of different features on the precision and recall values using 3-fold cross-validation on the CHEMDNER corpus, and finally on Section 5 we express our main conclusions.

2 BioCreative 2013 - CHEMDNER Task

2.1 CHEMDNER Corpus

The CHEMDNER corpus consists in 10,000 MEDLINE titles and abstracts and was originally partitioned randomly in three sets: training, development and test. The chosen articles were sampled from a list of articles published in 2013 by the top 100 journals of a list of categories related to the chemistry field. These articles were manually annotated according to the guidelines, by a team of curators with background in chemistry. Each annotation consisted in the article identifier, type of text (title or abstract), start and end indices, the text string and the type of the CEM which could be one of the following: trivial, formula, systematic, abbreviation, family and multiple. There was no limit for the number of words that could refer to a CEM but due to the annotation format, the sequence of words had to be continuous. There was a total of 59,004 annotations on the training and development sets, which consisted in 7,000 documents.

2.2 CEM and CDI Subtasks

There were two types of predictions the participants could submit for the CHEMDNER task: a ranked list of unique chemical entities described on each document (CDI task) and the start and end indices of each chemical entity mentioned on each document (CEM task). Each list should be ordered by how confident the system is that each prediction is a chemical entity. Using the CEM predictions, it was possible to generate results for the CDI subtask, by excluding multiple mentions of the same entity in a text.

A gold standard for both subtasks was included with the corpus, which could be used to calculate precision and recall of the results, with the evaluation script released by the organization. Each team was allowed to submit up to five different runs for each subtask.

3 Methods

3.1 Submission to BioCreative 2013

Our method uses Conditional Random Fields (CRFs) for building probabilistic models based on training datasets. We used the MALLET [6] implementation of CRFs, adapted to also output the probability of the most probable sequence. This probability was used as a confidence score for each prediction, making it possible to filter predictions with low confidence.

To train models and classify new text, it is necessary to tokenize the text and generate features from word tokens. Then, the corresponding label is added to the feature list. This label could be "Not Chemical", "Single", "Start", "Middle" or "End", to include chemical entities composed by more than one token. We have used a specifically adapted word tokenizer for chemical text adapted from an open source project [7]. Four features were being extracted from each word token by our system: Stem, Prefix and suffix (size 3) and a boolean which indicates if the token contains a number (Has number). We merged the training and development sets of the CHEMDNER corpus into one training set and generated one dataset for each type of CEM. With this method we expected to identify more correct chemical entities since we were including the results of classifiers focused on just one type of CEM. The confidence score used when more than one of the classifiers identified the same CEM was the average of the three best confidence scores. This system was then evaluated with 3-fold cross-validation.

With the terms identified as chemical entities, we employed an adaptation of FiGO, a lexical similarity method [8], to perform the search for the most likely ChEBI terms. Then, we were able to calculate the Gentleman's simUI [9] semantic similarity measure for each pair of entities identified in the text and successfully mapped to the ChEBI ontology. We used the maximum semantic similarity value for each entity as a feature for filtering and ranking. This value has shown to be crucial to achieve high precision results [10].

Since each team could submit up to five runs for each subtask, we generated three runs to achieve our best F-measure, precision and recall, based on the cross-validation results we obtained on the training set. For the other two runs, we filtered the predictions by semantic similarity only. The best results we obtained were with the run we submitted for best precision (run 2), achieving the second highest precision value in the competition. For this run, we excluded results with the classifier confidence score and the semantic similarity measure lower than 0.8. We now focused on keeping the precision of our system at high values, while improving the recall and F-measure.

3.2 New Features

After implementing thirteen new features, we studied the effect of adding one new feature at a time, while always keeping the four original features constant. These new features are based on orthographic and morphological properties of the words used to represent the entity, inspired by other CRF-based chemical NER systems [4, 5, 11–13]. We integrated the following features:

Prefix and Suffix sizes 1, 2 and 4: The first and last n characters of a word token.

Greek symbol: Boolean that indicates if the token contains greek symbols.

Non-alphanumeric character: Boolean that indicates if the token contains non-alphanumeric symbols.

Case pattern: "Lower" if all characters are lower case, "Upper" if all characters are upper case, "Title" if only the first character is upper case and "Mixed" if none of the others apply.

Word shape: Normalized form of the token by replacing every number with '0', every letter with 'A' or 'a' and every other character with 'x'.

Simple word shape: Simplified version of the word shape feature where consecutive symbols of the same kind are merged.

Periodic Table element: Boolean that indicates if the token matches a periodic table symbols or name.

Amino acid: Boolean that indicates if the token matches a 3 letter code amino acids.

For example, for the sentence fragment "Cells exposed to α-MeDA showed an increase in intracellular glutathione (GSH) levels", the list of tokens obtained by the tokenizer and some possible features are shown on Table 1.

After applying the same methods described on Section 3.1 for each new feature, we were able to compare the effect of each one on the results. Then, we selected the features that achieved a higher precision, recall and F-measure, creating three sets of features for each metric and a fourth set with all the features tested.

4 Results

4.1 BioCreative 2013

Using 3-fold cross-validation on the training and development sets, we obtained the results presented in Table 2. The first three runs were aimed at achieving a high F-measure, precision and recall, respectively. On runs 4 and 5 we filtered only by semantic similarity. We used as reference the results of run 2 since the precision value obtained with the test set was the second highest in the CHEMDNER task. Our objective was to improve recall and F-measure values with minimal effect on the precision.

Table 1. Example of a sequence of some the new features, and the corresponding label, derived from a sentence fragment (PMID 23194825)

Token	Prefix 4	Suffix 4	Case pattern	Word shape	Label
Cells	Cell	ells	titlecase	Aaaaa	Not Chemical
exposed	expo	osed	lowercase	aaaaaaa	Not Chemical
to	to	to	lowercase	aa	Not Chemical
α-MeDA	α-Me	MeDA	mixed	xxAaAA	Chemical
showed	show	owed	lowercase	aaaaaa	Not Chemical
an	an	an	lowercase	aa	Not Chemical
increase	incr	ease	lowercase	aaaaaaaa	Not Chemical
in	in	in	lowercase	aa	Not Chemical
intracellular	intr	ular	lowercase	aaaaaaaaaaaaa	Not Chemical
glutathione	glut	ione	lowercase	aaaaaaaaaaa	Chemical
(	(	(	-	x	Not Chemical
GSH	GSH	GSH	uppercase	AAA	Chemical
)	)	)	-	x	Not Chemical
levels	leve	vels	lowercase	aaaaaa	Not Chemical

Table 2. Precision, Recall and F-measure estimates for each run submitted to BioCreative 2013, obtained with cross-validation on the training and development dataset for the CDI and CEM subtasks

	CDI			CEM		
	P	R	F_1	P	R	F_1
Run 1	84.8%	71.2%	77.4%	87.3%	70.2%	77.8%
Run 2	95.0%	6.5%	12.2%	95.0%	6.0%	11.1%
Run 3	52.1%	80.4%	63.3%	57.1%	76.6%	65.4%
Run 4	87.9%	22.7%	36.1%	89.7%	21.2%	34.3%
Run 5	87.9%	22.7%	36.1%	79.9%	22.6%	35.3%

4.2 New Features

The precision, recall and F-measure values obtained using our four original features plus one new one are presented in Table 3 For each metric, we added a shaded column which compares that value with the corresponding one on Table 2, for the run with best precision.

The features that returned the best recall and F-measure were the simple word shape and prefix and suffix with size=2. Using prefix and suffix with size=1 and the alphanumeric boolean decreased our precision the most, without improving the other metrics as much as other features. The periodic table feature, which was one of our two domain-specific features, achieved a recall value of 16.4%, while maintaining the precision at 94%. Our other domain-specific feature, amino acid, achieved our highest precision in this work. The general effect of using five features instead of the original four was a decrease in precision by 0.8%-4.5% and increase in recall and F-measure by 0.4%-19.5%.

For each subtask, we performed another cross-validation run with the original four features to use as baseline values. We created three feature sets composed by the original features we used for BioCreative and the features that improved precision, recall or F-measure on any subtask, compared to the baseline. The three feature sets created were:

Best precision: Stem, Prefix/suffix 3, Has number, Prefix/suffix 4, Has greek symbol, Has periodic table element, Has amino acid.

Best recall: Stem, Prefix/suffix 3, Has number, Prefix/suffix 1, Prefix/suffix 2, Has greek symbol, Has periodic table element, Case pattern, Word shape, Simple word shape.

Best F-measure: Stem, Prefix/suffix 3, Has number, Prefix/suffix 1, Prefix/ suffix 2, Has greek symbol, Has periodic table element, Has amino acid, Case pattern, Word shape, Simple word shape.

The results obtained with these sets are presented in Table 4 Although there was a decrease in precision in every case, the difference in recall and F-measure values was always much higher. The feature set with best F-measure was able to improve the recall by 21.0% while taking only 3.2% of the precision.

To determine the statistical significance of the improvement between the expanded feature set and the original, we ran a bootstrap resampling simulation similar to the BioCreative II gene mention task [14] and BioCreative CHEMDNER task evaluations. We picked 1000 PMIDs from the train and development sets and computed the recall and F-measure for this subset of documents. Then we repeated this process 10,000 times, and estimated the average recall and F-measure, and respective standard deviation for each feature set. With the original features, the average recall was 8.00% (SD=0.53%) and the average F-measure was 14.74% (SD=0.90%) while using the expanded feature set, the average recall was 27.20% (SD=0.92%) and the average F-measure was 42.02% (SD=1.13%).

Table 3. Precision, Recall and F-measure estimates for each new features used with the original set, obtained with cross-validation on the training and development dataset for the CDI subtask

	CDI						CEM					
Feature set	P	ΔP	R	ΔR	F_1	ΔF_1	P	ΔP	R	ΔR	F_1	ΔF_1
Prefix/suffix 1	91.0%	-4.0%	14.0%	+7.5%	24.3%	+12.1%	92.4%	-2.6%	13.4%	+7.4%	23.4%	+12.3%
Prefix/suffix 2	92.4%	-2.6%	19.1%	+12.6%	31.6%	+19.4%	93.5%	-1.5%	18.3%	+12.3%	30.6%	+19.5%
Prefix/suffix 4	93.3%	-1.7%	6.9%	+0.4%	12.9%	+0.7%	94.2%	-0.8%	6.6%	+0.6%	12.2%	+1.1%
Greek letter	93.4%	-1.6%	12.0%	+5.5%	21.2%	+9.0%	94.2%	-0.8%	11.8%	+5.8%	20.9%	+9.8%
Periodic table	94.0%	-1.0%	16.3%	+9.8%	27.8%	+15.6%	94.7%	-0.3%	16.4%	+10.4%	28.0%	+16.9%
Amino acid	95.0%	0.0%	9.0%	+2.5%	16.4%	+4.2%	95.1%	+0.1%	8.7%	+2.7%	16.0%	+4.9%
Alphanumeric	90.4%	-4.6%	5.3%	-1.2%	10.0%	-2.2%	92.0%	-3.0%	4.4%	-1.6%	8.4%	-2.7%
Case pattern	93.0%	-2.0%	15.7%	+9.2%	26.9%	+14.7%	93.5%	-1.5%	14.9%	+8.9%	25.6%	+14.5%
Word shape	93.9%	-1.1%	11.8%	+5.3%	20.9%	+8.7%	93.3%	-1.7%	12.7%	+6.7%	22.4%	+11.3%
Simple word shape	92.2%	-2.8%	17.1%	+10.6%	28.9%	+16.7%	92.4%	-2.6%	16.9%	+10.9%	28.7%	+17.6%

Table 4. Precision, Recall and F-measure estimates for each feature set used with the original set, obtained with cross-validation on the training and development dataset for the CDI and CEM subtasks

	CDI						CEM					
Feature set	P	ΔP	R	ΔR	F_1	ΔF_1	P	ΔP	R	ΔR	F_1	ΔF_1
Precision	93.7%	-1.3%	15.4%	+8.9%	26.5%	+14.3%	94.1%	-0.9%	15.0%	+9.0%	25.9%	+14.8%
Recall	91.5%	-3.5%	24.7%	+18.2%	38.9%	+26.7%	92.0%	-3.0%	23.9%	+17.9%	37.9%	+26.8%
F-measure	91.7%	-3.3%	28.3%	+21.8%	43.2%	+31.0%	92.3%	-2.7%	28.0%	+22.0%	43.0%	+31.9%
All features	91.5%	-3.5%	24.5%	+18.0%	38.7%	+26.5%	93.0%	-2.0%	24.2%	+18.2%	38.4%	+27.3%

5 Conclusion

Our participation in the CHEMDNER task of BioCreative 2013 achieved high precision values for both subtasks, but at the expense of a low recall. This manuscript shows how ICE improved its recall and F-measure maintaining the same levels of precision, by using a more comprehensive feature set. The effect of adding each new feature to ICE was evaluated by cross-validation on the CHEMDNER corpus. We then evaluated feature sets composed by the features that achieved the best precision, recall and F-measure, using the same method.

Individually, the features that were specific to chemical compounds achieved the best balance between precision and recall. Adding only the prefixes and suffixes with size 2, we were able to increase the recall and F-measure by 12.3% and 19.5%, while decreasing the precision by 1.5%. Using a combination of the features that achieved the best results individually, we were able to increase the recall and F-measure by 21.2% and 31.0% respectively while decreasing the precision by 2.6% (Table 4).

Considering the run that achieved the highest precision in the official BioCreative results for the CDI task, our precision is 6.9% lower, but the recall and F-measure are 10.9% and 13.9% higher, respectively. Considering the run with best precision in the CEM task, our precision is 5.7% lower, but the recall and F-measure are 9.3% and 11.8% higher. Our precision values would be the third and sixth highest in the CDI and CEM subtasks, respectively. However, notice that the results presented here were not obtained with CHEMDNER test set, and for the competition, using the official test set, our results were higher than the cross-validation estimates we obtained.

In the future we intend to use more domain-specific features, and filter predictions with a more powerful semantic similarity measure [15].

Acknowledgments. The authors want to thank the Portuguese Fundação para a Ciência e Tecnologia through the financial support of the SPNet project (PTDC/EBB-EBI/113824/2009), the SOMER project (PTDC/EIA-EIA/119119/2010) and through funding of LaSIGE Strategic Project, ref. PEst-OE/EEI/UI0408/2014.

References

1. Krallinger, M., Leitner, F., Rabal, O., Vazquez, M., Oyarzabal, J., Valencia, A.: Overview of the chemical compound and drug name recognition (CHEMDNER) task. In: BioCreative Challenge Evaluation Workshop, vol. 2, p. 2 (2013)
2. Grego, T., Pęzik, P., Couto, F.M., Rebholz-Schuhmann, D.: Identification of chemical entities in patent documents. In: Omatu, S., Rocha, M.P., Bravo, J., Fernández, F., Corchado, E., Bustillo, A., Corchado, J.M. (eds.) IWANN 2009, Part II. LNCS, vol. 5518, pp. 942–949. Springer, Heidelberg (2009)
3. Lamurias, A., Grego, T., Couto, F.M.: Chemical compound and drug name recognition using CRFs and semantic similarity based on ChEBI. In: BioCreative Challenge Evaluation Workshop vol. 2, 489, p. 75 (2013)
4. Huber, T., Rocktäschel, T., Weidlich, M., Thomas, P., Leser, U.: Extended feature set for chemical named entity recognition and indexing. In: BioCreative Challenge Evaluation Workshop, vol. 2, p. 88 (2013)
5. Leaman, R., Wei, C.H., Lu, Z.: NCBI at the biocreative IV CHEMDNER task: Recognizing chemical names in PubMed articles with tmChem. In: BioCreative Challenge Evaluation Workshop, vol. 2, p. 34 (2013)
6. McCallum, A.K.: Mallet: A machine learning for language toolkit (2002)
7. Corbett, P., Batchelor, C., Teufel, S.: Annotation of chemical named entities. In: Proceedings of the Workshop on BioNLP 2007: Biological, Translational, and Clinical Language Processing, Association for Computational Linguistics, pp. 57–64 (2007)
8. Couto, F.M., Silva, M.J., Coutinho, P.M.: Finding genomic ontology terms in text using evidence content. BMC Bioinformatics 6(suppl. 1), 21 (2005)
9. Gentleman, R.: Visualizing and distances using GO (2005), `http://www.bioconductor.org/docs/vignettes.html`
10. Grego, T., Couto, F.M.: Enhancement of chemical entity identification in text using semantic similarity validation. PloS One 8(5), e62984 (2013)
11. Batista-Navarro, R.T., Rak, R., Ananiadou, S.: Chemistry-specific features and heuristics for developing a CRF-based chemical named entity recogniser. In: BioCreative Challenge Evaluation Workshop, vol. 2, p. 55 (2013)
12. Usié, A., Cruz, J., Comas, J., Solsona, F., Alves, R.: A tool for the identification of chemical entities (CheNER-BioC). In: BioCreative Challenge Evaluation Workshop, vol. 2, p. 66 (2013)
13. Campos, D., Matos, S., Oliveira, J.L.: Chemical name recognition with harmonized feature-rich conditional random fields. In: BioCreative Challenge Evaluation Workshop, vol. 2, p. 82 (2013)
14. Smith, L., Tanabe, L.K., Ando, R.J., Kuo, C.J., Chung, I.F., Hsu, C.N., Lin, Y.S., Klinger, R., Friedrich, C.M., Ganchev, K., et al.: Overview of BioCreative II gene mention recognition. Genome Biology 9(suppl. 2), 2 (2008)
15. Couto, F., Pinto, H.: The next generation of similarity measures that fully explore the semantics in biomedical ontologies. Journal of Bioinformatics and Computational Biology 11(5 (1371001), 1–12 (2013)

Bringing Named Entity Recognition on Drupal Content Management System

José Ferrnandes[1] and Anália Lourenço[1,2]

[1] ESEI - Escuela Superior de Ingeniería Informática, University of Vigo, Edificio Politécnico, Campus Universitario As Lagoas s/n, 32004 Ourense, Spain
[2] IBB - Institute for Biotechnology and Bioengineering, Centre of Biological Engineering, University of Minho, Campus de Gualtar, 4710-057 Braga, Portugal
jose@bloomidea.com, analia@{ceb.uminho.pt,uvigo.es}

Abstract. Content management systems and frameworks (CMS/F) play a key role in Web development. They support common Web operations and provide for a number of optional modules to implement customized functionalities. Given the increasing demand for text mining (TM) applications, it seems logical that CMS/F extend their offer of TM modules. In this regard, this work contributes to Drupal CMS/F with modules that support customized named entity recognition and enable the construction of domain-specific document search engines. Implementation relies on well-recognized Apache Information Retrieval and TM initiatives, namely Apache Lucene, Apache Solr and Apache Unstructured Information Management Architecture (UIMA). As proof of concept, we present here the development of a Drupal CMS/F that retrieves biomedical articles and performs automatic recognition of organism names to enable further organism-driven document screening.

Keywords: Drupal, text mining, named entity recognition, Apache Lucene, Apache Solr, Apache UIMA.

1 Introduction

The number of generic Text Mining (TM) software tools available now is considerable [1, 2], and almost every computer language has some module or package dedicated to natural language processing [3]. Notably, Biomedical TM (BioTM), i.e. the area of TM dedicated to applications in the Biomedical domain, has grown considerably [4, 5].

One of the main challenges in BioTM is achieving a good integration of TM tools with tools that are already part of the user workbench, in particular data curation pipelines [4, 6]. Many TM products (especially, commercial products) are built in a monolithic way and often, their interfaces are not disclosed and open standards are not fully supported [7]. Also, it is important to note that biomedical users have grown dependent of Web resources and tools, such as online data repositories, online and downloadable data analysis tools, and scientific literature catalogues [8]. Therefore, it

J. Sáez-Rodríguez et al. (eds.), *8th International Conference on Practical Appl. of Comput. Biol. & Bioinform. (PACBB 2014)*, Advances in Intelligent Systems and Computing 294,
DOI: 10.1007/978-3-319-07581-5_31, © Springer International Publishing Switzerland 2014

is desirable to integrate TM tools with these resources and tools, and it seems logical to equip Web development frameworks, such as Content Management Systems and Frameworks (CMS/F), with highly customizable TM modules.

Drupal, which is one of the most common open source CMS/Fs (http://trends.builtwith.com/cms), already presents some contributions to Bioinformatics and TM: the GMOD Drupal Bioinformatic Server Framework [9], which aims to speed up the development of Drupal modules for bioinformatics applications; the OpenCalais project that integrates Drupal with the Thomson Reuters' Calais Web service (http://www.opencalais.com), a service for annotating texts with URIs from the Linked Open Data cloud; and, RDF/RDFa support so to enable the use of this ontology language in Web knowledge exchange and facilitate the development of document-driven applications, and promote the availability of knowledge resources [10].

The aim of this work was to extend further Drupal TM capabilities, notably to enable the incorporation of third-party specialized software and the development of customized applications. The proof of concept addressed Named Entity Recognition (NER), i.e. the identification of textual references to entities of interest, which is an essential step in automatic text processing pipelines [11, 12]. There are many open-source and free NER tools available, covering a wide range of bio-entities and approaches. So, our efforts were focused on implementing a Drupal module that would support customised NER and, in particular, to equip Drupal with the necessary means to construct domain-specific document search engines. For this purpose, we relied on Apache Information Retrieval (IR) and TM initiatives, namely Apache Lucene, Apache Solr and Apache Unstructured Information Management Architecture (UIMA).

The next sections describe the technologies used and their integration in the new Drupal model. The recognition of species names in scientific papers using the state-of-the-art and open source Linnaeus tool [13] is presented as an example of application.

2 Apache Software Foundation Information Retrieval and Extraction Initiatives

Apache organization supports some of the most important open source projects for the Web [14]. The Web server recommended to run Drupal CMS/F is the Apache HTTP Server [15]. Now, we want to take advantage of Apache Lucene, Apache Solr and Apache UIMA to incorporate IR and TM capabilities in Drupal CMS/F.

The Apache Lucene and Apache Solr are two distinct Java projects that have joined forces to provide a powerful, effective, and fully featured search tool. Solr is a standalone enterprise search server with a REST-like API [16] and Lucene is a high-performance and scalable IR library [17]. Due to its scalability and performance, Lucene is one of the most popular, free IR libraries [17, 18]. Besides the inverted index for efficient document retrieval, Lucene provides search enhancing features, namely: a rich set of chainable text analysis components, such as tokenizers and language-specific stemmers; a query syntax with a parser and a variety of query types that support from simple term lookup to fuzzy term matching; a scoring algorithm, with flexible means to affect the scoring; and utilities such as the highlighter, the query spell-checker, and "more like this" [16].

Apache Solr can be seen as an enabling layer for Apache Lucene that extends its capabilities in order to support, among others: external configuration via XML; advanced full-text search capabilities, standard-based open interfaces (e.g. XML, JSON and HTTP); extensions to the Lucene Query Language; and, Apache UIMA integration for configurable metadata extraction (http://lucene.apache.org/solr/features.html).

Originally, the Apache UIMA started as an IBM Research project with the aim to deliver a powerful infrastructure to store, transport, and retrieve documents and annotation knowledge accumulated in NLP pipeline systems [19]. Currently, Apache UIMA supports further types of unstructured information besides text, like audio, video and images and is a *de facto* industry standard and software framework for content analysis [20]. Its main focus is ensuring interoperability between the processing components and thus, allowing a stable data transfer through the use of common data representations and interfaces.

3 New Supporting Drupal NER Module

We looked for a tight integration of the aforementioned Apache technologies in order to provide the basic means to deploy any NER task, namely those regarding basic natural language processing and text annotation (Fig.1). Using XML as common document interchange format, the new module allows the incorporation of third-party NER tools through pre-existent or newly developed Apache UIMA wrappers.

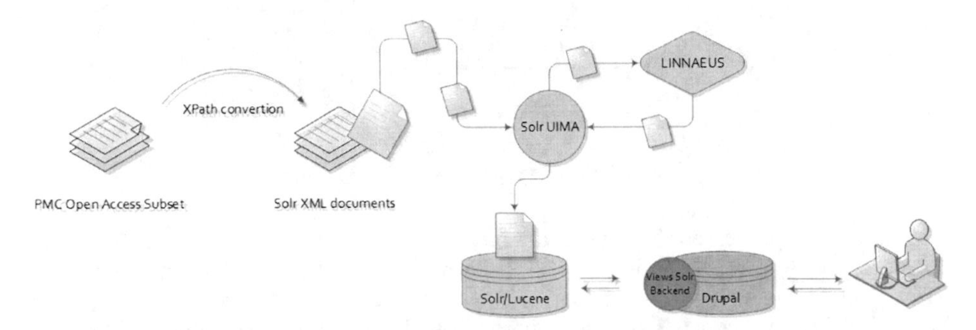

Fig. 1. Interoperation of Apache technology and third-party NER tools in the Drupal NER module

The proof of concept was the development of a Drupal CMS/F that retrieves scientific articles from the PubMed Central Open Access subset (PMC-OA) [21] and performs automatic recognition of organism mentions to enable further organism-driven document screening. The next subsections detail the interoperation of the different technologies and the integration of the Linnaeus UIMA NER wrapper as means to deploy such CMS/F.

3.1 Document Retrieval and Indexing

Documents are retrieved from the PMC-OA through the FTP service. An XSLT stylesheet is used to specify the set of rules that guide document format transformation. Notably, the XML Path (XPath) language is used to identify matching nodes and navigate through the elements and attributes in the PMC-OA XML documents.

After that, the Apache Solr engine is able to execute the UIMA-based NER pipeline to identify textual references of interest (in this case, organism names) and produce a list of named entities. The textual references are included in the metadata of the documents, and the entities recognized are added to the Apache Lucene index as means to enable further organism-specific document retrieval by the Drupal application.

3.2 Document Processing and Annotation

Apache UIMA supports the creation of highly customized document processing pipelines [22]. At the beginning of any processing pipeline is the Collection Reader component (Fig. 2), which is responsible for document input and interaction. Whenever a document is processed by the pipeline, a new object-based data structure, named Common Analysis Structure (CAS), is created. UIMA associates a Type System (TS), like an object schema for the CAS, which defines the various types of objects that may be discovered in documents. The TS can be extended by the developer, permitting the creation of very rich type systems.

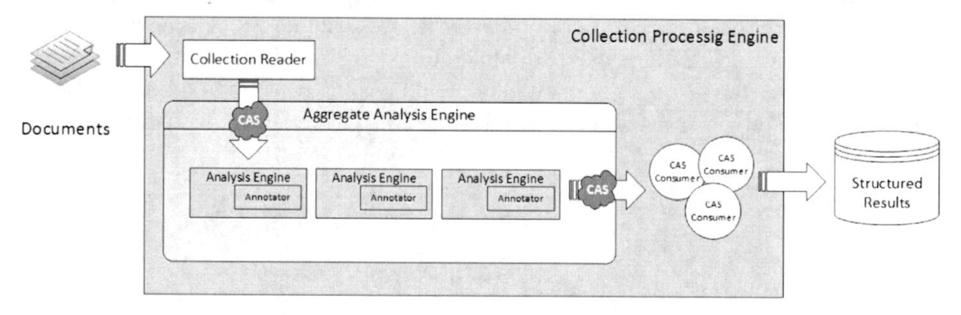

Fig. 2. High-Level UIMA Component Architecture

This CAS is processed throughout the pipeline and information can be added to the object by the Analysis Engine (AE) at different stages. The UIMA framework treats AEs as pluggable, compatible, discoverable, managed objects that analyze documents as needed. An AE consists of two components: Java classes, typically packaged as one or more JAR files, and AE descriptors, consisting of one or more XML files. The simplest type of AE is the primitive type, which contains a single annotator at its core (e.g. a tokenizer), but AEs can be combined together into an Aggregate Analysis Engine (AAE). The basic building block of any AE is the Annotator, which comprises the analysis algorithms responsible for the discovery of the desired types and the CAS update for upstream processing.

At the end of the processing pipeline are the CAS Consumers, which receive the CAS objects, after they have been analyzed by the AE/AAE, and conduct the final CAS processing.

3.3 Integration of Third-Party NER Tools

Apache UIMA supports seamless integration of third-party TM tools such as NER tools. Indeed, there already exist UIMA wrappers for several state-of-the-art NER tools, such as the organism tagger Linnaeus [23].

The first step to create an UIMA annotator wrapper is to define the AE Descriptor, i.e. the XML file that contains the information about the annotator, such as the configuration parameters, data structures, annotator input and output data types, and the resources that the annotator uses. The UIMA Eclipse plug-ins help in this creation by auto-generating this file based on the options configured in a point and click window (Fig. 3 - A).

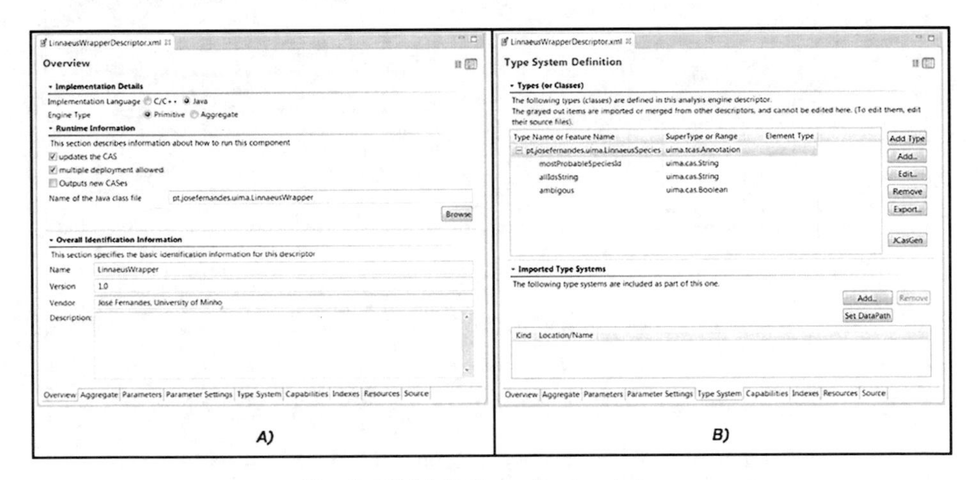

Fig. 3. UIMA Eclipse plug-in windows

The AE is then able to load the annotator in the UIMA pipeline, and the next step is to define the TS, namely the output types produced by the annotator, as described in the AE descriptor file (Fig. 3 - B).

```
    public void process(JCas cas) throws AnalysisEngineProcessException
{
        String text = cas.getDocumentText();
        List<Mention> mentions = matcher.match(text);
        for (Mention mention : mentions) {
            String mostProbableID = mention.getMostProbableID();
            String idsToString = mention.getIdsToString();
            LinnaeusSpecies species = new LinnaeusSpecies(cas);
            species.setBegin(mention.getStart());
            species.setEnd(mention.getEnd());
            species.setMostProbableSpeciesId(mostProbableID);
            species.setAllIdsString(idsToString);
            species.setAmbigous(mention.isAmbigous());
            species.addToIndexes();
        }
    }
```

Fig. 4. process() method of the LinnaeusWrapper.java class

The implementation of AE's Annotator is based on the standard interface AnalysisComponent. Basically, the wrapping of third-party tools implies the implementation of the annotator process( ) method, i.e. the desired Annotator logic (Fig. 4).

The CAS Visual Debugger is useful while implementing and testing the UIMA wrappers, in particular the annotators (Fig. 5). After wrappers are fully functioning the pipeline is ready to be used.

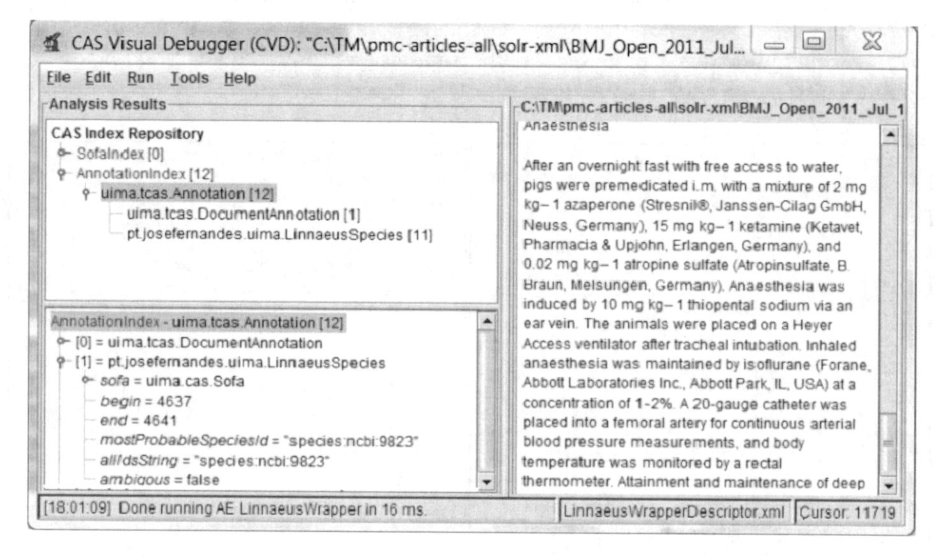

Fig. 5. Linnaeus UIMA wrapper running on a PMC Open Access article

3.4 Integration between Drupal and Apache Solr

Drupal allows developers to alter and customize the functionality of almost all of its components. Here, we developed a new Drupal module, named Views Solr Backend (https://drupal.org/project/views_solr_backend), to allow the easy and flexible querying of the Apache Solr index. The module is written in PHP and uses the APIs available for Drupal Views and Solarium, an Apache Solr client library for PHP applications [24]. This Drupal module can be easily configured and is even able to support different configurations for multiple Solr hosts (Fig. 6). Notably, it enables the administration of network parameters, the path to connect to the Solr host, and other parameters regarding the presentation of the search results in Drupal.

After configuration, it is possible to query any Solr schema, i.e. all indexed documents and the corresponding annotations. Therefore, our Drupal's Views module simplifies custom query display while increasing the interoperability with the CMS.

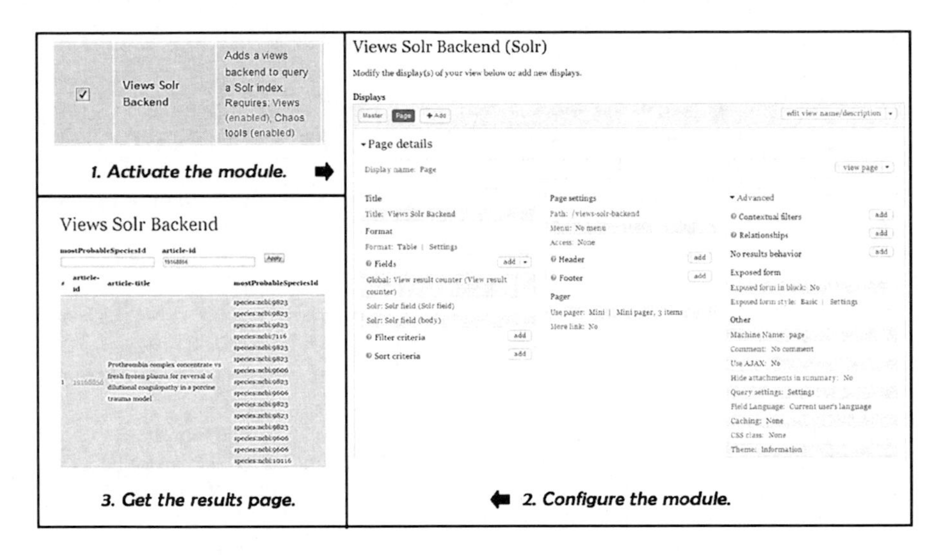

Fig. 6. Drupal module setup and presentation

4 Conclusions and Future Work

Drupal is a powerful and agile CMS/F that suits a number of development efforts in the Biomedical domains. Given the increasing demand for automatic document processing in support of the population of biomedical knowledge systems, BioTM has become almost a required module in such a framework.

This work addressed this need through the exploitation of major open source IR and IE initiatives. The new Drupal Views Solr Backend module, available at Drupal.org website (https://drupal.org/project/views_solr_backend), integrates Apache Solr with Drupal and thus, enables the implementation of customized search engines in Drupal applications. Moreover, the Apache Solr UIMA plug-in developed here for the Linnaeus NER tool exemplifies the integration of third-party NER tools in the document analysis processes of Apache Solr, granting a powerful and seamlessly means of specialized document annotation and indexing.

Acknowledgements. This work was supported by the IBB-CEB, the Fundação para a Ciência e Tecnologia (FCT) and the European Community fund FEDER, through Program COMPETE [FCT Project number PTDC/SAU-SAP/113196/2009/FCOMP-01-0124-FEDER-016012], and the Agrupamento INBIOMED from DXPCTSUG-FEDER unha maneira de facer Europa (2012/273). The research leading to these results has received funding from the European Union's Seventh Framework Programme FP7/REGPOT-2012-2013.1 under grant agreement n° 316265, BIOCAPS. This document reflects only the author's views and the European Union is not liable for any use that may be made of the information contained herein.

References

1. Kano, Y., Baumgartner, W.A., McCrohon, L., et al.: U-Compare: share and compare text mining tools with UIMA. Bioinformatics 25, 1997–1998 (2009), doi:10.1093/bioinformatics/btp289
2. Fan, W., Wallace, L., Rich, S., Zhang, Z.: Tapping the power of text mining. Commun. ACM 49, 76–82 (2006), doi:10.1145/1151030.1151032
3. Gemert, J.: Van Text Mining Tools on the Internet An overview. Univ. Amsterdam 25, 1–75 (2000)
4. Lourenço, A., Carreira, R., Carneiro, S., et al.: @Note: A workbench for biomedical text mining. J. Biomed. Inform. 42, 710–720 (2009), doi:10.1016/j.jbi.2009.04.002
5. Hucka, M., Finney, A., Sauro, H.: A medium for representation and exchange of biochemical network models (2003)
6. Lu, Z., Hirschman, L.: Biocuration workflows and text mining: overview of the BioCreative, Workshop Track II. Database (Oxford) 2012:bas043 (2012), doi:10.1093/database/bas043
7. Feinerer, I., Hornik, K., Meyer, D.: Text Mining Infrastructure in R. J. Stat. Softw. 25, 1–54 (2008), doi:citeulike-article-id:2842334
8. Fernández-Suárez, X.M., Rigden, D.J., Galperin, M.Y.: The 2014 Nucleic Acids Research Database Issue and an updated NAR online Molecular Biology Database Collection. Nucleic Acids Res. 42, 1–6 (2014), doi:10.1093/nar/gkt1282
9. Papanicolaou, A., Heckel, D.G.: The GMOD Drupal bioinformatic server framework. Bioinformatics 26, 3119–3124 (2010), doi:10.1093bioinformatics/btq599
10. Decker, S., Melnik, S., van Harmelen, F., et al.: The Semantic Web: the roles of XML and RDF. IEEE Internet Comput. 4, 63–73 (2000), doi:10.1109/4236.877487
11. Rebholz-Schuhmann, D., Kafkas, S., Kim, J.-H., et al.: Monitoring named entity recognition: The League Table. J. Biomed Semantics 4, 19 (2013), doi:10.1186/2041-1480-4-19
12. Rzhetsky, A., Seringhaus, M., Gerstein, M.B.: Getting started in text mining: Part two. PLoS Comput. Biol. 5, e1000411 (2009), doi:10.1371/journal.pcbi.1000411
13. Gerner, M., Nenadic, G., Bergman, C.M.: LINNAEUS: A species name identification system for biomedical literature. BMC Bioinformatics 11, 85 (2010), doi:10.1186/1471-2105-11-85
14. Fielding, R.T., Kaiser, G.: The Apache HTTP Server Project. IEEE Internet Comput. (1997), doi:10.1109/4236.612229
15. Web server | Drupal.org., `https://drupal.org/requirements/webserver`
16. Smiley, D., Pugh, E.: Apache Solr 3 Enterprise Search Server, p. 418 (2011)
17. McCandless, M., Hatcher, E., Gospodnetic, O.: Lucene in Action, Second Edition: Covers Apache Lucene 3.0, p. 475 (2010)
18. Konchady, M.: Building Search Applications: Lucene, LingPipe, and Gate, p. 448 (2008)
19. Ferrucci, D., Lally, A.: UIMA: An architectural approach to unstructured information processing in the corporate research environment. Nat. Lang. Eng. (2004)
20. Rak, R., Rowley, A., Ananiadou, S.: Collaborative Development and Evaluation of Text-processing Workflows in a UIMA-supported Web-based Workbench. In: LREC (2012)
21. Lin, J.: Is searching full text more effective than searching abstracts? BMC Bioinformatics 10, 46 (2009), doi:10.1186/1471-2105-10-46
22. Baumgartner, W.A., Cohen, K.B., Hunter, L.: An open-source framework for large-scale, flexible evaluation of biomedical text mining systems. J. Biomed. Discov. Collab. 3, 1 (2008), doi:10.1186/1747-5333-3-1
23. Móra, G.: Concept identification by machine learning aided dictionary-based named entity recognition and rule-based entity normalisation. Second CALBC Work
24. Kumar, J.: Apache Solr PHP Integration, p. 118 (2013)

Marky: A Lightweight Web Tracking Tool for Document Annotation

Martín Pérez-Pérez[1], Daniel Glez-Peña[1], Florentino Fdez-Riverola[1], and Anália Lourenço[1,2]

[1] ESEI - Escuela Superior de Ingeniería Informática, Edificio Politécnico, Campus Universitario As Lagoas s/n, Universidad de Vigo, 32004 Ourense, Spain
[2] IBB - Institute for Biotechnology and Bioengineering, Centre of Biological Engineering, University of Minho, Campus de Gualtar, 4710-057 Braga, Portugal
mpperez3@esei.uvigo.es, {dgpena,riverola}@uvigo.es, analia@{ceb.uminho.pt,uvigo.es}

Abstract. Document annotation is an elementary task in the development of Text Mining applications, notably in defining the entities and relationships that are relevant to a given domain. Many annotation software tools have been implemented. Some are particular to a Text Mining framework while others are typical stand-alone tools. Regardless, most development efforts were driven to basic functionality, i.e. performing the annotation, and to interface, making sure operation was intuitive and visually appellative. The deployment of large-scale annotation jamborees and projects showed the need for additional features regarding inter- and intra-annotation management. Therefore, this paper presents Marky, a new Web-based document annotation tool that integrates a highly customisable annotation environment with a robust project management system. Novelty lays on the annotation tracking system, which supports *per* user and *per* round annotation change tracking and thus, enables automatic annotation correction and agreement analysis.

Keywords: Text mining, document annotation, annotation guidelines, inter-annotator agreement, Web application.

1 Introduction

Text Mining (TM) has a wide range of applications that require differentiated processing of documents of various natures [1]. Overall, the goal is to be able to recognise and contextualise information of relevance, notably named entities and relationships among them. Language knowledge plays a key role characterising meaningful elements in sentence composition, such as nouns and verbs. Domain implementation implies to be generally familiar with the written language and specifically aware of the terminology and "writing structure" employed in the context under analysis. For example, TM practitioners of written English are required to learn about the structure of scientific papers, and the specificities of the terminology used, in order to apply TM methods and algorithms to biomedical research documents.

J. Sáez-Rodríguez et al. (eds.), *8th International Conference on Practical Appl. of Comput. Biol. & Bioinform. (PACBB 2014)*, Advances in Intelligent Systems and Computing 294,
DOI: 10.1007/978-3-319-07581-5_32,

Ontologies and controlled vocabularies are crucial in capturing the semantics of a domain, and machine learning models have proven successful in employing these resources to automatically recognise and extract information of interest. Currently, there are many commercial and free TM frameworks and software tools available. Apache Unstructured Information Management Architecture (UIMA) [2] and General Architecture for Text Engineering (GATE) [3] are two meaningful examples of open source initiatives. Apart from the natural language processors and machine learning recognisers, the most sophisticated components of TM tools are the document annotator and the document viewer. Typically, user-system interaction relies on these components and therefore, attractiveness, intuitiveness, ergonomics and flexibility are major development directives. UIMA's U-Compare [4] and GATE's Teamware [5], as others alike, are offered as an integrated framework option. Solutions not bound to TM frameworks also exist. For example, MyMiner [6], EGAS [7] and PubTator [8] offer free Web-based solutions, benefiting from feedback on user experience collected at jamborees and annotation evaluations. Arguably, the data staging area is the component of the annotation life-cycle less developed so far. Namely, existing tools come short in features such as: monitoring intra-annotator and inter-annotator annotation patterns, assessing the suitability of annotation guidelines, and identifying unanticipated semantics, or other annotation issues, while still conducting annotation rounds. These features are equally important to large-scale annotation projects and smaller, more application-specific projects. Notably, they are quite important when the annotators involved in the project present different levels of domain expertise and/or are not so familiar with the concept and implications of document annotation.

This paper presents Marky, a freely accessible Web-based annotation tool that aims to provide for customised document annotation while supporting project management. Notably, the novelty lays on the annotation tracking system, which ensures that all actions occurring within the annotation project are recorded and may be reverted at any point. This ability is crucial to assess inter-annotator agreement and observe intra-annotator patterns and thus, this tracking system is expected to improve the overall quality of project's results.

The next sections detail Marky design and its main functionalities. Attention is called to the following key activities: the creation of annotation projects, which involves the definition of the entities of interest and the main guidelines of annotation; the deployment of annotation rounds, which includes intra-annotator and inter-annotator statistics analysis; and the use of the annotation tracking system.

2 Marky Web Application

Marky is a Web-based multi-purpose document annotation application. The application was developed using the CakePHP framework (http://cakephp.org/) [9], which follows the Model–View–Controller (MVC) software pattern. Crafting application tasks into separate models, views, and controllers has made Marky lightweight, maintainable and modular. Notably, the modular design separates back-end development (e.g. the inclusion of natural language tools) from front-end development (e.g.

documents and annotations visual representation), and allows developers to make changes in one part of the application without affecting the others.

Marky reaches for state-of-the-art and free Web technologies to offer the best possible user experience and provide for efficient project management. The HTML5 (http://www.w3.org/TR/html5/) and CSS3 (http://www.css3.info/) technologies support the design of intuitive interfaces whereas Ajax and JQuery (http://jquery.com/) technologies account for user-system interaction, notably document traversal and manipulation, event handling, animation, and efficient use of the network layer. Additionally, the Rangy library (http://code.google.com/p/rangy/) is used in common DOM range and selection tasks to abstract from the different browser implementations of these functionalities (namely, Internet Explorer versus DOM-compliant browsers). MySQL database engine supports data management.

This section describes the annotation life-cycle and the core management and analysis functionalities currently provided by the application.

2.1 Project Life-Cycle

A project accounts for the following main components: documents or corpus, species or concepts of interest, annotations and users (administrator and annotators). The project administrator and the team of annotators have one shared goal: to carry out the work adequately to meet the project's objectives.

At initiation, the annotation goal of the project is established and the team is defined. The documents to be annotated are automatically retrieved from an online source (e.g. PubMed Central) or uploaded by the administrator. The concepts of interest, in particular the different types of concepts and their association, are identified manually (Fig. 1) and their semantics is formalised in a set of annotation guidelines.

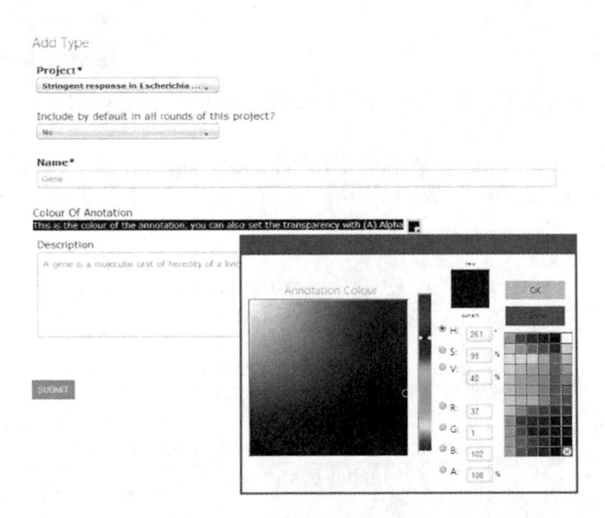

Fig. 1. Defining the types of concepts to be annotated in the project

This formalisation typically occurs after the preliminary meetings with annotators, and is meant to guide the annotation process, help annotators decide on (or disambiguate) textual mentions, and leverage annotators' domain expertise.

Often, the annotation project involves several annotators and is conducted in multiple rounds to guarantee the quality of the final annotations (Fig 2). Marky keeps track of the work done by each annotator at every round. Annotation rounds may prompt unanticipated issues, which may lead to changes in annotation guidelines, and even in the set of annotation types. Therefore, each round of annotation has associated its own set of guidelines and concept types.

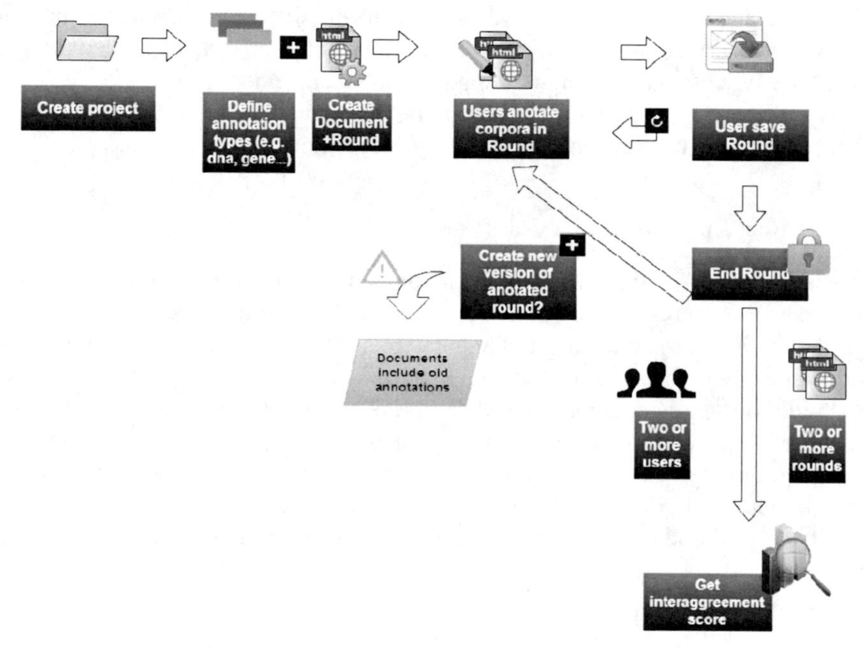

Fig. 2. The life-cycle of an annotation project in Marky

Round and user results are compared for the concept types in common. The improvement in the rates of agreement is quantified using the F-score, a common metric in IAA evaluations [10] presented below:

$$\text{F} - \text{score} = 2 \times \frac{\text{precision} \times \text{recall}}{\text{precision} + \text{recall}}$$

$$\text{Precision} = \frac{\text{number of identical entities in set A and set B}}{\text{number of entities in set A}}$$

$$\text{Recall} = \frac{\text{number of identical entities in set A and set B}}{\text{number of entities in set B}}$$

such that Set A refers to the set of annotations produced by annotator A and set B refers to the set of annotations produced by annotator B, and recall(set A, set B)=precision(set B, set A) [11].

2.2 Annotation Function

Marky offers an interactive interface allowing annotators to identify various kinds of concepts or entities within documents, according to task definition. The annotation component handles both plain text and HTML documents, and relies on state-of-the-art Web technologies, such as HTML5, CSS3, Ajax and JQuery, to offer an intuitive What-You-See-Is-What-You-Get editor.

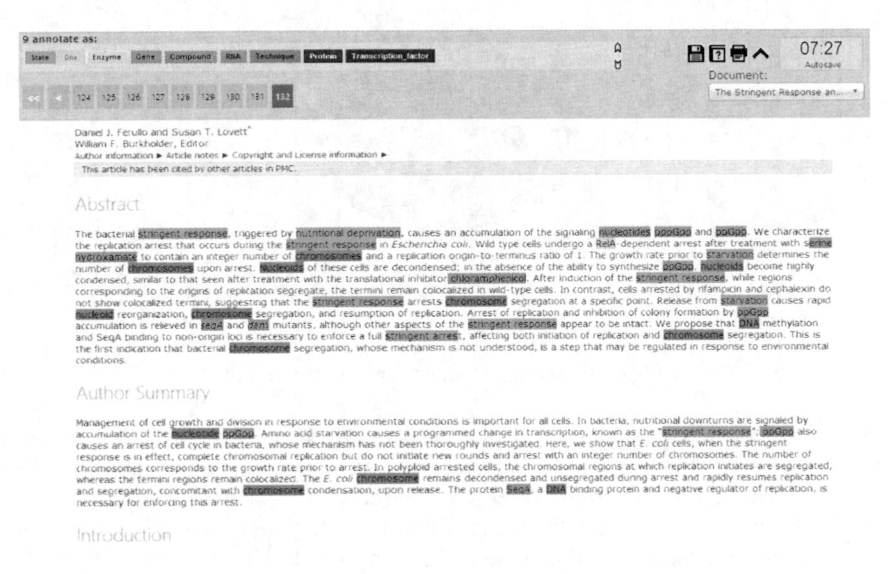

Fig. 3. Document annotation in action

At each round, the annotator has a list of documents to annotate. Documents are rendered in a Web form that supports term annotation as well as annotation visual presentation (Fig 3). By right-clicking on one or more words, the annotator marks a term or concept of interest (coloured in accordance to its type). While the annotation round is open, any annotation can be edited or removed. After the administrator closes the round, annotation statistics are calculated and round assessment is conducted, to evaluate the quality of the current version of annotations and decide upon launching a new round or not.

2.3 Annotation Tracking Function

By monitoring annotation changes, the project administrator may supervise the compliance of annotators with annotation guidelines and thus, adjust these guidelines and alert annotators about incorrect or dubious curation patterns. Notably, it is highly unlikely that two annotators annotate the very same text fragments, or completely agree on text-concept associations. Mostly, variability arises from differences in domain expertise, annotation skills, and interpretation of annotation criteria.

Per round, annotator statistics describe the volume of annotations achieved by the annotator regarding the different types of concepts in analysis (Fig 4). These data are useful to assess what concepts are most annotated and are rarely annotated by a given annotator, or within the round.

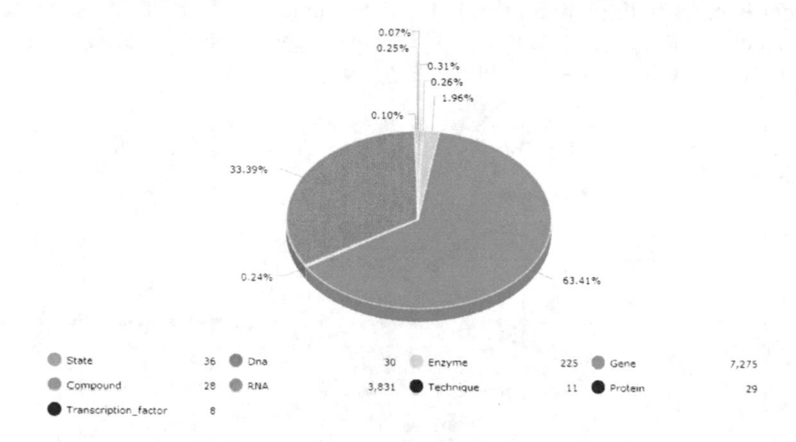

Fig. 4. Reporting annotator statistics in round

By assessing the agreement between annotators in the round, the project administrator evaluates the annotation discrepancies regarding the present set of guidelines (Fig 5). Typically, it is important to see whether certain concepts are being systematically missed or only some annotators are able to identify them. Moreover, annotators may not agree on term classification, i.e. terms are classified differently (and accounted in the statistics of different concept types), or term boundaries, i.e. annotations may only match partially.

Annotation guidelines may thus be revised so to contemplate any new semantics contexts and help to solve/minimise the annotation discrepancies observed. After deploying a new round, and besides analysing intra-round behaviour, the project administrator may compare the results between rounds to assess whether the revised set of guidelines was successful or new corrections/additions are still in need. Typically, the number of rounds performed depends considerably on time and cost constraints, leading the team to commit to a satisfactory score of agreement.

3 Conclusions

Marky is a free Web-based generic annotation tool that aims to provide highly customised annotation while supporting project management life-cycle. Indeed, Marky detaches from existing annotation tools in that it incorporates an annotation tracking system to monitor compliance with annotation guidelines and inter-annotator agreement. The ability to redo or undo annotations automatically is of help while consolidating annotation guidelines and minimises the manual work required from the annotators. Notably, the active monitoring of annotation patterns helps the administrator to "leverage" (at some extent) the expertise of the annotators by pin pointing interpretation/semantics issues that require further discussion and contextualisation.

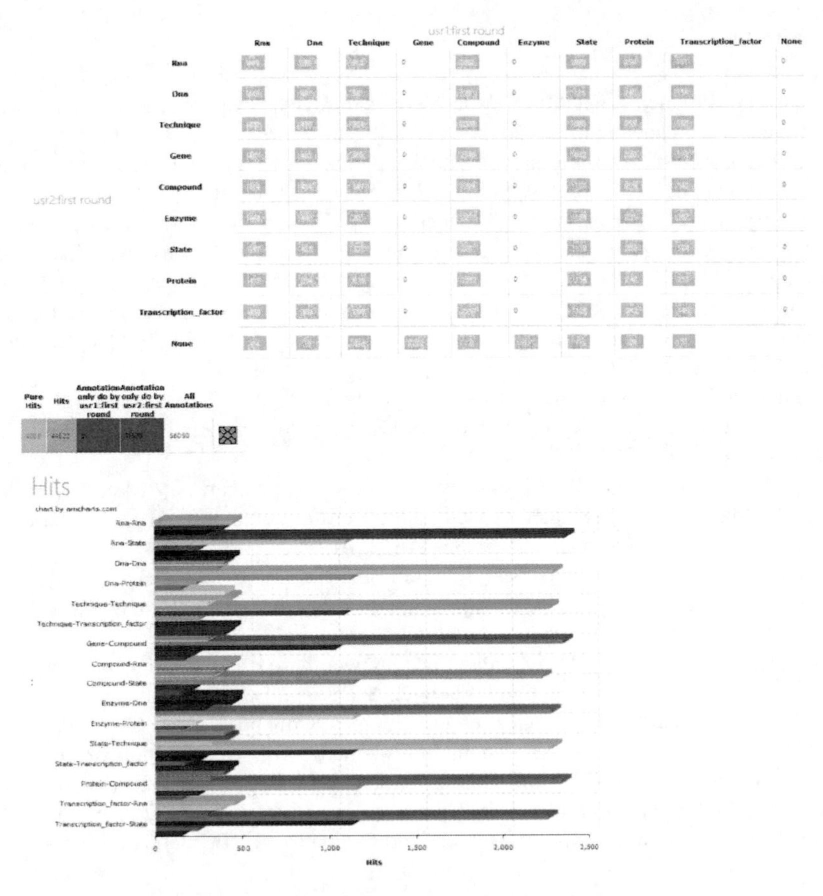

Fig. 5. Reporting inter-annotator agreement

Regarding annotation functionalities, Marky design has favoured the use of state-of-the-art Web technologies as means to ensure wide user-system interaction and tool interoperability. Currently, Marky offers the same extent of manual operation of other tools. The ability to plug in named entity recognisers, or deploy the automatic recognition of dictionary entries will be sought after in the near future.

Acknowledgements. This work was supported by the IBB-CEB, the Fundação para a Ciência e Tecnologia (FCT) and the European Community fund FEDER, through Program COMPETE [FCT Project number PTDC/SAU-SAP/113196/2009/FCOMP-01-0124-FEDER-016012], and the Agrupamento INBIOMED from DXPCTSUG-FEDER unha maneira de facer Europa (2012/273). The research leading to these results has received funding from the European Union's Seventh Framework Programme FP7/REGPOT-2012-2013.1 under grant agreement n° 316265, BIOCAPS. This document reflects only the author's views and the European Union is not liable for any use that may be made of the information contained herein.

References

1. Miner, G.: Practical Text Mining and Statistical Analysis for Non-structured Text Data Applications. Academic Press (2012)
2. Ferrucci, D., Lally, A.: Building an example application with the Unstructured Information Management Architecture. IBM Syst. J. 43, 455–475 (2004)
3. Cunningham, H., Maynard, D., Bontcheva, K., Tablan, V., Aswani, N., Roberts, I., Gorrell, G., Funk, A., Roberts, A., Damljanovic, D., Heitz, T., Greenwood, M.A., Saggion, H., Petrak, J., Li, Y., Peters, W.: Text Processing with GATE, Version 6 (2011)
4. Kano, Y., Baumgartner, W.A., McCrohon, L., Ananiadou, S., Cohen, K.B., Hunter, L., Tsujii, J.: U-Compare: share and compare text mining tools with UIMA. Bioinformatics 25, 1997–1998 (2009)
5. Bontcheva, K., Cunningham, H., Roberts, I., Roberts, A., Tablan, V., Aswani, N., Gorrell, G.: GATE Teamware: A web-based, collaborative text annotation framework. Lang. Resour. Eval. 47, 1007–1029 (2013)
6. Salgado, D., Krallinger, M., Depaule, M., Drula, E., Tendulkar, A.V., Leitner, F., Valencia, A., Marcelle, C.: MyMiner: A web application for computer-assisted biocuration and text annotation. Bioinformatics 28, 2285–2287 (2012)
7. Campos, D., Lourenço, J., Nunes, T., Vitorino, R., Domingues, P.S.M., Oliveira, J.L.: Egas - Collaborative Biomedical Annotation as a Service. In: Fourth BioCreative Challenge Evaluation Workshop, pp. 254–259 (2013)
8. Wei, C.-H., Kao, H.-Y., Lu, Z.: PubTator: A web-based text mining tool for assisting biocuration. Nucleic Acids Res. 41, W518–22 (2013)
9. Iglesias, M.: CakePHP 1.3 Application Development Cookbook. Packt Publishing (2011)
10. Thompson, P., Iqbal, S.A., McNaught, J., Ananiadou, S.: Construction of an annotated corpus to support biomedical information extraction. BMC Bioinformatics 10, 349 (2009)
11. Brants, T.: Inter-annotator agreement for a German newspaper corpus. In: Second International Conference on Language Resources and Evaluation, LREC 2000 (2000)

A Nanopublishing Architecture for Biomedical Data

Pedro Sernadela[1,*], Eelke van der Horst[2], Mark Thompson[2], Pedro Lopes[1], Marco Roos[2], and José Luís Oliveira[1]

[1] DETI/IEETA, University of Aveiro, Aveiro, Portugal
{sernadela,pedrolopes,jlo}@ua.pt
[2] Human Genetics Department, Leiden University Medical Center, Leiden, Netherlands
{e.van_der_horst,m.thompson,m.roos}@lumc.nl

Abstract. The massive production of data in the biomedical domain soon triggered a phenomenon known as information overload. The majority of these data are redundant without linked statements or associations, which hinders research methods. In this work, we describe an innovative and automated approach to integrate scientific results into small RDF-based data snippets called nanopublications. A nanopublication enhances attribution and ownership of specific data elements, representing the smallest unit of publishable information. It is particularly relevant for the scientific domain, where controlled publication, validation and ownership of data are essential. This proposal extends an existing semantic data integration framework by enabling the generation of nanopublications. Furthermore, we explore a streamlined integration and retrieval pipeline, empowered by current Semantic Web standards.

Keywords: Nanopublications, COEUS, Semantic Web, Data Integration.

1 Introduction

Peer-reviewed publications remain the main means for exchanging biomedical research information. However, there are several ways, apart from publishing and sharing scientific articles, in which researchers can contribute to scientific community, for example, the submission or curation of biological databases [1]. In both cases, most part of the information is actually growing at high levels [2] and it is increasingly difficult to find scientific data that are linked or associated, including provenance details. For example, if one researcher decides to investigate if the gene APP has some specific association with Alzheimer's disease, he may spent several days searching and analyzing the current scientific information available on the Web. This scenario will worsen if he wants to analyze multiple gene combinations in complex diseases, one of the most challenging domains of biomedical research.

In addition, few initiatives specify how academic credit is established for biomedical data sharing. Traditionally, the evaluation measure of a researcher's scientific career relies on his publication record in international peer-reviewed scientific journals. As stated above, there is a multitude of ways to contribute to the scientific community such as the submission and curation of databases entries and

J. Sáez-Rodríguez et al. (eds.), *8th International Conference on Practical Appl. of Comput. Biol. & Bioinform. (PACBB 2014)*, Advances in Intelligent Systems and Computing 294,
DOI: 10.1007/978-3-319-07581-5_33,

records. In these specific cases, there is no successful way to attribute the credits of this work.

In an effort to tackle these challenges and with the dawn of the Semantic Web, a new strategy arises to interconnect and share data – nanopublications. With this approach for relating atomic data with its authors, accessing and exchanging knowledge becomes a streamlined process. The idea is that nanopublications are suitable to represent relationships between research data and efficient exchange of knowledge [3]. With the nanopublications format, most of experimental data or negative studies can be published in a standard format, such as RDF triples, instead of archived as supplemental information in an arbitrary format or independent databases. Researchers also need to access supporting data to make progress with their investigation. Analyzing only data content is not enough to fulfill most research studies requirements, becoming essential to analyze all the associated metadata. For these reasons, publishing this type of information as nanopublications will benefit similar studies saving time and unnecessary costs.

Additionally, even with the adoption of standards, some data sharing problems persist. The main reason for this is the lack of expertise by institutions or authors to transform local data into accepted data standards [4]. In this way, it is evident that researchers need an easy-setup mechanism that allows them to publish and share their scientific results through a reliable system.

In this paper, we propose to follow this idea presenting an innovative architecture to integrate automatically several data studies into a reusable format - the nanopublication. With this system, we make the transition from several common data formats to the Semantic Web paradigm, "triplifying" the data and making it publicly available as nanopublications. The main goal is to exploit the nanopublication format to efficiently share the information produced by the research community, assigning the respective academic credit to its authors. The proposed approach is provided as an extension of the COEUS[1] framework [5], a semantic data integration system. This framework includes advanced data integration and triplication tools, base ontologies, a web-oriented engine with interoperability features, such as REST (Representational State Transfer) services, a SPARQL (SPARQL Protocol and RDF Query Language) endpoint and LinkedData publication. Moreover, the resources can be integrated from heterogeneous data sources, including CSV and XML files or SQL and SPARQL query results, which will benefit our final solution.

This document is organized in 4 sections. Section 2 introduces the nanopublications standard and some related projects. Section 3 describes the system architecture. Finally, Section 4 discusses ongoing work and future research perspectives.

2 Background

Nanopublications make it possible to report individualized knowledge assertions in a more efficient way. Due to the schema extensibility, it allows a useful aggregation

[1] http://bioinformatics.ua.pt/coeus

alternative to manage disparate data. Next, supported by the actual model, we analyze some uses cases demonstrating nanopublications' current potential.

2.1 The Nanopublication Model

The basic Semantic Web (SW) knowledge unit is built through the union of two concepts (subject and object) through a predicate, a triple statement, which formulates the assertion about something that can be uniquely identified. Nanopublications are also built on this SW strategy, allowing knowledge summarization to a set of thoroughly individualized list of assertions - the nanopublication [6]. It standardizes how one can attribute provenance, authorship, publication information and further relationships, always with the intention to stimulate information reuse. It is serializable through the interoperable RDF format, opening the door to new knowledge exchange possibilities and fostering their retrieval and use. Moreover, with universal nanopublications identifiers, each nanopublication can be cited and their impact tracked, encouraging compliance with open SW standards. Various efforts are under way to create guidelines and recommendations for the final schema [6]. Nowadays, the standard is being developed as an incremental process by the nanopublication community[2]. Figure 1 represents the basic model according to nanopublications schema[3].

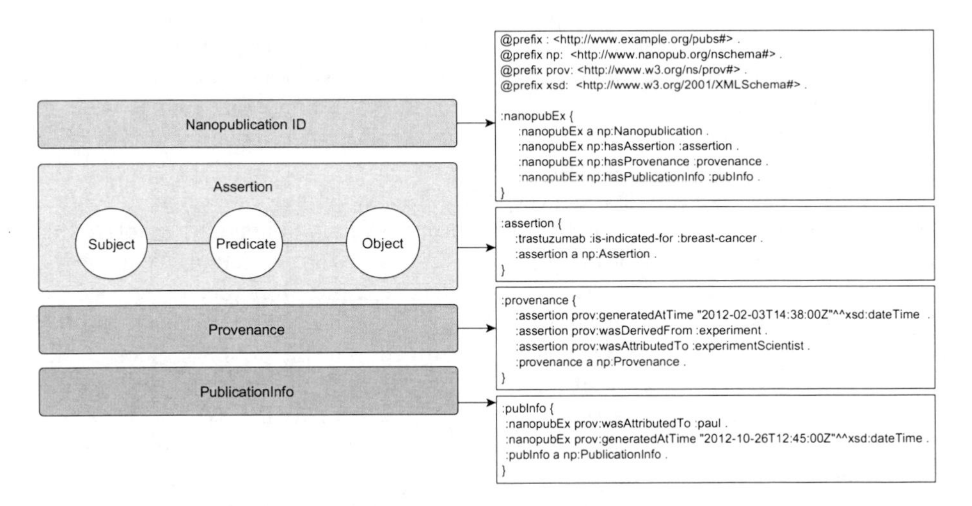

Fig. 1. Basic nanopublication structure (left) with corresponding example (right)

The unique nanopublication identifier is connected to Assertion, Provenance and Publication Information objects. Each of these contains a set of axioms representing the nanopublication metadata. The Assertion graph must contain, at least, one assertion comprised by one or more RDF triples. Supporting information about these

[2] `http://nanopub.org`

[3] `http://nanopub.org/nschema`

assertions is included in the Provenance scope, where DOIs, URLs, timestamps or associated information can be described. Additional information, such as attribution, generated time, keywords or tags can be added too in the Publication Information graph to offer provenance information regarding the nanopublication itself.

In a sense, nanopublications are a natural response to the exploding number and complexity behind scientific communications. In this way, it offers not only a great opportunity to improve and publish conventional papers' research data, but also to explore experimental or negative data studies. Studies of this type are rarely published. Moreover, deploying data as nanopublications allows authors to receive the rightful credit for the shared content.

2.2 Related Projects

In 2008, the scientific journal Nature Genetics, was the first to introduce the concept of "microattribution" to enable an alternative reward system for scientific contributions [1]. Nevertheless, the first practical demonstration was only achieved in 2011, with a series of interrelated locus-specific databases to store all published and unpublished genetic variation related to hemoglobinopathies and thalassemia [7]. At the same time, some approaches emerged to deposit scientific results as nanopublications due to recent SW initiatives empowerment. Some of them are outlined next.

The Open PHACTS project[4] provides a nanopublications use case in their semantic knowledge infrastructure for public and commercial drug discovery research [8]. The nanopublications are used to store information as individual assertions from drug databases and from generated individuals through annotation tools. With the nanopublication format they promote data citation and provide credit to those producing important scientific assertions.

The LOVD nanopublication tool [1] encourages the submission of human genomic variant data for scientific community sharing. This application enables first-generation nanopublications from the Leiden Open-Access Variation Database[5] [9]. From the local database, the system populates a triple store and aggregates all different triples into nanopublications. The content can also be retrieved in XML format. Another module has also been developed for this tool to specify allele frequency data [5]. In this case, the data is submitted by uploading a pre-formatted Excel spreadsheet template in order to extract the data to the system, creating a nanopublication per record. To attribute work recognition the system uses the ResearcherID[6] user unique identity.

The Prizms approach [4] enables the creation of nanopublication data by providing an automated RDF conversion tool. The input data can be in any format, including CSV, XML, JSON and others formats. Making use of an extension of a data management infrastructure (Comprehensive Knowledge Archive Network – CKAN) it can cite derived datasets using the nanopublication provenance standards.

---

[4] http://openphacts.org/

[5] http://lovd.nl

[6] http://researcherid.com

Essentially, it generates RDF data to describe the datasets as a "datapub", a nanopublication model for describing datasets, according to the authors. A public demonstration with 330 melanoma datasets is publicly available[7].

Other approach is related with exposing experimental data in life sciences. Mina *et al.* make use of the nanopublication model to create scientific assertions from the workflow analysis of Huntington's Disease data, making it machine-readable, interoperable, and citable [10]. Mainly, they present how the results of a specific case study can be represented as nanopublications and how this integration could facilitate the data search by means of SPARQL queries. Also, they include and connect the nanopublications provenance graph to Research Objects (RO) [11], an aggregation object that bundles experimental studies resources. This linkage allows a context description of the workflow process. In contrast to nanopublications, RO encapsulate elements for an entire investigation, as opposed to individual claims [12].

The Nanobrowser portal[8] is a different approach that lets users create interactive and manual statements through the nanopublication concept. The tool uses English sentences to represent informal and underspecified scientific claims [13]. These sentences follow a syntactic and semantic scheme that the authors call AIDA (Atomic, Independent, Declarative, Absolute), which provides a uniform and succinct representation of scientific assertions [14]. Essentially, authors and curators manually write AIDA sentences, and text mining approaches automatically extract the content to create nanopublications assertions.

3 Nanopublishing Architecture

The previous projects show how nanopublications can be used in real world scenarios. However, there are several issues and challenges that still have to be addressed. Most of the available solutions target a specific domain (e.g. Open PHACTS, LOVD, etc.), which limits the creation of nanopublications by researchers. Others miss the main functionality that is actually needed: an automated transition from several data formats to nanopublications. In this way, we believe that certain features must be employed for a nanopublication transition solution to be successfully implemented in practice, which are described next:

1. The solution must accept common input data types (databases, delimited or structured files, etc.) and be capable of generating new nanopublications automatically, assigning the respective credit to its authors.
2. The application content, i.e. all nanopublications, must be created with the goal to be publicly available, promoting data sharing.
3. A search engine, supported by a SPARQL endpoint for instance, must be developed to provide a mechanism to query nanopublications.
4. A query federation solution for users' information exchange must be available, according to LinkedData principles [15].
5. The solution must be easy to setup by researchers.

---

[7] `http://data.melagrid.org`

[8] `http://nanobrowser.inn.ac`

To tackle these requirements, we propose an extension to the COEUS' architecture to allow easy integration from several data formats to the nanopublications ontology graph. In the next section, we describe the main changes in the core system architecture to enable nanopublishing.

3.1 Integrating Nanopublications

The COEUS framework offers a good starting point to develop an architecture to support generic data loading and integration. Its main handicap is the data transformation process that must match the internal model ontology. Changing this strategy, COEUS' architecture will allow publishing universal scientific results automatically as nanopublications.

The COEUS engine provides a variety of connectors (CSV, XML, JSON, SQL, SPARQL, RDF, and TTL) to aggregate data from different sources. However, the "triplification" process is made through an organized ontology model. In this model, the data relationships are in an "Entity-Concept-Item" structure (e.g. Protein-Uniprot-P51587) to enable the integration of generic data into the Knowledge Base (KB). However, in the scenario addressed in this work, we know in advance the model of the data to be stored. Hence, we facilitate the user setup by completing automatically the nanopublications structure model. In this specific case, the user must only configure each "Resource" (data source properties) to integrate data as nanopublications. This introduces the first change to COEUS' internal setup.

The COEUS ontology translates the data elements into "Items" (coeus:Item), a basic representation of the produced data. As we are creating nanopublications, a specific data model, we associate a new predicate property: "Nanopublication" (np:Nanopublication). By adding this property to the internal ontology of the application, we split the core data transformation in 2 ways: the creation of COEUS concept data and the creation of nanopublication data. Attending to these modifications, the core application proceeds differently if the user wants to integrate data into COEUS' original model or into COEUS' nanopublication model.

To publish data as nanopublications, we also change the triples generation process. Due to the nanopublications schema, each nanopublication produced includes at least one Assertion along with the Provenance and PublicationInfo field. The creation of each field is an automatic and incremental process. The mapping between the data source and each nanopublication field content remains a manual user interaction process. This procedure is conducted through a specific COEUS web interface, facilitating the user interaction.

Figure 2 shows the new workflow diagram. The user starts by associating the data, creating one or more resources and their data endpoints (Figure 2, block 1). Linking the selected data to the nanopublications different fields will depend on the data type. This mapping process allows the engine to complete the nanopublication structure. Based on advanced Extract-Transform-and-Load (ETL) features, the engine generates, for each entry, a nanopublication record (Figure 2, block 2). Every nanopublication created is stored and made publicly accessible by several services (Figure 2, block 3).

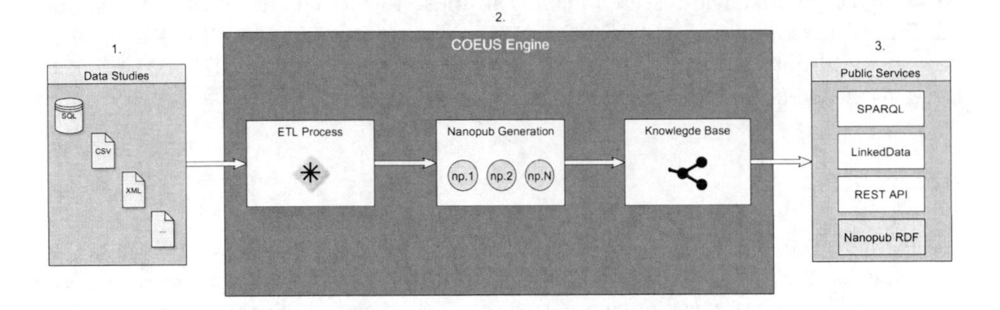

Fig. 2. Nanopublishing workflow: from generic data to nanopublications

To explore the data, COEUS has several interoperability features including REST services, a SPARQL endpoint and LinkedData interfaces. The creation of a nanopublication store forces this platform to adopt a new strategy to retrieve data. In this way, the system includes a RDF/XML exporting format option (represented in Figure 2 as "Nanopub RDF"), concordant with the nanopublication schema and accessible by a Uniform Resource Identifier (URI). We are also making collaborative efforts to maintain a compatibility interface with the nanopublication API, currently in development by the nanopublications community.

4 Discussion and Conclusions

The massive growth of scientific data generated year by year, including experimental data, begs for new strategies to grasp novel scientific outcomes. The nanopublication standard arises as a Semantic Web solution to this problem enabling researchers to synthesize and interconnect their results data. The appearance of this prominent solution, quickly triggered the provision of some tools. The majority are prototype solutions, each one targeting a specific domain. The approach described in this paper, intends to incentivize researches to publish and integrate their data as nanopublications in an easy way. Studies results can be generated in common formats to be submitted later to this framework. According to the workflow described, the user has the option to include the desired data into the engine, selecting and mapping the essential structured fields. Through the new design of ETL features on COEUS, we can integrate and deliver the data as nanopublications. This approach allows the reduction of redundancy and ambiguity of scientific statements contained in integrated studies. Also, it provides an attribution system with proper recognition to their authors, enabling appropriate data sharing mechanisms, according to LinkedData principles. With our expertise, we believe that such open source framework will benefit the research community and promote data sharing standards. We are also making efforts to offer an easy setup solution. By designing a new setup web interface we plan to make this framework more user-friendly and increase the researcher's application range. In a near future, this work in progress will deliver, in a package, all tools needed to help researchers publish, store and retrieve all their outcomes as nanopublications.

Acknowledgments. The research leading to these results has received funding from the European Community (FP7/2007-2013) under ref. no. 305444 – the RD-Connect project, and from the QREN "MaisCentro" program, ref. CENTRO-07-ST24-FEDER-00203 – the CloudThinking project.

References

1. Patrinos, G.P., Cooper, D.N., van Mulligen, E., Gkantouna, V., Tzimas, G., Tatum, Z., Schultes, E., Roos, M., Mons, B.: Microattribution and nanopublication as means to incentivize the placement of human genome variation data into the public domain. Hum. Mutat. 33, 1503–1512 (2012)
2. Velterop, J.: Nanopublications*: The future of coping with information overload. LOGOS J. World B. Community (2010)
3. Mons, B., Velterop, J.: Nano-Publication in the e-science era. Work. Semant. Web Appl. Sci. Discourse (2009)
4. McCusker, J., Lebo, T.: Next Generation Cancer Data Discovery, Access, and Integration Using Prizms and Nanopublications. Data Integr. Life Sci. 105–112 (2013)
5. Lopes, P., Oliveira, J.L.: COEUS: "semantic web in a box" for biomedical applications. J. Biomed. Semantics. 3, 11 (2012)
6. Groth, P., Gibson, A., Velterop, J.: The anatomy of a nanopublication. Inf. Serv. Use (2010)
7. Giardine, B., Borg, J., Higgs, D.R., Peterson, K.R., Philipsen, S., et al.: Systematic documentation and analysis of human genetic variation in hemoglobinopathies using the microattribution approach. Nat. Genet. 43, 295–301 (2011)
8. Harland, L.: Open PHACTS: A semantic knowledge infrastructure for public and commercial drug discovery research. Knowl. Eng. Knowl. Manag. (2012)
9. Fokkema, I.F.A.C., Taschner, P.E.M., Schaafsma, G.C.P., Celli, J., Laros, J.F.J., den Dunnen, J.T.: LOVD v.2.0: The next generation in gene variant databases. Hum. Mutat. 32, 557–563 (2011)
10. Mina, E., Thompson, M.: Nanopublications for exposing experimental data in the life-sciences: A Huntington's Disease case study. In: Proc. 6th Int. Semant. Web Appl. Tools Life Sci. Work 2013 (2013)
11. Belhajjame, K., Corcho, O., Garijo, D.: Workflow-centric research objects: First class citizens in scholarly discourse. In: Proc. ESWC 2012 Work. Futur. Sch. Commun. Semant. Web (2012)
12. Belhajjame, K., Zhao, J., Garijo, D., Hettne, K., Palma, R., Corcho, Ó., Gómez-Pérez, J.-M., Bechhofer, S., Klyne, G., Goble, C.: The Research Object Suite of Ontologies: Sharing and Exchanging Research Data and Methods on the Open Web. arXiv Prepr. arXiv 1401.4307. 20 (2014)
13. Kuhn, T., Krauthammer, M.: Underspecified scientific claims in nanopublications. arXiv Prepr. arXiv1209.1483 (2012)
14. Kuhn, T., Barbano, P.: Broadening the scope of nanopublications. Semant. Web Semant. Big Data. 487–501 (2013)
15. Bizer, C., Heath, T., Berners-Lee, T.: Linked data-the story so far. Int. J. Semant. Web Inf. Syst. (2009)

Retrieval and Discovery of Cell Cycle Literature and Proteins by Means of Machine Learning, Text Mining and Network Analysis

Martin Krallinger, Florian Leitner, and Alfonso Valencia

Structural Biology and Biocomputing Programme, Spanish National Cancer, Research Centre (CNIO), C/ Melchor Fernndez Almagro, 3. 28029 Madrid, Spain

Abstract. The cell cycle is one of the most important biological processes, being studied intensely by experimental as well as bioinformatics means. A considerable amount of literature provides relevant descriptions of proteins involved in this complex process. These proteins are often key to understand cellular alterations encountered in pathological conditions such as abnormal cell growth. The authors explored the use of text mining strategies to improve the retrieval of relevant articles and individual sentences for this topic. Moreover information extraction and text mining was used to detect and rank automatically Arabidopsis proteins important for the cell cycle. The obtained results were evaluated using independent data collections and compared to keyword-based strategies. The obtained results indicate that the use of machine learning methods can improve the sensitivity compared to term-co-occurrence, although with considerable differences when using abstracts and full text articles as input. At the level of document triage the recall ranges for abstracts from around 16% for keyword indexing, 37% for a sentence SVM classifier to 57% for SVM abstract classifier. In case of full text data, keyword and cell cycle phrase indexing obtained a recall of 42% and 55% respectively compared to 94% reached by a sentence classifier. In case of the cell cycle protein detection, the cell cycle keyword-protein co-occurrence strategy had a recall of 52% for abstracts and 70% for full text while a protein mentioning sentence classifier obtained a recall of over 83% for abstracts and 79% for full text. The generated cell cycle term co-occurrence statistics and SVM confidence scores for each protein were explored to rank proteins and filter a protein network in order to derive a topic specific subnetwork. All the generated protein cell cycle scores together with a global protein interaction and gene regulation network for Arabidopsis are available at: http://zope.bioinfo.cnio.es/cellcyle_addmaterial.

Keywords: text mining, natural language processing, cell cycle, machine learning, protein ranking.

1 Introduction

The cell cycle is characterised by a series of coordinated spatiotemporal events that involve transcription regulation, synchronised control of dynamic subcellular location changes and interactions of gene products. To better understand the

J. Sáez-Rodríguez et al. (eds.), *8th International Conference on Practical Appl. of Comput. Biol. & Bioinform. (PACBB 2014)*, Advances in Intelligent Systems and Computing 294,
DOI: 10.1007/978-3-319-07581-5_34,

cell cycle, model organisms are used, including the plant *Arabidopsis thaliana*. This organism provides relevant insights to understand commonalities during the cell cycle process of higher eukaryotes. It helps to determine connections between the cell cycle and plant growth, an issue of key importance for agricultural biotechnology [1]. To generate a collection of plant cell cycle-modulated genes a popular strategy is to examine periodically expressed genes during distinct cell cycle phases [2]. Therefore, genome-wide temporal expression studies of cell division were performed [3]. These studies only cover one level of association to the cell cycle. Another source that provides information covering various different types of associations between genes/proteins and the cell cycle is the literature.

Text mining methods have been used on a range of biological topics to improve information access [4] from scientific publications. In addition to the recognition of mentions of biologically relevant entities such as genes and proteins [5], the extraction of relationships between entities has attracted considerable attention [6]. Machine learning methods were applied not only to find entities and their interactions but also to detect articles of relevance for a variety of biological topics such as protein interactions [7], pharmacogenetics [8] or alternative transcripts [9]. Currently one of the most popular statistical learning methods to classify abstracts and sentences from the biomedical literature are Support Vector Machines (SVMs) [10,11]. Some general purpose classification systems of PubMed abstracts have been implemented, such as MedlineRanker [12] or MScanner [13]. The results of text classification methods have been explored to rank also bio-entities mentioned in documents [14,15] or to analyze list of genes [16,17].

Previous work did not sufficiently examine the combination of the detection individual textual items (documents and sentences) with particular bio-entities at the level of abstract and full text data. The authors examined the classification/ranking of textual evidences to facilitate a more targeted retrieval of cell cycle related information for Arabidopsis. The presented work goes beyond the use of abstracts, covering the retrieval of sentences and full text passages. The used strategy returned a core set of cell cycle genes for the model organism *A. thaliana* described in publications. The ranking of bio-entity relations (protein interactions and transcriptional gene regulation) using text mining was attempted. We foresee that such results might be potentially useful in the analysis of interaction networks for the characterisation of cell cycle proteins. The presented methodology could in principle be adapted to other organisms or topics of interest.

2 Materials and Methods

Document Selection and Preprocessing. The cell cycle text mining system integrated three main modules: (1) a cell cycle keyword recognition module, (2) several cell cycle text classifiers and (3) the bio-entity recognition module. Figure 1 provides a general overview of the cell cycle text mining system. The collection of abstracts and full text articles relevant for *A. thaliana* was collected

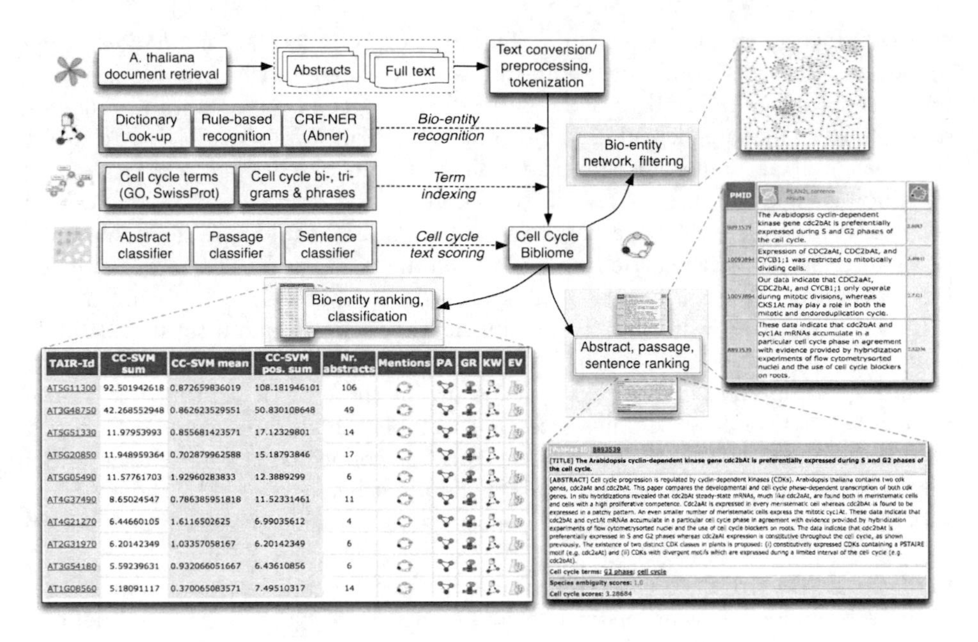

Fig. 1. Cell cycle classifier. This figure shows a simplified flowchart of the cell cycle protein and literature text mining system. The lower part of the figure shows the protein ranking by the abstract cell cycle classifier while on the right side example sentences and abstracts returned by the classifier are shown.

by considering citations for this species from various databases[1] together with a hand-crafted Boolean PubMed query that covered the various aliases and relevant journals for this species[2]. Abstracts and full text papers were tokenized using an in-house sentence boundary recognition script. The resulting set of documents contained 11,636 full text articles and 16,536 abstracts.[3]

Gene/Protein Mention Recognition and Normalisation. On the literature collection we then performed bio-entity mention recognition[4]. The approach was based on the construction and look-up of a specific gene/protein lexicon, followed by a protein normalisation scoring/disambiguation step. The lexicon had two entry types: a) a database-derived gene lexicon containing names that can be linked to database records, and b) the NER-derived lexicon containing names that cannot be normalised directly to any database record but corresponded

[1] TAIR (Rhee et al. 2003), SwissProt (Boeckmann et al. 2003) and GOA (Camon et al. 2004).

[2] Query: ((thale AND cress) OR (arabidopsis) OR (thaliana) OR (thale-cress) OR (mouse-ear AND cress) OR (mouse ear AND cress)) OR (a.thaliana) OR (arabidopsis thaliana) OR (mouse-ear AND rock cress) OR ("Pilosella siliquosa") OR ("thale-cress") OR ("The Arabidopsis book /American Society of Plant Biologists"[Journal]).

[3] Not all abstracts had an associated full text article and vice versa.

[4] We refer to genes and proteins as bio-entities in this article.

to Arabidopsis gene/protein names. The first type integrated *A. thaliana* gene names and symbols from multiple resources. It covered all the Arabidopsis gene names contained in TAIR and SwissProt[5]. The second type contained names detected by a machine learning named entity recognition program (i.e. ABNER [18][6]) as well as names recognised through a rule based approach exploiting morphological word cues (organism source gene prefixes and suffixes like At or AT[7]) and name length characteristics for potential Arabidopsis gene symbols. Lexicon expansion was carried out using manually crafted rules to account for typographical name variations. Ambiguous names were removed using a specially compiled stop word list. The resulting gene lexicon had a total of 919,994 unique name entries, out of which 14,294 could be detected in abstracts and 27,497 were matched to full text articles. From the 6,214 UniProt *A. thaliana* records, a total of 1,908 unique accession numbers could be linked to PubMed abstract sentences (30.70%) while 4,732 could be connected to full text sentences (76.15%). From the initial 10,287 TAIR gene identifiers (covered by the lexicon), 4,741 could be linked to abstract sentences (46.09%) while 7,163 (69.63%) were associated to full text sentences.

Cell Cycle Term Detection. A simple strategy to retrieve documents for a topic is by indexing mentions of relevant terms or keywords from a predefined vocabulary. This article selection criterion is often used as a simple baseline for more sophisticated retrieval and classification approaches. In addition of document indexing, those mentions can be used to retrieve relevant entities by considering co-occurrence. Therefore a list of 303 keywords related to the cell cycle topic was selected from Gene Ontology (GO)[8] and SwissProt keywords. Child terms and synonyms of the GO cell cycle term were also included. This original list was refined (highly ambiguous terms were deleted) and some additional terms were added. 4,430 term-sentence associations were recovered from abstracts and 37,892 from full text articles using case insensitive look-up. 1,806 abstracts and 7,275 full text documents had at least a single term. From the list of 303 keywords, 43 were mentioned in abstracts and 80 in full text. 978 co-occurrences with Arabidopsis proteins were found in abstracts and 11,199 in full text.

The authors also extracted semi-automatically candidate cell cycle terms directly from the literature. Using a PubMed search, candidate abstracts related to plant cell cycle were retrieved[9]. From those abstracts noun phrases, bigrams and trigrams were obtained and ranked based on their raw frequencies. These were extracted using the NLTK toolkit. Bigrams and trigrams were processed using the Justeson-Katz filter. The ranked list was manually examined to select

[5] TAIR: `http://www.arabidopsis.org`; SwissProt: `http://www.uniprot.org`

[6] `http://pages.cs.wisc.edu/~bsettles/abner`

[7] AT consists of the acronym for Arabidopsis thaliana. It is often used as an affix attached to gene symbols.

[8] `http://www.geneontology.org`

[9] Search query used: mitotic[All Fields] OR "mitosis"[MeSH Terms] AND ("plants"[MeSH Terms] OR "plants"[All Fields] OR "plant"[All Fields]) AND hasabstract[text] AND hasabstract[text].

phrases associated with cell cycle or mitosis. From the initial noun phrase list, those cases that contained cell cycle terms either as substrings or sub-tokens were incorporated as well. From the resulting list of phrases 3,479 could be mapped to documents (1,319 in abstracts and 3129 in full text). A total of 4,430 mentions were detected in abstracts and 37,892 in will text. For this phrase list 5,917 co-occurrences with Arabidopsis proteins were found in abstracts and 59,288 in full text.

Machine Learning Classifiers. To complement the term-indexing approach we explored a supervised machine learning strategy based on SVMs. The SVM packages scikit-learn [19] and SVMlight were used[10] for this purpose. Two distinct classifiers were trained; one for classifying abstracts (and full text passages) and another more granular for individual sentences. As the cell cycle topic is fairly general with a less uniform word usage selecting suitable training instances was challenging. Initially we explored using literature citations from GO annotations, but as they were rather noisy (many corresponding abstracts did only indirectly refer to this topic), we finally selected the positive training abstracts through a specific PubMed query[11]. From these hits, a random sample of 4,800 recent abstracts served as positive training collection. The negative abstract training data was composed of 4,800 randomly selected PubMed abstracts. Samples of 50 records from both sets were manually checked to make sure that the training data was correct.

For the sentence classifier training data, we manually examined sentences mentioning within the corresponding abstracts both cell cycle terms and proteins. A total of 5,840 cell cycle related sentences were selected as positive training set, while 5,840 randomly selected sentences from the Arabidopsis abstracts constituted the negative training collection. We used in both cases a linear support vector machine with the following parameter settings: whitespace tokenisation pattern, stripped accents, minimum token frequency of 2, UTF-8 encoding, word n-gram range of 1-4, lowercase tokens, parameter C of 0.05, L2 penalty and term frequency-inverse document frequency term weighting. Overall the linear SVM classifier was slightly better when compared to a Naïve Bayes multinomial or Bernoulli classifier both at the abstract and sentence classification tasks (data not shown). We also analysed the effect of masking protein mentions in the training data sets. A total of 159,523 features were extracted from the abstract and 24,970 from the sentence training sets. Using a 5-fold cross validation the abstract classifier obtained a precision of 90.3 and a recall of 92.6 with a corresponding F1-score of 91.5. In case of the sentence classifier the obtained precision was of 90.2 with a recall of 97.5 (F1-score of 93.7).

[10] scikit-learn: `http://scikit-learn.org`; SVMlight: `http://svmlight.joachims.org`

[11] Query: "cell cycle"[MeSH Terms] AND ("plants"[MeSH Terms] OR "plants"[All Fields] OR "plant"[All Fields])

3 Discussion and Results

Carrying out an exhaustive manual validation of the entire collection of Arabidopsis documents was not viable. Manual database annotations are commonly used to evaluate bioinformatics predictions. We therefore validated our methods using documents annotated as cell cycle relevant by the UniProt database for Arabidopsis proteins[12]: 915 abstracts and 504 full text articles. Table 1 summarises the results for the triage of cell cycle documents and for detecting cell cycle genes/proteins. It is important to note that these documents are cell cycle protein annotation significant and not just cell cycle related[13].

Table 1. Evaluation of the cell cycle document triage (top) and protein detection (lower)

DOCUMENT TRIAGE RESULTS

	Method	Recall %	Precision %	f-score %	Accuracy
ABSTRACTS	Nr. GO sent. mentions	15.52	21.52	18.03	0.92194
ABSTRACTS	Nr. Phrase sent. mentions	17.81	17.14	17.47	0.90689
ABSTRACTS	Nr. relevant SVM sentences	37.92	10.84	16.86	0.79310
ABSTRACTS	Perc. relevant SVM sentences	37.92	10.84	16.86	0.79310
ABSTRACTS	Sum rel. sent. conf. scores	37.92	10.84	16.86	0.79310
ABSTRACTS	Sum all sent. conf. scores	12.90	23.69	16.70	0.92883
ABSTRACTS	Abstract classifier SVM conf.	57.27	5.64	10.26	0.44586
FULL TEXT	Nr. GO sent. mentions	42.16	11.62	*18.22*	0.78552
FULL TEXT	Nr. Phrase sent. mentions	55.51	10.03	16.99	0.69245
FULL TEXT	Sum rel. sent. conf. scores	*93.86*	6.46	12.09	0.22601
FULL TEXT	Nr. relevant SVM sentences	*93.86*	6.46	12.09	0.22601
FULL TEXT	Perc. relevant SVM sentences	*93.86*	6.46	12.09	0.22601
FULL TEXT	Sum all sent. conf. scores	1.06	*27.78*	2.04	*0.94236*

PROTEIN DETECTION RESULTS

	Method	Recall %	Precision %	f-score %	AP	FAP
ABSTRACTS	χ^2 protein-phrase	81.7	12.73	22.03	0.3875	0.2809
ABSTRACTS	χ^2 protein-GO	51.63	30.04	37.98	0.2975	0.3336
ABSTRACTS	Total protein-GO coocur.	51.63	29.48	37.53	0.3181	0.3444
ABSTRACTS	Total protein-phrase coocur.	82.35	12.51	21.72	0.3217	0.2594
ABSTRACTS	Perc. relevant SVM abstracts	83.66	4.51	8.56	0.0654	0.0742
ABSTRACTS	Perc. relevant SVM sentences	83.01	10.18	18.13	0.2468	0.2091
ABSTRACTS	Mean rel. SVM abs. conf. scores	83.66	4.51	8.56	0.1315	0.1037
ABSTRACTS	Mean rel. SVM sent. conf. scores	83.01	10.18	18.13	0.1657	0.1732
ABSTRACTS	Mean all SVM abs. conf. scores	87.58	3.46	6.66	0.0978	0.0792
ABSTRACTS	Total SVM relevant sent.	100	3.32	6.42	0.0455	0.0533
ABSTRACTS	Total SVM relevant abstracts	83.66	4.51	8.56	0.0527	0.0653
FULL TEXT	χ^2 protein-phrase	81.66	4.85	9.16	0.2756	0.1375
FULL TEXT	χ^2 protein-GO	70.41	10.95	18.95	0.2543	0.2171
FULL TEXT	Total protein-GO coocur.	70.41	10.76	18.67	0.298	0.2296
FULL TEXT	Total protein-phrase coocur.	82.84	4.86	9.18	0.2771	0.1379
FULL TEXT	Perc. relevant SVM sentences	78.7	5.86	10.91	0.1841	0.137
FULL TEXT	Total SVM relevant sent.	78.7	5.86	10.91	0.2142	0.1446

Literature biocuration is particularly interested in high recall and reasonable ranking of the text mining results, making sure that as many relevant instances are captured as possible for further manual validation. From the results obtained for the cell cycle document triage it is clear that the recall was higher when using full text data as compared to abstracts (associated with a larger number of potential false positive hits). The recall of the cell cycle phrase co-occurrence

[12] UniProt online version 20th March 2013.

[13] Database annotations are generally incomplete, covering only a fraction of the information contained in the literature, thus the evaluation metrics that depend on the precision have to be taken with care.

approach worked better for full text data when compared to the initial term co-occurrence. Overall, the machine learning approach had a much higher recall then term indexing, indicating that it is more appropriate for exhaustive literature curation. Under the used evaluation setting it seems that using full text data can more than double the recall for document triage, but it remains unclear how much time is *de facto* saved by human curators when using text mining results for abstracts as opposed to full text. Providing a ranked list of relevant sentences is key to find annotation relevant text passages for manual curation. The highest recall for abstracts was obtained by the abstract classifier (57.27) followed by the sentence classifier method (37.92). Although the provided precision scores have to be taken with caution, it looks as if co-occurrence statistics between entities and terms can be more reliable for retrieving protein annotations. To evaluate the detection of cell-cycle proteins, all proteins annotated in the TAIR database as cell cycle relevant were used as the Gold Standard set. A total of 174 of these proteins had detected literature mentions (153 in abstracts and 169 in full text sentences). The text mining results were compared to these proteins separately for abstracts and full text. Table 1 (lower part) illustrates the protein results. These were generally better than the document-based results. The use of cell cycle phrase-protein co-occurrence had a greater boost in recall over the initial cell cycle term list and could reach a recall comparable to the SVM classifier, of over 80%. All three methods generated a list of additional potential cell cycle related proteins that lack such annotations in the UniProt database.

One exploratory application of these results is the use of literature derived protein scores to filter interaction networks. Such a network was assembled for Arabidopsis derived from in multiple databases[14] together with results from a rule based interaction and gene regulation information extraction method [20]. We kept only interactions between proteins that where scored as cell cycle relevant. Manual inspection of the resulting interaction network showed that highly connected nodes in this network corresponded to the key players of various central cell cycle events. Some issues that still need further examination include, a more thorough analysis of precision related issues, the use of a combined system integrating the various techniques presented here and the transfer of generated scored to homologue proteins in other plant genomes.

References

1. Lenhard, M.: Plant growth: Jogging the cell cycle with JAG. Curr. Biol. 22(19), R838–840 (2012)
2. Menges, M., Hennig, L., Gruissem, W., Murray, J.A.: Cell cycle-regulated gene expression in Arabidopsis. J. Biol. Chem. 277(44), 41987–4(2002)
3. Breyne, P., Zabeau, M.: Genome-wide expression analysis of plant cell cycle modulated genes. Curr. Opin. Plant Biol. 4(2), 136–142 (2001)

[14] Protein-protein interaction annotations were assembled from the following databases: BIND, BioGRID, MINT, TAIR and DIP (see additional materials).

4. Jensen, L.J., Saric, J., Bork, P.: Literature mining for the biologist: from information retrieval to biological discovery. Nature Reviews Genetics 7(2), 119–129 (2006)
5. Leser, U., Hakenberg, J.: What makes a gene name? named entity recognition in the biomedical literature. Briefings in Bioinformatics 6(4) (2005)
6. Zhou, D., He, Y.: Extracting interactions between proteins from the literature. Journal of Biomedical Informatics 41(2), 393–407 (2008)
7. Krallinger, M., Vazquez, M., Leitner, F., Salgado, D., Chatr-aryamontri, A., Winter, A., Perfetto, L., Briganti, L., Licata, L., Iannuccelli, M., et al.: The protein-protein interaction tasks of biocreative iii: classification/ranking of articles and linking bio-ontology concepts to full text. BMC Bioinformatics 12(suppl. 8), S3 (2011)
8. Rubin, D.L., Thorn, C.F., Klein, T.E., Altman, R.B.: A statistical approach to scanning the biomedical literature for pharmacogenetics knowledge. Journal of the American Medical Informatics Association 12(2), 121–129 (2005)
9. Shah, P.K., Jensen, L.J., Boué, S., Bork, P.: Extraction of transcript diversity from scientific literature. PLoS Computational Biology 1(1), e10 (2005)
10. Cortes, C., Vapnik, V.: Support-vector networks. Machine learning 20(3), 273–297 (1995)
11. Joachims, T.: Text categorization with support vector machines: Learning with many relevant features. In: Nédellec, C., Rouveirol, C. (eds.) ECML 1998. LNCS, vol. 1398, pp. 137–142. Springer, Heidelberg (1998)
12. Fontaine, J.F., Barbosa-Silva, A., Schaefer, M., Huska, M.R., Muro, E.M., Andrade-Navarro, M.A.: Medlineranker: flexible ranking of biomedical literature. Nucleic Acids Research 37(suppl. 2), W141–W146 (2009)
13. Poulter, G.L., Rubin, D.L., Altman, R.B., Seoighe, C.: Mscanner: A classifier for retrieving medline citations. BMC Bioinformatics 9(1), 108 (2008)
14. Fontaine, J.F., Priller, F., Barbosa-Silva, A., Andrade-Navarro, M.A.: Genie: literature-based gene prioritization at multi genomic scale. Nucleic Acids Research 39(suppl. 2), W455–W461(2011)
15. Krallinger, M., Rojas, A.M., Valencia, A.: Creating reference datasets for systems biology applications using text mining. Annals of the New York Academy of Sciences 1158(1), 14–28 (2009)
16. Soldatos, T.G., O'Donoghue, S.I., Satagopam, V.P., Barbosa-Silva, A., Pavlopoulos, G.A., Wanderley-Nogueira, A.C., Soares-Cavalcanti, N.M., Schneider, R.: Caipirini: Using gene sets to rank literature. BioData Mining 5(1), 1 (2012)
17. Soldatos, T.G., Pavlopoulos, G.A.: Mining cell cycle literature using support vector machines. In: Maglogiannis, I., Plagianakos, V., Vlahavas, I. (eds.) SETN 2012. LNCS (LNAI), vol. 7297, pp. 278–284. Springer, Heidelberg (2012)
18. Settles, B.: Abner: an open source tool for automatically tagging genes, proteins and other entity names in text. Bioinformatics 21(14), 3191–3192 (2005)
19. Pedregosa, F., Varoquaux, G., Gramfort, A., Michel, V., Thirion, B., Grisel, O., Blondel, M., Prettenhofer, P., Weiss, R., Dubourg, V., Vanderplas, J., Passos, A., Cournapeau, D., Brucher, M., Perrot, M., Duchesnay, E.: Scikit-learn: Machine learning in Python. Journal of Machine Learning Research 12, 2825–2830 (2011)
20. Krallinger, M., Rodriguez-Penagos, C., Tendulkar, A., Valencia, A.: PLAN2L: A web tool for integrated text mining and literature-based bioentity relation extraction. Nucleic Acids Res. 37, W160–165 (2009)

Author Index